D0526474

The Construction of Buildings

Volume 1

SEVENTH EDITION

By the same author

Volume Two
Fifth Edition
Windows - Doors – Stairs – Fires, Stoves and Chimneys – Internal Finishes and External Rendering

Volume Three
Fourth Edition
Lattice Truss, Beam, Portal Frame and Flat Roof Structures – Roof and Wall Cladding, Decking and Flat Roof Weathering – Rooflights – Diaphragm, Fin Wall and Tilt-up Construction – Shell Structures

Volume Four
Fourth Edition
Multi-storey Buildings – Foundations – Steel Frames – Concrete Frames – Floors – Wall Cladding

Volume Five
Third Edition
Water – Electricity and Gas Supplies – Foul Water Discharge – Refuse Storage

THE CONSTRUCTION OF BUILDINGS

Volume 1

SEVENTH EDITION

R. BARRY
Architect

FOUNDATIONS and OVERSITE CONCRETE –
WALLS – FLOORS – ROOFS

Blackwell
Science

Copyright © R. Barry 1958, 1962, 1969, 1980, 1989, 1996, 1999

Blackwell Science Ltd
Editorial Offices:
Osney Mead, Oxford OX2 0EL
25 John Street, London WC1N 2BL
23 Ainslie Place, Edinburgh EH3 6AJ
350 Main Street, Malden
MA 02148 5018, USA
54 University Street, Carlton
Victoria 3053, Australia
10, rue Casimir Delavigne
75006 Paris, France

Other Editorial Offices:

Blackwell Wissenschafts-Verlag GmbH
Kurfürstendamm 57
10707 Berlin, Germany

Blackwell Science KK
MG Kodenmacho Building
7–10 Kodenmacho Nihombashi
Chuo-ku, Tokyo 104, Japan

The right of the Author to be identified as the Author of this Work has been asserted in accordance with the Copyright, Designs and Patents Act 1988.

All rights reserved. No part of this publication may be reproduced, stored in a retrieval system, or transmitted, in any form or by any means, electronic, mechanical, photocopying, recording or otherwise, except as permitted by the UK Copyright, Designs and Patents Act 1988, without the prior permission of the publisher.

First Edition published by Crosby Lockwood & Son Ltd 1958
Second Edition, published 1962
Reprinted 1964, 1965, 1968
Third Edition (Metric) 1969
Reprinted 1971
Reprinted 1972, 1974, 1975 by Crosby Lockwood Staples
Fourth Edition published by Granada Publishing 1980
Reprinted 1982, 1984
Reprinted by Collins Professional and Technical Books 1985, 1987
Fifth Edition published by BSP Professional Books 1989
Fifth Edition revised 1993
Sixth Edition published by Blackwell Science Ltd 1996
Reprinted 1997
Seventh Edition 1999

Set in 11/14pt Times
by DP Photosetting, Aylesbury, Bucks
Printed and bound in Great Britain by
MPG Books Ltd, Bodmin, Cornwall

The Blackwell Science logo is a trade mark of Blackwell Science Ltd, registered at the United Kingdom Trade Marks Registry

DISTRIBUTORS

Marston Book Services Ltd
PO Box 269
Abingdon
Oxon OX14 4YN
(*Orders:* Tel: 01235 465500
Fax: 01235 465555)

USA
Blackwell Science, Inc.
Commerce Place
350 Main Street
Malden, MA 02148 5018
(*Orders:* Tel: 800 759 6102
781 388 8250
Fax: 781 388 8255)

Canada
Login Brothers Book Company
324 Saulteaux Crescent
Winnipeg, Manitoba R3J 3T2
(*Orders:* Tel: 204 837 2987
Fax: 204 837 3116)

Australia
Blackwell Science Pty Ltd
54 University Street
Carlton, Victoria 3053
(*Orders:* Tel: 03 9347 0300
Fax: 03 9347 5001)

A catalogue record for this title is available from the British Library

ISBN 0-632-05261-9

Library of Congress Cataloging-in-Publication Data
Barry, R. (Robin)
The construction of buildings / R. Barry. — 7th ed.
p. cm.
Includes index.
Contents: v. 1. Foundations and oversite concrete, walls, floors, roofs
ISBN 0-632-05261-9 (v. 1)
1. Building. I. Title.
TH146.B3 1999
690—dc21 99-32090
CIP

For further information on
Blackwell Science, visit our website:
www.blackwell-science.com

Contents

Preface

The initial concept on which the series was prepared was that of principles of building under the headings functional requirements, common to all building, with diagrams to illustrate the application of the requirements to the elements of building. Subsequent changes in the use of traditional materials and the use of new materials in novel forms of construction have illustrated the value of the concept of functional requirements as a measure of the suitability of materials and construction for both traditional and novel forms of construction.

The text has been revised and rearranged to improve the sequence of subject matter to more clearly follow principles of building, with notes on the history of such changes in use of materials, largely dictated by economics and fashion, and the consequences of such changes.

A new page layout has been adopted for the series which is more suited to setting diagrams next to the relevant text than the old format. The text is set in a wide right hand column with smaller diagrams set in a left hand column which also contains headings for quick reference. These changes to the text and layout have helped to underline more clearly the original concept on which the series was based.

Notes on the properties and uses of both traditional and new materials are included in each chapter. Such notes on the changes in Building Regulations that have occurred over the years that are relevant to principles have been included without extensive use of reference to standards and the use of tables.

As appropriate to the sense of the material, diagrams have been altered, rearranged and augmented with new diagrams to update the series.

The basis for the series is an explanation of the principles of building through an understanding of the nature, properties and uses of materials in the construction of buildings adequate to the work of designing and construction.

R. Barry

Acknowledgements

My thanks are due to my friend Ross Jamieson who redrafted all of my original diagrams for the five volume series; to Mrs Sue Moore for advice and help in the new page layout and to Polly Andrews who is now three fifths of the way through typing my drafts of the five volumes of the revised series.

Table 3 is Crown © and is reproduced with the permission of the Controller of HMSO.

Tables 2 and 8 are extracted from British Standards and are reproduced with the permission of BSI under licence number PD/1999 1050. Complete copies of the standards can be obtained by post from BSI Customer Services, 389 Chiswick High Road, London, W4 4AL.

1: Foundations and Oversite Concrete

HISTORY

Up to the latter part of the nineteenth century, when Portland cement first came into general use for making concrete, the majority of buildings were built directly off the ground. Walls of stone or brick were built on a bed of rough stones or brick footings and timber framed buildings on a base of rough stones or brick. As walls were built their weight gradually compressed soils such as clay, sand or gravel to form a sound, adequate foundation.

Local experience of the behaviour of soils and rocks, under the load of buildings, generally provided sufficient information to choose a foundation of the required depth and spread by this method of construction.

Where a small variation of the degree of compression of soils under buildings occurred the natural arching effect of the small, bonded units of stone and brick and the flexibility of lime mortar would allow a transfer of load to the sound foundation without damage to the building.

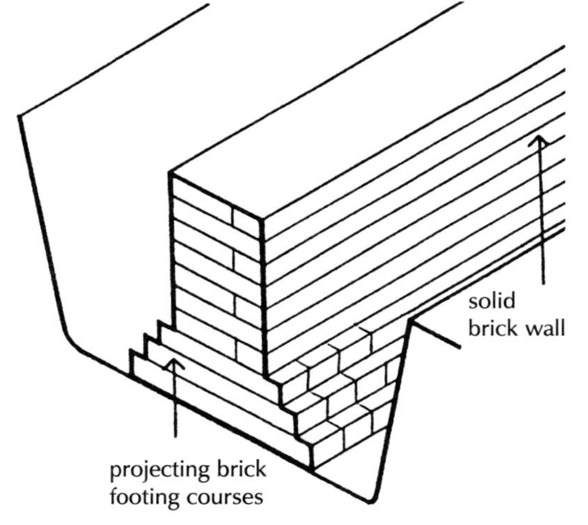

Fig. 1 Brick footings.

From the beginning of the twentieth century concrete was increasingly used as a foundation base for walls. Initially concrete bases were used for the convenience of a solid, level foundation on which to lay and bond stone and brick walls. Brick walls which, prior to the use of concrete, had been laid as footings, illustrated in Fig. 1, to spread the load, were built on a concrete base wider than the footings for the convenience of bricklaying below ground. This massive and unnecessary form of construction was accepted practice for some years.

With the introduction of local and, more recently, general building regulations in this century, standard forms of concrete foundations have become accepted practice in this country along with more rigorous investigation of the nature and bearing capacity of soils and rocks.

The move from the practical, common sense approach of the nineteenth century to the closely regulated systems of today has to an extent resulted in some foundations so massive as to exceed the weight of the entire superstructure above and its anticipated loads. This tendency to over design the foundations of larger buildings has been exacerbated by the willingness of building owners to seek compensation for damage, caused by the claimed negligence of architects, engineers and builders who, in order to control the amount of premium they pay for insurance against such claims, have tended to over design as an insurance.

FOUNDATIONS

The foundation of a building is that part of walls, piers and columns in direct contact with and transmitting loads to the ground. The building foundation is sometimes referred to as the artificial, and the ground on which it bears as the natural foundation.

Ground is the general term for the earth's surface, which varies in composition within the two main groups, rocks and soils. Rocks include hard, strongly cemented deposits such as granite and soils the loose, uncemented deposits such as clay. Rocks suffer negligible compression and soils measurable compression under the load of buildings.

The size and depth of a foundation is determined by the structure and size of the building it supports and the nature and bearing capacity of the ground supporting it.

ROCKS

Rocks may be divided into three broad groups as igneous, sedimentary and metamorphic.

Igneous rocks

Igneous rocks, such as granite, dolerite and basalt, are those formed by the fusion of minerals under great heat and pressure. Beds of strong igneous rock occur just below or at the surface of ground in Scotland and Cornwall as Aberdeen and Cornish granite. The nature and suitability of such rocks as a foundation may be distinguished by the need to use a pneumatic drill to break up the surface of sound, incompressible rock to form a roughly level bed for foundations.

Because of the density and strength of these rocks it would be sufficient to raise walls directly off the rock surface. For convenience it is usual to cast a bed of concrete on the roughly levelled rock surface as a level surface on which to build. The concrete bed need be no wider than the wall thickness it supports.

Sedimentary rocks

Sedimentary rocks, such as limestone and sandstone, are those formed gradually over thousands of years by the settlement of particles of calcium carbonate or sand to the bottom of bodies of water where the successive layers of deposit have been compacted as beds of rock by the weight of water above. Because of the irregular and varied deposit of the sediment, these rocks were formed in layers or laminae. In dense rock beds the layers are strongly compacted and in others the layers are weakly compacted and may vary in the nature of the layers and so have poor compressive strength. Because of the layered nature of these rocks the material should be laid as a building stone with the layers at right angles to the loads.

Many of the beds of sound limestone and sandstone in this country have been quarried for the production of natural building stones such as Portland and Bath limestones and Darly Dale and Crosland Hill sandstones. The suitability of sound limestone and sandstone as a foundation may be determined by the need to use a pneumatic drill to level the material ready for use as a foundation. As with igneous rock it is usual to cast a concrete base on the roughly levelled rock for the convenience of building.

Metamorphic rocks

Metamorphic rocks such as slates and schists are those changed from igneous, sedimentary or from soils into metamorphic by pressure or heat or both. These rocks vary from dense slates in which the layers of the material are barely visible to schists in which the layers of various minerals are clearly visible and may readily split into thin plates. Because of the mode of the formation of these rocks the layers or planes rarely lie horizontal in the ground and so generally provide an unsatisfactory or poor foundation.

SOILS

Soil is the general term for the upper layer of the earth's surface which consists of various combinations of particles of disintegrated rock such as gravel, sand or clay with some organic remains of decayed vegetation generally close to the surface.

Top soil

The surface layer of most of the low lying land in this country, which is most suited to building, consists of a mixture of loosely compacted particles of sand, clay and an accumulation of decaying vegetation. This layer of top soil, which is about 100 to 300 mm deep, is sometimes referred to as vegetable top soil. It is loosely compacted, supports growing plant life and is unsatisfactory as a foundation. It should be stripped from the site of buildings because of its poor bearing strengths and its ability to retain moisture and support vegetation which might adversely affect the health of occupants of buildings.

Subsoil

Subsoil is the general term for soil below the top soil.

It is unusual for a subsoil to consist of gravel, sand or clay by itself. The majority of subsoils are mixes of various soils. Gravel, sand and clay may be combined in a variety of proportions. To make a broad assumption of the behaviour of a particular soil under the load on foundations it is convenient to group soils such as gravel, sand and clay by reference to the size and nature of the particles.

The three broad groups are coarse grained non-cohesive, fine grained cohesive and organic. The nature and behaviour under the load on foundations of the soils in each group are similar.

Coarse grained non-cohesive soils

Soils which are composed mainly of, or combinations of, sand and gravel consist of largely siliceous, unaltered products of rock weathering. They have no plasticity and tend to lack cohesion, especially when dry. Under pressure of the loads on foundations the soils in this group compress and consolidate rapidly by some rearrangement of the coarse particles and the expulsion of water.

A foundation on coarse grained non-cohesive soils settles rapidly by consolidation of the soil, as the building is erected, so that there is no further settlement once the building is completed.

Gravel

Gravel consists of particles of a natural coarse grained deposit of rock fragments and finer sand. Many of the particles are larger than 2 mm.

Sand

Sand is a natural sediment of granular, mainly siliceous, products of rock weathering. Particles are smaller than 2 mm, are visible to the naked eye and the smallest size is 0.06 mm. Sand is gritty, has no real plasticity and can be easily powdered by hand when dry.

Dense, compact gravel and sand requires a pick to excavate for foundation trenches. A test of the suitability of these soils as a foundation is that it is difficult to drive a 5 mm wooden peg more than some 150 mm into compact gravel or sand.

As a foundation for small buildings, such as a house, it is sufficient to spread and level a continuous strip of concrete in the excavated trenches as a level base for load bearing walls.

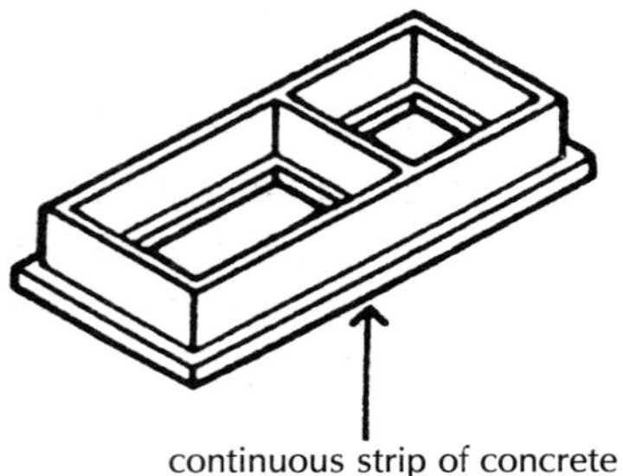

Fig. 2 Strip foundation.

Figure 2 is a diagram illustrating a strip foundation. The continuous strip of concrete is spread in the trenches excavated down to an undisturbed level of compact soil. The strip of concrete may well need to be no wider than the thickness of the wall. In practice the concrete strip will generally be wider than the thickness of the wall for the convenience of covering the whole width of the trench and to provide a wide enough level base for bricklaying below ground. A continuous strip foundation of concrete is the most economic form of foundation for small buildings on compact soils.

Fine grained cohesive soils

Fine grained cohesive soils, such as clays, are a natural deposit of the finest siliceous and aluminous products of rock weathering. Clay is smooth and greasy to the touch, shows high plasticity, dries slowly and shrinks appreciably on drying. Under the pressure of the load on foundations clay soils are very gradually compressed by the expulsion of water through the very many fine capillary paths, so that buildings settle gradually during building work and this settlement may continue for some years after the building is completed.

The initial and subsequent small settlement by compression during and after building on clay subsoils will generally be uniform under most small buildings, such as houses, to the extent that no damage is caused to the structure and its connected services.

Volume change

Firm, compact shrinkable clays suffer appreciable vertical and horizontal shrinkage on drying and expansion on wetting due to seasonal changes. Seasonal volume changes under grass extend to about 1 m below the surface in Great Britain and up to depths of 4 m or more below large trees.

The extent of volume changes, particularly in firm clay soils, depends on seasonal variations and the proximity of trees and shrubs. The greater the seasonable variation, the greater the volume change. The more vigorous the growth of shrubs and trees in firm clay soils, the greater the depth below surface the volume change will occur.

As a rough guide it is recommended that buildings on shallow foundations should not be closer to single trees than the height of the tree at maturity, and one-and-a-half times the height at maturity of groups of trees, to reduce the risk of damage to buildings by seasonal volume changes in clay subsoils.

When shrubs and trees are removed to clear a site for building on firm clay subsoils there will, for some years after the clearance, be ground recovery as the clay gradually recovers moisture previously withdrawn by the shrubs and trees. This gradual recovery of water by the clay and consequent expansion may take several years. The depth at which the recovery and expansion is appreciable will be roughly proportional to the height of the trees and shrubs removed, and the design and depth of foundations of buildings must allow for this gradual expansion to limit damage by differential settlement. Similarly, if vigorous shrub or tree growth is stopped by removal, or started by planting, near to a building on firm clay subsoil with foundations at a shallow depth, it is most likely that gradual expansion or contraction of the soil will cause damage to the building by differential movement.

At the recommended depth of at least 0.9 m it is not generally economic to use the traditional strip foundation and hence the narrow strip or trench fill foundation (Fig. 9) has been used. A narrow trench 400 mm wide is excavated by machine and filled with concrete to just below the surface. If the concrete is placed immediately after the excavation there is no need to support the sides of the trench in stiff clays, the sides of the trench will not be washed away by rain and the exposed clay will not suffer volume change.

The foundations of buildings sited adjacent to past, present or future deep-rooted vegetation can be affected at a considerable depth below the surface by the gain or removal of ground moisture and consequent expansion or shrinkage. Appreciable expansion, following the removal of deep-rooted vegetation, may continue for some years as the subsoil gains moisture. Significant seasonal volume change, due to deep-rooted vegetation, will be pronounced during periods of drought and heavy continuous rainfall.

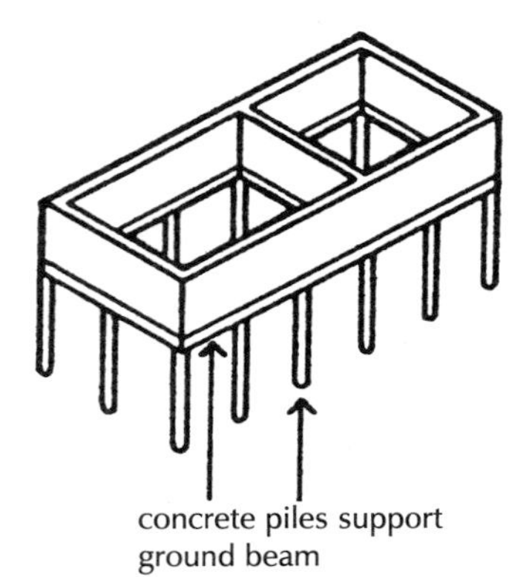

Fig. 3 Pile foundation.

The vigorous growth of newly planted deep-rooted vegetation adjacent to buildings may cause continuous shrinkage in clay soils for some years. The most economical and effective foundation for low rise buildings on shrinkable clays close to deep-rooted vegetation is a system of short-bored piles and ground beams (Fig. 3). The piles should be taken down to a depth below which vegetation roots will not cause significant volume changes in the subsoil. Single deep-rooted vegetation such as shrubs and trees as close as their mature height to buildings, and groups of shrubs and trees one-and-a-half times their mature height to buildings, can affect foundations on shrinkable clay subsoils.

Many beds of clay consist of combinations of clay with sand or silt in various proportions. The mix of sand or silt to clay will affect the behaviour of these soils as a foundation. In general where the proportion of sand or silt to clay is appreciable the less dense the soil will be. Because of variations in the proportion of clay to sand or silt and the general loose or soft nature of the soil it is practice to assume that their bearing capacity is less than that of clay.

Frost heave

Where the water table is high, that is near the surface, soils, such as silts, chalk, fine gritty sands and some lean clays, near the surface may expand when frozen. This expansion, or frost heave, is due to crystals of ice forming and expanding in the soil and so causing frost heave. In this country, ground water near the surface rarely freezes at depths of more than 0.5 m, but in exposed positions on open ground during frost it may freeze up to a depth of 1 m. Even in exposed positions during severe frost it is most unlikely that ground water under and adjacent to the foundations of heated buildings will freeze because of the heat stored in the ground under and around the building. There is, therefore, no need to consider the possibility of ground movement due to frost heave under and around heated buildings.

For unheated buildings and heated buildings with insulated ground floors, a foundation depth of 450 mm is generally sufficient against the possibility of damage by ground movement due to frost heave.

Made up ground

Areas of low lying ground near the coast and around rivers close to towns and cities have been raised by tipping waste, refuse and soil from excavations. Over the years the fill will have settled and consolidated to some extent. Areas of made up ground are often used for buildings as the towns and cities expand. Because of the varied nature of the materials tipped to fill and raise ground levels and the uncertainty of the bearing capacity of the fill, conventional foundations may well be unsatisfactory as a foundation.

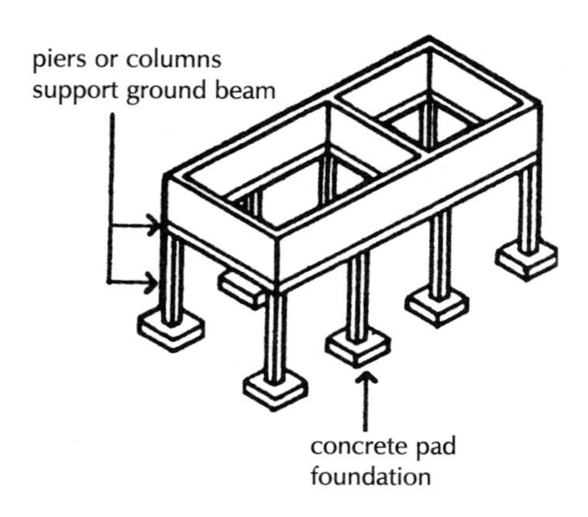

Fig. 4 Pad foundation.

An example of made up ground is the area of Westminster now known as Pimlico where the soil excavated during the construction of the London docks was transported by barge to what was low lying land that was usually flooded when high tides and heavy rainfall caused the Thames river to overflow. The raised land was subsequently heavily built on.

A uniformly stable, natural, sound foundation may well be some 3 or more metres below the surface of made up ground. To excavate to that level below the surface for conventional strip foundations would be grossly uneconomic. A solution is the use of piers on isolated pad foundations supporting reinforced concrete ground beams on which walls are raised, as illustrated in Fig. 4.

Unstable ground

There are some extensive areas of ground in this country where mining and excavations for coal and excavations for taking out chalk for use as a fertiliser and making lime and others for extracting sand and gravel may have made the ground unstable. The surface of the ground under deep and shallow excavations below the surface may well be subject to periodic, unpredictable subsidence.

Where it is known that ground may be unstable and there is no ready means of predicting the possibility of mass movement of the subsoil and it is expedient to build, a solution is to use some form of reinforced concrete raft under the whole of the buildings, as illustrated in Fig. 5.

The concrete raft, which is cast on or just below the surface, is designed to spread the load of the building over the whole of the underside of the raft so that in a sense the raft floats on the surface.

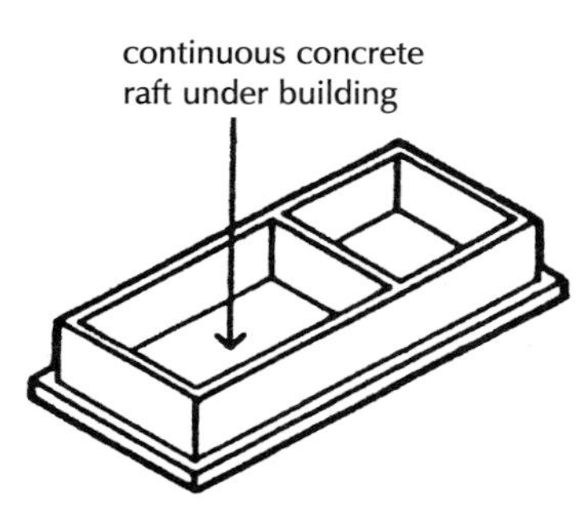

Fig. 5 Raft foundation.

SITE INVESTIGATION

To select a foundation from tables, or to design a foundation, it is necessary to calculate the loads on the foundation and determine the nature of the subsoil, its bearing capacity, its likely behaviour under seasonal and ground water level changes and the possibility of ground movement. Where the nature of the subsoil is known from geological surveys, adjacent building work or trial pits or borings and the loads on foundations are small, as for single domestic buildings, it is generally sufficient to excavate for foundations and confirm, from the exposed subsoil in the trenches, that the soil is as anticipated.

Under strip and pad foundations there is a significant pressure on the subsoil below the foundations to a depth and breadth of about one-and-a-half-times the width of the foundation. If there were, in this area below the foundation, a soil with a bearing capacity less than that below the foundation, then appreciable settlement of the foundation might occur and damage the building. It is important, therefore, to know or ascertain the nature of the subsoil both at the level of the foundation and for some depth below.

Where the nature of the subsoil is uncertain or there is a possibility of ground movement or a need to confirm information on subsoils, it is wise to explore the subsoil over the whole of the site of the building.

As a first step it is usual to collect information on soil and subsoil conditions from the County and Local Authority, whose local knowledge from maps, geological surveys, aerial photography and works for buildings and services adjacent to the site may in itself give an adequate guide to subsoil conditions. In addition geological maps from the British Geological Survey, information from local geological societies, Ordnance Survey maps, mining and river and coastal information may be useful.

Site visit

A visit to the site and its surroundings should always be made to record everything relevant from a careful examination of the nature of the subsoil, vegetation, evidence of marshy ground, signs of ground water and flooding, irregularities in topography, ground erosion and ditches and flat ground near streams and rivers where there may be soft alluvial soil. A record should be made of the foundations of old buildings on the site and cracks and other signs of movement in adjacent buildings as evidence of ground movement.

Trial pits

To make an examination of the subsoil on a building site, trial pits or boreholes are excavated. Trial pits are usually excavated by machine or hand to depth of 2 to 4 m and at least the anticipated depth of the foundations. The nature of the subsoil is determined by examination of the sides of the excavations. Boreholes are drilled by hand auger or by machine to withdraw samples of soil for examination. Details of the subsoil should include soil type, consistency or strength, soil structure, moisture conditions and the presence of roots at all depths. From the nature of the subsoil the bearing capacity, seasonal volume changes and other possible ground movements are assumed. To determine the nature of the subsoil below the foundation level it is either necessary to excavate trial pits some depth below the foundation or to bore in the base of the trial hole to withdraw samples. Whichever system is adopted will depend on economy and the nature of the subsoil. Trial pits or boreholes should be sufficient in number to determine the nature of the subsoil over and around the site of the building and should be at most say 30 m apart.

Ground movements that may cause settlement are:

(1) compression of the soil by the load of the building
(2) seasonal volume changes in the soil

(3) mass movement in unstable areas such as made up ground and mining areas where there may be considerable settlement
(4) ground made unstable by adjacent excavations or by dewatering, for example, due to an adjacent road cutting.

It is to anticipate and accommodate these movements that site investigation and exploration is carried out. For further details of site investigation and exploration see Volume 4.

FUNCTIONAL REQUIREMENT

The functional requirement of a foundation is: strength and stability.

Strength and stability

The requirements from the Building Regulations are, as regards 'Loading', that 'The building shall be so constructed that the combined, dead, imposed and wind loads are sustained and transmitted to the ground safely and without causing such deflection or deformation of any part of the building, or such movement of the ground, as will impair the stability of any part of another building' and as regards 'ground movement' that 'The building shall be so constructed that movements of the subsoil caused by swelling, shrinkage or freezing will not impair the stability of any part of the building'.

A foundation should be designed to transmit the loads of the building to the ground so that there is, at most, only a limited settlement of the building into the ground. A building whose foundation is on sound rock will suffer no measurable settlement whereas a building on soil will suffer settlement into the ground by the compression of the soil under the foundation loads.

Foundations should be designed so that settlement into the ground is limited and uniform under the whole of the building. Some settlement of a building on a soil foundation is inevitable as the increasing loads on the foundation, as the building is erected, compress the soil. This settlement should be limited to avoid damage to service pipes and drains connected to the building. Bearing capacities for various rocks and soils are assumed and these capacities should not be exceeded in the design of the foundation to limit settlement.

In theory, if the foundation soil were uniform and foundation bearing pressure were limited, the building would settle into the ground uniformly as the building was erected, and to a limited extent, and there would be no possibility of damage to the building or its connected services or drains. In practice there are various possible ground movements under the foundation of a building that may cause one part of the foundation to settle at a different rate and to a different extent than another part of the foundation.

This different or differential settlement must be limited to avoid damage to the superstructure of the building. Some structural forms can accommodate differential or relative foundation movement without damage more than others. A brick wall can accommodate limited differential movement of the foundation or the structure by slight movement of the small brick units and mortar joints, without affecting the function of the wall, whereas a rigid framed structure with rigid panels cannot to the same extent. Foundations are designed to limit differential settlement, the degree to which this limitation has to be controlled or accommodated in the structure depends on the nature of the structure supported by the foundation.

FOUNDATION CONSTRUCTION

Strip foundations

Strip foundations consist of a continuous strip, usually of concrete, formed centrally under load bearing walls. This continuous strip serves as a level base on which the wall is built and is of such a width as is necessary to spread the load on the foundations to an area of subsoil capable of supporting the load without undue compaction. Concrete is the material principally used today for foundations as it can readily be placed, spread and levelled in foundation trenches, to provide a base for walls, and it develops adequate compressive strength as it hardens to support the load on foundations. Before Portland cement was manufactured, strip foundations of brick were common, the brick foundation being built directly off firm subsoil or built on a bed of natural stones.

The width of a concrete strip foundation depends on the bearing capacity of the subsoil and the load on the foundations. The greater the bearing capacity of the subsoil the less the width of the foundation for the same load.

A table in Approved Document A to the Building Regulations sets out the recommended minimum width of concrete strip foundations related to six specified categories of subsoil and calculated total loads on foundations as a form of ready reckoner. The widths vary from 250 mm for a load of not more than 20 kN/linear metre of wall on compact gravel or sand through 450 mm for loads of 40 kN/linear metre on firm clay, to 850 mm for loads not exceeding 30 kN/linear metre on soft silt, clay or sandy clay.

The dimensions given are indicative of what might be acceptable in the conditions specified rather than absolutes to be accepted regardless of the conditions prevailing on individual sites.

The strip foundation for a cavity external wall and a solid internal, load bearing wall illustrated in Fig. 6 would be similar to the width recommended in the Advisory Document for a firm clay subsoil when the load on the foundations is no more than 50 kN/linear metre. In practice the linear load on the foundation of a house would be appreciably less than 50 kN/linear metre and the foundation may well be made wider than the minimum requirement for the convenience of

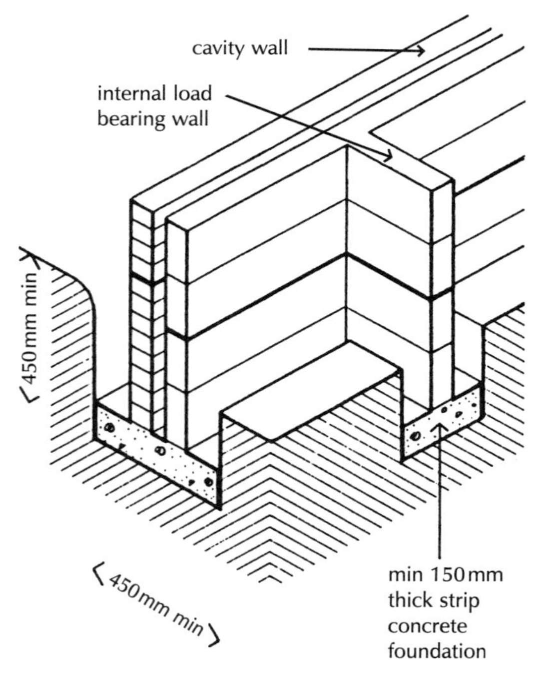

Fig. 6 Strip foundation.

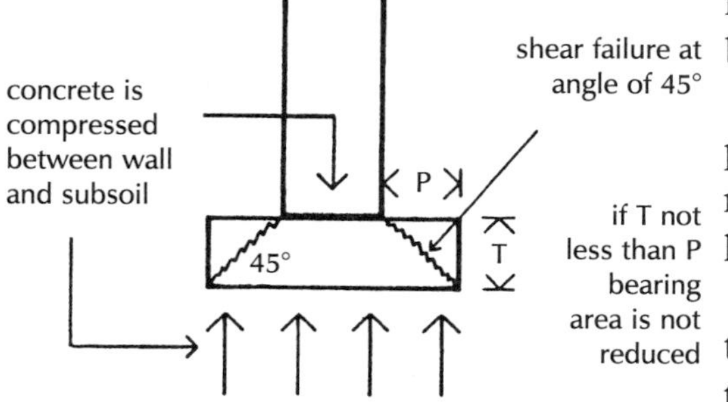

Fig. 7 Shear failure.

filling a wider trench with concrete for the convenience of laying brick below ground.

The least thickness of a concrete strip foundation is determined in part by the size of the aggregate used in the concrete, the need for a minimum thickness of concrete so that it does not dry too quickly and lose strength and to avoid failure of the concrete by shear.

If the thickness of a concrete strip foundation were appreciably less than its projection each side of a wall the concrete might fail through the development of shear cracks by the weight of the wall causing a 45° crack as illustrated in Fig. 7. If this occurred the bearing surface of the foundation on the ground would be reduced to less than that necessary for stability.

Shear is caused by the two opposing forces of the wall and the ground acting on and tearing or shearing the concrete as scissors or shears cut or shear materials apart.

Wide strip foundation

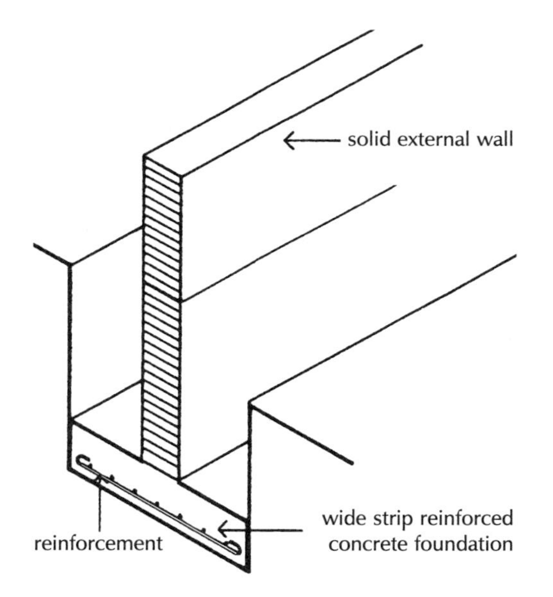

Fig. 8 Wide strip foundation.

Strip foundations on subsoils with poor bearing capacity, such as soft sandy clays, may need to be considerably wider than the wall they support to spread the load to a sufficient area of subsoil for stability. The concrete strip could be as thick as the projection of the strip each side of the wall which would result in concrete of considerable uneconomic thickness to avoid the danger of failure by shear.

The alternative is to form a strip of reinforced concrete, illustrated in Fig. 8, which could be no more than 150 mm thick.

The reason for the use of reinforcement of steel in concrete is that concrete is strong in compression but weak in tension. The effect of the downward pressure of the wall above and the supporting pressure of the soil below is to make the concrete strip bend upwards at the edges, creating tensile stress in the bottom and compressive stress under the wall. These opposing pressures will tend to cause the shear cracking illustrated in Fig. 7. It is to reinforce and strengthen concrete in tension that steel reinforcing bars are cast in the lower edge because steel is strong in tension. There has to be a sufficient cover of concrete below the steel reinforcing rods to protect them from rusting and losing strength.

Narrow strip (trench fill) foundation

Stiff clay subsoils have good bearing strength and are subject to seasonal volume change. Because of seasonal changes and the withdrawal of moisture by deep rooted vegetation it is practice to adopt a foundation depth of at least 0.9 m to provide a stable foundation.

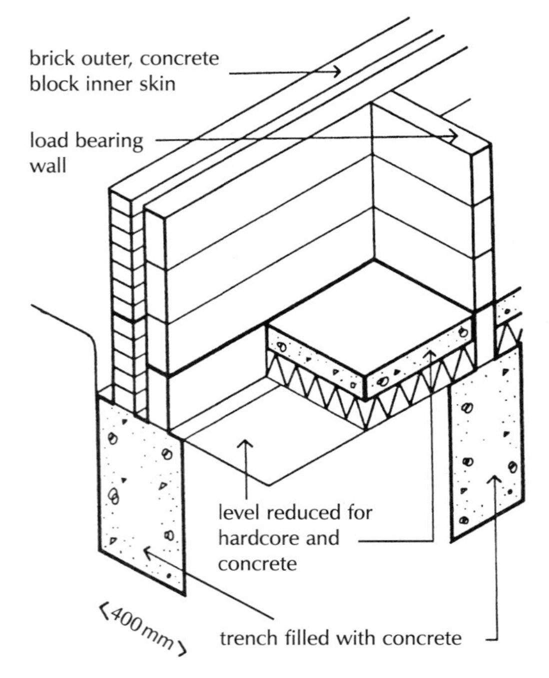

Fig. 9 Narrow trench fill foundation.

Because of the good bearing capacity of the clay the foundation may need to be little wider than the thickness of the wall to be supported. It would be laborious and uneconomic to excavate trenches wide enough for laying bricks down to the required level of a strip foundation.

Practice today is to use a mechanical excavator to take out the clay down to the required depth of at least 0.9 m below surface and immediately fill the trenches with concrete up to a level just below finished ground level, as illustrated in Fig. 9. The width of the trench is determined by the width of the excavator bucket available, which should not be less than the minimum required width of foundation.

The trench is filled with concrete as soon as possible so that the clay bed exposed does not dry out and shrink and against the possibility of the trench sides falling in, particularly in wet weather.

With the use of mechanical excavating equipment to dig the trenches and to move the excavated soil and spread it over other parts of the site or cart it from site, and the use of ready mixed concrete to fill the trenches this is the most expedient, economic and satisfactory method of making foundations on stiff, shrinkage subsoils for small buildings.

Short bored pile foundation

Where the subsoil is of firm, shrinkable clay which is subject to volume change due to deep rooted vegetation for some depth below surface and where the subsoil is of soft or uncertain bearing capacity for some few metres below surface, it may be economic and satisfactory to use a system of short bored piles as a foundation.

Piles are concrete columns which are either precast and driven (hammered) into the ground or cast in holes that are augered (drilled) into the ground down to a level of a firm, stable stratum of subsoil.

The piles that are used as a foundation down to a level of some 4 m below the surface for small buildings are termed short bore, which refers to the comparatively short length of the piles as compared to the much longer piles used for larger buildings. Short bored piles are generally from 2 to 4 m long and from 250 to 350 mm diameter.

Holes are augered in the ground by hand or machine. An auger is a form of drill comprising a rotating shaft with cutting blades that cuts into the ground and is then withdrawn, with the excavated soil on the blades that are cleared of soil. The auger is again lowered into the

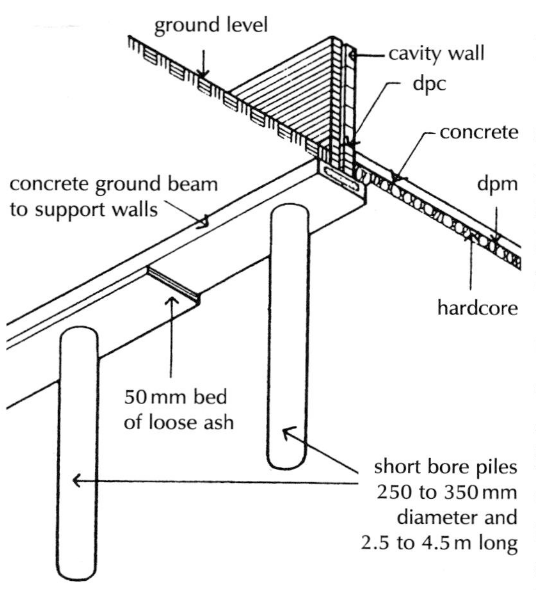

Fig. 10 Short bored pile foundation.

ground and withdrawn, cleared of soil and the process repeated until the required depth is reached.

The advantage of this system of augered holes is that samples of the subsoil are withdrawn, from which the bearing capacity of the subsoil may be assessed. The piles may be formed of concrete by itself or, more usually, a light, steel cage of reinforcement is lowered into the hole and concrete poured or pumped into the hole and compacted to form a pile foundation.

The piles are cast below angles and intersection of load bearing walls and at intervals between to reduce the span and depth of the reinforced ground beam they are to support. A reinforced concrete ground beam is then cast over the piles as illustrated in Fig. 10. The ground beam is cast in a shallow trench on a 50 mm bed of ash with the reinforcement in the piles linked to that in the beams for continuity. The spacing of the piles depends on the loads to be supported and on economic sections of ground beam.

Pad foundations

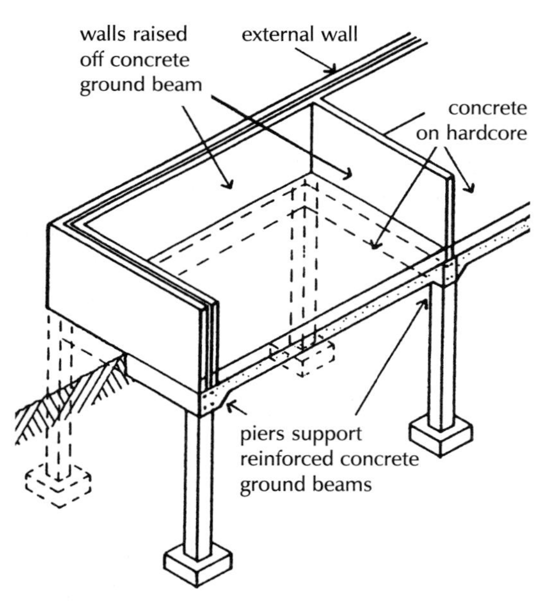

Fig. 11 Pad foundations.

On made up ground and ground with poor bearing capacity where a firm, natural bed of, for example, gravel or sand is some few metres below the surface, it may be economic to excavate for isolated piers of brick or concrete to support the load of buildings of some four storeys in height. The piers will be built at the angles, intersection of walls and under the more heavily loaded wall such as that between windows up the height of the building.

Pits are excavated down to the necessary level, the sides of the excavation temporarily supported and isolated pads of concrete are cast in the bottom of the pits. Brick piers or reinforced concrete piers are built or cast on the pad foundations up to the underside of the reinforced concrete beams that support walls as illustrated in Fig. 11. The ground beams or foundation beams may be just below or at ground level, the walls being raised off the beams.

The advantage of this system of foundation is that pockets of tipped stone or brick and concrete rubble that would obstruct bored piling may be removed as the pits are excavated and that the nature of the subsoil may be examined as the pits are dug to select a level of sound subsoil. This advantage may well be justification for this labour intensive and costly form of construction.

Raft foundations

A raft foundation consists of a raft of reinforced concrete under the whole of a building. This type of foundation is described as a raft in the sense that the concrete raft is cast on the surface of the ground which supports it, as water does a raft, and the foundation is not fixed by foundations carried down into the subsoil.

Raft foundations may be used for buildings on compressible ground such as very soft clay, alluvial deposits and compressible fill material where strip, pad or pile foundations would not provide a stable foundation without excessive excavation. The reinforced concrete raft is designed to transmit the whole load of the building from the raft to the ground where the small spread loads will cause little if any appreciable settlement.

The two types of raft foundation commonly used are the flat raft and the wide toe raft.

The flat slab raft is of uniform thickness under the whole of the building and reinforced to spread the loads from the walls uniformly over the under surface to the ground. This type of raft may be used under small buildings such as bungalows and two storey houses where the comparatively small loads on foundations can be spread safely and economically under the rafts.

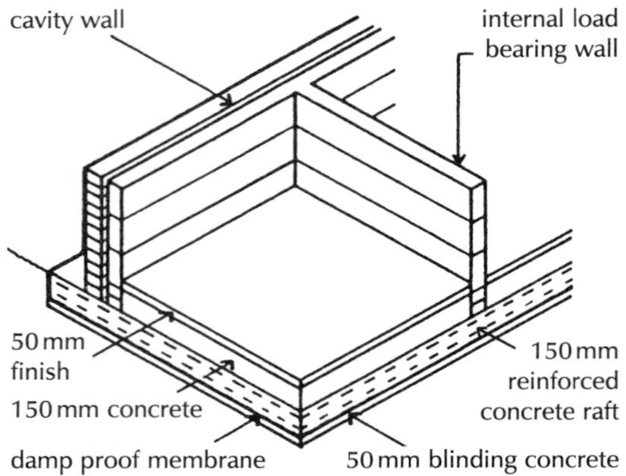

Fig. 12 Flat slab raft.

The concrete raft is reinforced top and bottom against both upward and downward bending. Vegetable top soil is removed and a blinding layer of concrete 50 mm thick is spread and levelled to provide a base on which to cast the concrete raft. A waterproof membrane is laid, on the dry concrete blinding, against moisture rising into the raft. The top and bottom reinforcement is supported and spaced preparatory to placing the concrete which is spread, consolidated and finished level.

When the reinforced concrete raft has dried and developed sufficient strength the walls are raised as illustrated in Fig. 12. The concrete raft is usually at least 150 mm thick.

The concrete raft may be at ground level or finished just below the surface for appearance sake. Where floor finishes are to be laid on the raft a 50 mm thick layer of concrete is spread over the raft, between the walls, to raise the level and provide a level, smooth finish for floor coverings. As an alternative a raised floor may be constructed on top of the raft to raise the floor above ground.

A flat slab recommended for building in areas subject to mining subsidence is similar to the flat slab, but cast on a bed of fine granular material 150 mm thick so that the raft is not keyed to the ground and is therefore unaffected by horizontal ground strains.

Where the ground has poor compressibility and the loads on the foundations would require a thick, uneconomic flat slab, it is usual to cast the raft as a wide toe raft foundation. The raft is cast with a reinforced concrete, stiffening edge beam from which a reinforced concrete toe extends as a base for the external leaf of a cavity wall as shown in Fig. 13. The slab is thickened under internal load bearing walls.

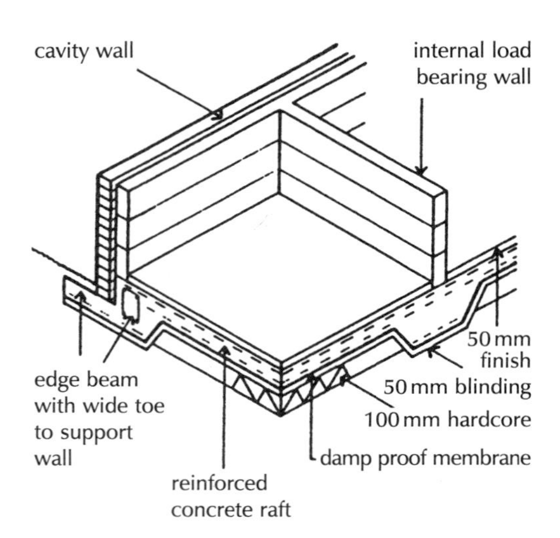

Fig. 13 Edge beam raft.

Vegetable top soil is removed and the exposed surface is cut away to roughly form the profile of the underside of the slab. As necessary

100 mm of hardcore or concrete is spread under the area of the raft and a 50 mm layer of blinding concrete is spread, shaped and levelled as a base for the raft and toes. A waterproof membrane is laid on the dried concrete blinding and the steel reinforcement fixed in position and supported preparatory to placing, compacting and levelling the concrete raft.

The external cavity and internal solid walls are raised off the concrete raft once it has developed sufficient strength. The extended toe of the edge beam is shaped so that the external brick outer leaf of the cavity wall is finished below ground for appearance sake. A floor finish is laid on 50 mm concrete finish or a raised floor constructed.

Raft foundation on sloping site

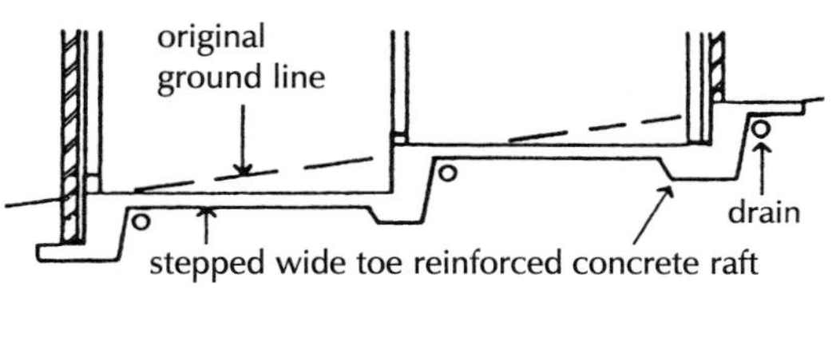

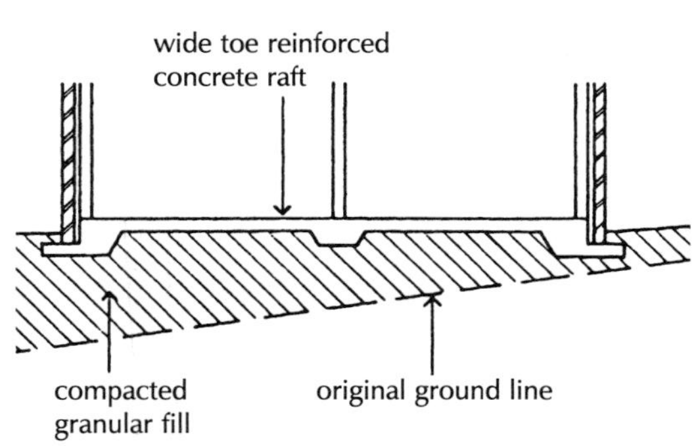

Fig. 14 Raft on sloping site.

On sites where the slope of the ground is such that there is an appreciable fall in the surface across the width or length of a building, and a raft foundation is to be used, because of the poor bearing capacity of subsoil, it is necessary either to cut into the surface or provide additional fill under the building or a combination of both to provide a level base for the raft.

It is advisable to minimise the extent of disturbance of the soft or uncertain subsoil. Where the slope is shallow and the design and use of the building allows, a stepped raft may be used down the slope, as illustrated in Fig. 14.

A stepped, wide toe, reinforced concrete raft is formed with the step or steps made at the point of a load bearing internal wall or at a division wall between compartments or occupations. The drains under the raft are to relieve and discharge surface water running down the slope that might otherwise be trapped against steps and promote dampness in the building.

The level raft illustrated in Fig. 14 is cast on imported granular fill that is spread, consolidated and levelled as a base for the raft. The disadvantage of this is the cost of the additional granular fill and the advantage a level bed of uniform consistency under the raft.

As an alternative the system of cut and fill may be used to reduce the volume of imported fill.

Raft foundations are usually formed on ground of soft subsoil or made up ground where the bearing capacity is low or uncertain, to minimise settlement. There is some possibility of there being some slight movement of the ground under the building which would fracture drains and other service pipes entering the building through the raft. Service pipes rising through the raft should run through collars, cast in the concrete, which will allow some movement of the raft without fracturing service pipes.

Foundations on sloping sites

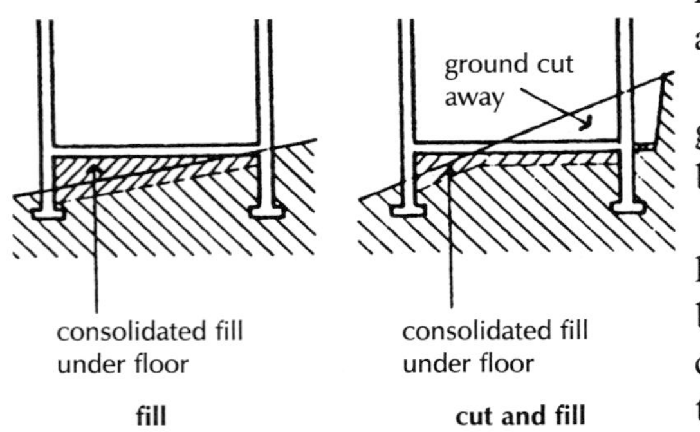

Fig. 15 Fill and cut and fill.

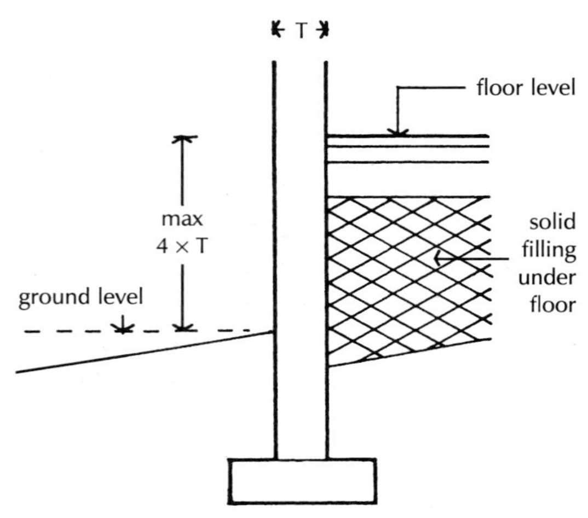

Fig. 16 Solid filling.

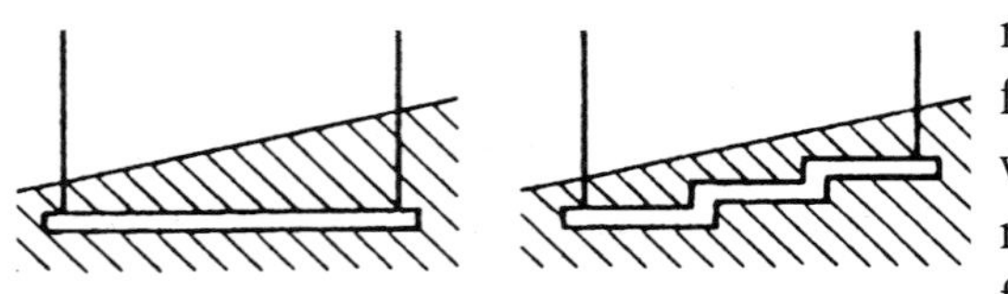

Fig. 17 Foundation on sloping site.

The natural surface of ground is rarely level to the extent that there may be an appreciable slope either across or along or both across and along the site of most buildings.

On sloping sites an initial decision to be made is whether the ground floor is to be above ground at the highest point or partly sunk below ground as illustrated in Fig. 15.

Where the ground floor is to be at or just above ground level at the highest point, it is necessary to import some dry fill material such as broken brick or concrete hardcore to raise the level of the oversite concrete and floor. This fill will be placed, spread and consolidated up to the external wall once it has been built.

The consolidated fill will impose some horizontal pressure on the wall. To make sure that the stability of the wall is adequate to withstand this lateral pressure it is recommended practice that the thickness of the wall should be at least a quarter of the height of the fill bearing on it as illustrated in Fig. 16. The thickness of a cavity wall is taken as the combined thickness of the two leaves unless the cavity is filed with concrete when the overall thickness is taken.

To reduce the amount of fill necessary under solid floors on sloping sites a system of cut and fill may be used as illustrated in Fig. 15. The disadvantage of this arrangement is that the ground floor is below ground level at the highest point and it is necessary to form an excavated dry area to collect and drain surface water that would otherwise run up to the wall and cause problems of dampness.

To economise in excavation and foundation walling on sloping sites where the subsoil, such as gravel and sand, is compact it is practice to use a stepped foundation as illustrated in Fig. 17, which contrasts diagrammatically the reduction in excavation and foundation walling of a level and a stepped foundation.

Figure 18 is an illustration of the stepped foundation for a small building on a sloping site where the subsoil is reasonably compact near the surface and will not be affected by volume changes. The foundation is stepped up the slope to minimise excavation and walling below ground. The foundation is stepped so that each step is no higher than the thickness of the concrete foundation and the foundation at the higher level overlaps the lower foundation by at least 300 mm.

The load bearing walls are raised and the foundation trenches around the walls backfilled with selected soil from the excavation. The concrete oversite and solid ground floor may be cast on granular fill no more than 600 mm deep or cast or placed as a suspended reinforced concrete slab. The drains shown at the back of the trench fill are laid to collect and drain water to the sides of the building.

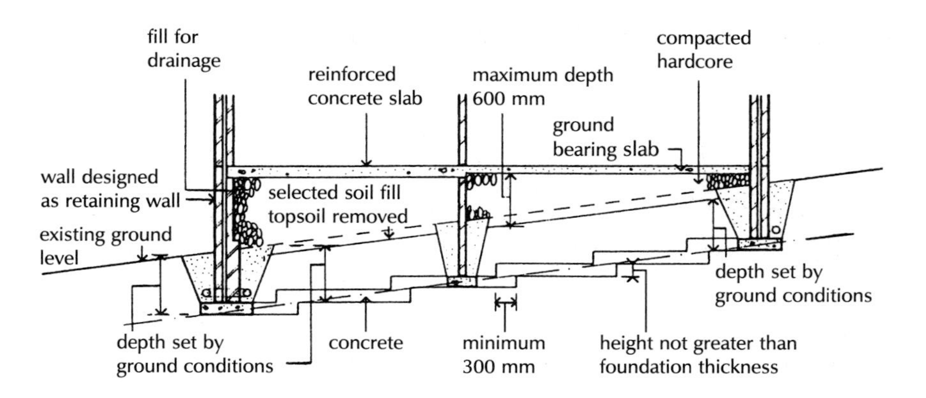

Fig. 18 Stepped foundation.

SITE PREPARATION

Turf and vegetable top soil should be removed from the ground to be covered by a building, to a depth sufficient to prevent later growth. Tree and bush roots, that might encourage later growth, are grubbed up and any pockets of soft compressible material, that might affect the stability of the building, are removed. The reasons for removing this vegetable soil are firstly to prevent plants, shrubs or trees from attempting to grow under the concrete. In growing, even the smallest of plant life exerts considerable pressure, which would quite quickly rupture the concrete oversite. The second reason for removing the vegetable top soil is that it is generally soft and compressible and readily retains moisture which would cause concrete over it to be damp at all times. The depth of vegetable top soil varies and on some sites it may be necessary to remove 300 mm or more vegetable top soil.

In practice most of the vegetable top soil over a building site is effectively moved by excavations for foundations, levelling and drain and other service pipes to the extent that it may be necessary to remove top soil that remains within or around the confines of a building.

Contaminants

In Approved Document C to the Building Regulations is a list of possible contaminants in or on ground to be covered by a building, that may be a danger to health or safety. Building sites that may be likely to contain contaminants can be identified from planning records or local knowledge of previous uses. Sites previously used as asbestos, chemical or gas works, metal works, munitions factories, nuclear installations, oil stores, railway land, sewage works and land fill are some examples given.

Site drainage

Surface water (stormwater) is the term used for natural water, that is rainwater that falls on the surface of the ground including open ground such as fields, paved areas and roofs. Rainwater that falls on paved areas and from roofs generally drains to surface water (stormwater) drains and thence to soakaways (see Volume 5), rivers, streams or the sea. Rainwater falling on natural open ground will in part lie on the surface of impermeable soils, evaporate to air, run off to streams and rivers and soak into the ground. On permeable soils much of the rainwater will soak into the ground as ground water.

Ground water is that water held in soils at and below the water table (which is the depth at which there is free water below the surface). The level of the water table will vary seasonally, being closest to the surface during rainy seasons and deeper during dry seasons when most evaporation to air occurs.

In Part C of the Building Regulations is a requirement for subsoil drainage, to avoid passage of ground moisture to the inside of a building or to avoid damage to the fabric of the building.

In Approved Document C to the Regulations are provisions for the need for subsoil drainage where the water table can rise to within 0.25 m of the lowest floor and where the water table is high in dry weather and the site of the building is surrounded by higher ground.

Paved areas are usually laid to falls to channels and gullies that drain to surface water drains (Volume 5).

Subsoil drains

Subsoil drains are used to improve the run off of surface water and the drainage of ground water to maintain the water table at some depth below the surface for the following reasons:

(1) to improve the stability of the ground
(2) to avoid surface flooding
(3) to alleviate or avoid dampness in basements
(4) to reduce humidity in the immediate vicinity of buildings.

Ground water, or land or field, drains are either open jointed or jointed, porous or perforated pipes of clayware, concrete, pitch fibre or plastic (see Volume 5). The pipes are laid in trenches to follow the fall of the ground, generally with branch drains discharging to a ditch, stream or drain.

On impervious subsoils, such as clay, it may be necessary to form a system of drains to improve the run off of surface water and drain subsoil to prevent flooding. Some of the drain systems used are natural, herring bone, grid, fan and moat or cut-off.

Natural system

This system, which is commonly used for field drains, uses the natural contours of the ground to improve run off of surface ground water to spine drains in natural valleys that fall towards ditches or streams. The drains are laid in irregular patterns to follow the natural contours as illustrated in Fig. 19A.

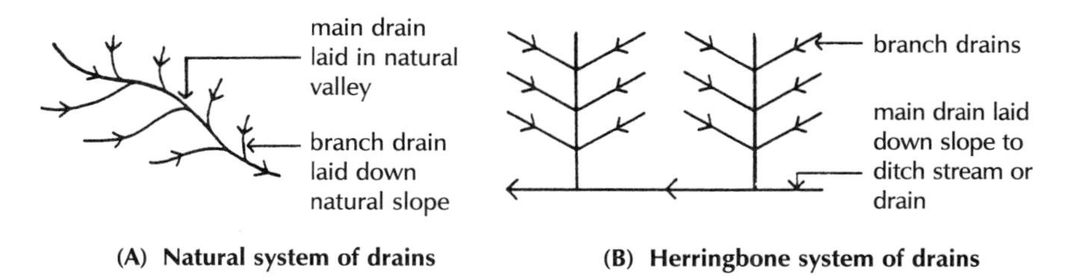

Fig. 19 (A) Natural system. (B) Herring bone system.

Herring bone system

In this system, illustrated in Fig. 19B, fairly regular runs of drains connect to spine drains that connect to a ditch or main drain. This system is suited to shallow, mainly one way slopes that fall naturally towards a ditch or main drain and can be laid to a reasonably regular pattern to provide a broad area of drainage.

Grid system

This is an alternative to the herring bone system for draining one way slopes where branch drains are fed by short branches that fall towards a ditch or main drain, as illustrated in Fig. 20A. This system may be preferred to the herring bone system, where the run off is moderate, because there are fewer drain connections that may become blocked.

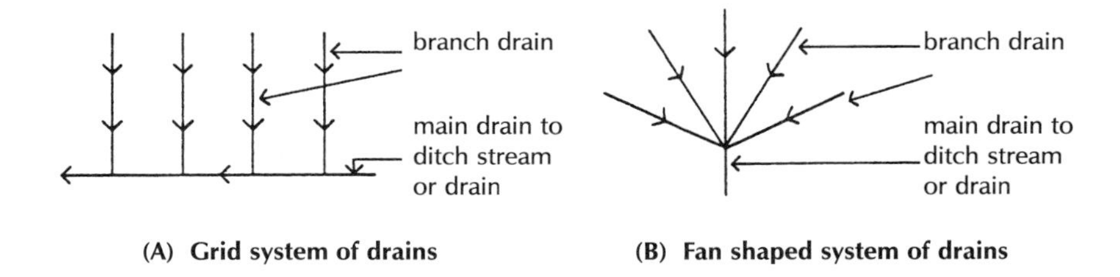

Fig. 20 (A) Grid system. (B) Fan systems.

Fan system

A fan shaped layout of short branches, illustrated in Fig. 20B, drains to spine drains that fan towards a soakaway, ditch or drain on narrow sites. A similar system is also used to drain the partially purified outflow from a septic tank, (see Volume 5), to an area of subsoil where further purification will be effected.

On sloping building sites on impervious soil where an existing system of land drains is already laid and where a new system is laid to prevent flooding a moat or cut off system is used around the new building to isolate it from general land drains, as illustrated in Fig. 21.

The moat or cut off system of drains is laid some distance from and around the new building to drain the ground between it and the new building and to carry water from the diverted land drains down the slope of the site. Plainly the moat drains should be clear of paved areas around the house.

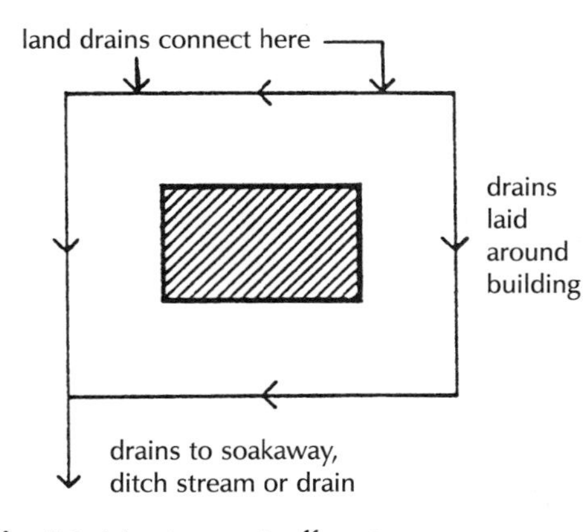

Fig. 21 Moat or cut off system.

Laying drains

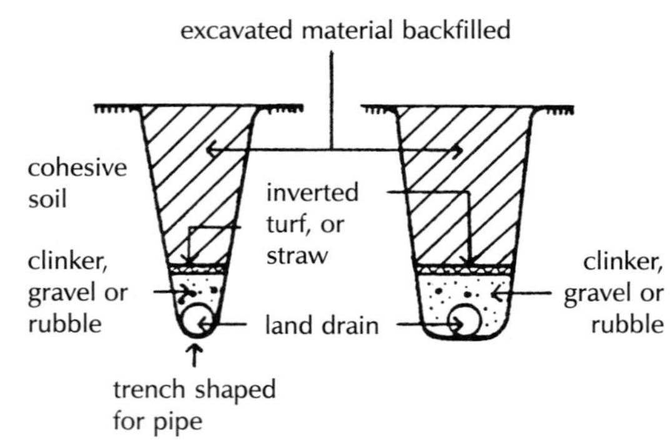

Fig. 22 Land drains.

Ground water (land) drains are laid in trenches at depths of 0.6 and 0.9 m in heavy soils and 0.9 to 1.2 m in light soils. The nominal bore of the pipes is usually 75 and 100 mm for main drains and 65 or 75 mm for branches.

The drain pipes are laid in the bed of the drain trench and surrounded with clinker, gravel or broken pervious rubble which is covered with inverted turf, brushwood or straw to separate the back fill from the pipes and their surround. Excavated material is backfilled into the drain trench up to the natural ground level.

The drain trench bottom may be shaped to take and contain the pipe or finished with a flat bed as illustrated in Fig. 22, depending on the nature of the subsoil and convenience in using a shaping tool.

Where drains are laid to collect mainly surface water the trenches are filled with clinker, gravel or broken rubble to drain water either to a drain or without a drain as illustrated in Fig. 23 in the form known as a French drain. Whichever is used will depend on the anticipated volume of water and the economy of dispensing with drainpipes.

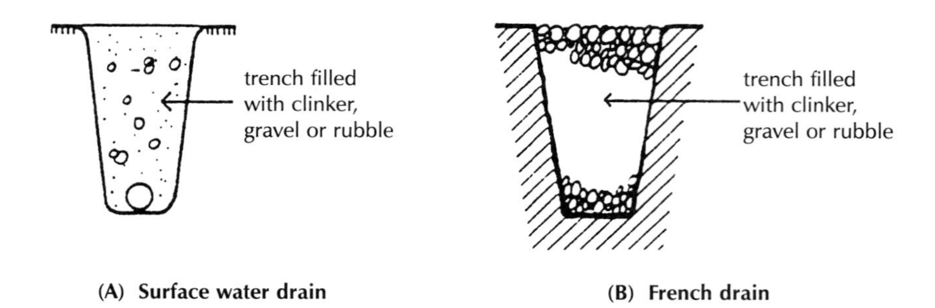

Fig. 23 (A) Surface water drain. (B) French drain.

RESISTANCE TO GROUND MOISTURE

Up to about the middle of the nineteenth century the ground floor of most buildings was formed on compacted soil or dry fill on which was laid a surface of stone flagstones, brick or tile or a timber boarded floor nailed to battens bedded in the compacted soil or fill. In lowland areas and on poorly drained soils most of these floors were damp and cold underfoot.

A raised timber ground floor was sometimes used to provide a comparatively dry floor surface of boards, nailed to timber joists, raised above the packed soil or dry fill. To minimise the possibility of the joists being affected by rising damp it was usual to ventilate the space below the raised floor. The inflow of cold outside air for ventilation tended to make the floor cold underfoot.

OVERSITE CONCRETE

When Portland cement was first continuously produced, towards the end of the nineteenth century, it became practical to cover the site of buildings with a layer of concrete as a solid level base for floors and as a barrier to rising damp. From the early part of the twentieth century

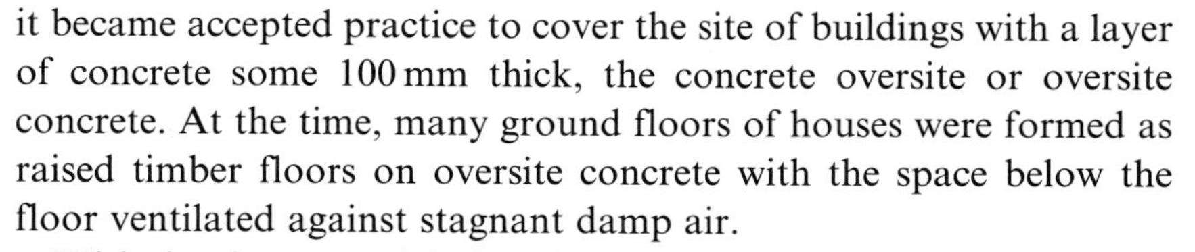

it became accepted practice to cover the site of buildings with a layer of concrete some 100 mm thick, the concrete oversite or oversite concrete. At the time, many ground floors of houses were formed as raised timber floors on oversite concrete with the space below the floor ventilated against stagnant damp air.

With the shortage of timber that followed the Second World War, the raised timber ground floor was abandoned and the majority of ground floors were formed as solid, ground supported floors with the floor finish laid on the concrete oversite. At the time it was accepted practice to form a continuous horizontal damp-proof course, some 150 mm above ground level, in all walls with foundations in the ground.

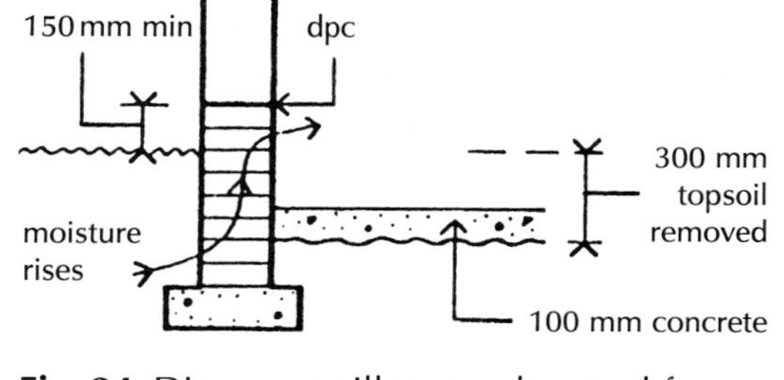

Fig. 24 Diagram to illustrate the need for hardcore.

With the removal of vegetable top soil the level of the soil inside the building would be from 100 to 300 mm below the level of the ground outside. If a layer of concrete were then laid oversite its finished level would be up to 200 mm below outside ground level and up to 350 mm below the horizontal dpc in walls. There would then be considerable likelihood of moisture rising through the foundation walls, to make the inside walls below the dpc damp, as illustrated in Fig. 24.

It would, of course, be possible to make the concrete oversite up to 450 mm thick so that its top surface was level with the dpc and so prevent damp rising into the building. But this would be unnecessarily expensive. Instead, a layer of what is known as hardcore is spread oversite, of sufficient thickness to raise the level of the top of the concrete oversite to that of the dpc in walls. The purpose of the hardcore is primarily to raise the level of the concrete oversite for solid, ground supported floors.

The layer of concrete oversite will serve as a reasonably effective barrier to damp rising from the ground by absorbing some moisture from below. The moisture retained in the concrete will tend to make solid floor finishes cold underfoot and may adversely affect timber floor finishes. During the second half of the twentieth century it became accepted practice to form a waterproof membrane under, in or over the oversite concrete as a barrier to rising damp, against the cold underfoot feel of solid floors and to protect floor finishes. Having accepted the use of a damp-proof membrane it was then logical to unite this barrier to damp, to the damp-proof course in walls, by forming them at the same level or by running a vertical dpc up from the lower membrane to unite with the dpc in walls.

Even with the damp-proof membrane there is some appreciable transfer of heat from heated buildings through the concrete and hardcore to the cold ground below. In Approved Document L to the Building Regulations is the inclusion of provision for insulation to ground floors for the conservation of fuel and power. The requirement can be met by a layer of insulating material under the site concrete, under a floor screed or under boarded or sheet floor finishes to provide a maximum U value of 0.45 W/m^2K for the floor.

The requirement to the Building Regulations for the resistance of the passage of moisture to the inside of the building through floors is met if the ground is covered with dense concrete laid on a hardcore bed and a damp-proof membrane. The concrete should be at least 100 mm thick and composed of 50 kg of cement to not more than 0.11 m^3 of fine aggregate and 0.16 m^3 of coarse aggregate of BS 5328 mix ST2. The hardcore bed should be of broken brick or similar inert material, free from materials including water soluble sulphates in quantities which could damage the concrete. A damp-proof membrane, above or below the concrete, should ideally be continuous with the dpc in the walls.

It is practice on building sites to first build external and internal load bearing walls from the concrete foundation up to the level of the dpc, above ground, in walls. The hardcore bed and the oversite concrete are then spread and levelled within the external walls.

If the hardcore is spread over the area of the ground floor and into excavations for foundations and soft pockets of ground that have been removed and the hardcore is thoroughly consolidated by ramming, there should be very little consolidation settlement of the concrete ground supported floor slab inside walls. Where a floor slab has suffered settlement cracking, it has been due to an inadequate hardcore bed, poor filling of excavation for trenches or ground movement due to moisture changes. It has been suggested that the floor slab be cast into walls for edge support. This dubious practice, which required edge formwork support of slabs at cavities, will have the effect of promoting cracking of the slab, that may be caused by any slight consolidation settlement. Where appreciable settlement is anticipated it is best to reinforce the slab and build it into walls as a suspended reinforced concrete slab.

Hardcore

Hardcore is the name given to the infill of materials such as broken bricks, stone or concrete, which are hard and do not readily absorb water or deteriorate. This hardcore is spread over the site within the external walls of the building to such thickness as required to raise the finished surface of the site concrete. The hardcore should be spread until it is roughly level and rammed until it forms a compact bed for the oversite concrete. This hardcore bed is usually from 100 to 300 mm thick.

The hardcore bed serves as a solid working base for building and as a bed for the concrete oversite. If the materials of the hardcore are hard and irregular in shape they will not be a ready path for moisture to rise by capillarity. Materials for hardcore should, therefore, be clean and free from old plaster or clay which in contact with broken brick or gravel would present a ready narrow capillary path for moisture to rise.

The materials used for hardcore should be chemically inert and not appreciably affected by water. Some materials used for hardcore, for example colliery spoil, contain soluble sulphate that in combination with water combine with cement and cause concrete to disintegrate. Other materials such as shale may expand and cause lifting and cracking of concrete. A method of testing materials for soluble sulphate is described in Building Research Station (BRS) Digest 174.

The materials used for hardcore are:

Brick or tile rubble

Clean, hard broken brick or tile is an excellent material for hardcore. Bricks should be free of plaster. On wet sites the bricks should not contain appreciable amounts of soluble sulphate.

Concrete rubble

Clean, broken, well-graded concrete is another excellent material for hardcore. The concrete should be free from plaster and other building materials.

Gravel and crushed hard rock

Clean, well-graded gravel or crushed hard rock are both excellent, but somewhat expensive materials for hardcore.

Chalk

Broken chalk is a good material for hardcore providing it is protected from expansion due to frost. Once the site concrete is laid it is unlikely to be affected by frost.

Blinding

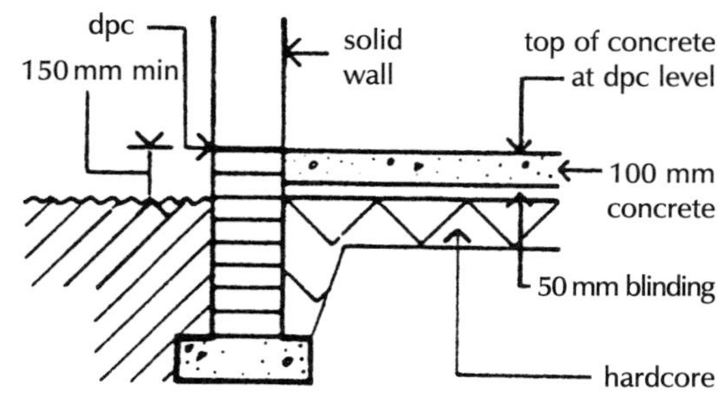

Fig. 25 Hardcore and blinding.

Before the oversite concrete is laid it is usual to blind the top surface of the hardcore. The purpose of this is to prevent the wet concrete running down between the lumps of broken brick or stone, as this would make it easier for water to seep up through the hardcore and would be wasteful of concrete. To blind, or seal, the top surface of the hardcore a thin layer of very dry coarse concrete can be spread over it, or a thin layer of coarse clinker or ash can be used. This blinding layer, or coat, will be about 50 mm thick, and on it the site concrete is spread and finished with a true level top surface. Figure 25 is an illustration of hardcore, blinding and concrete oversite. Even with a good hardcore bed below the site concrete a dense hard floor finish, such as tiles, may be slightly damp in winter and will be cold underfoot. To reduce the coldness experienced with some solid ground floor finishes it is good practice to form a continuous damp-proof membrane in the site concrete.

CONCRETE

Concrete (see also Volume 4) is the name given to a mixture of particles of stone bound together with cement. Because the major part of concrete is of particles of broken stones and sand, it is termed the aggregate. The material which binds the aggregate is cement and this is described as the matrix.

Aggregate

The materials commonly used as the aggregate for concrete are sand and gravel. The grains of natural sand and particles of gravel are very hard and insoluble in water and can be economically dredged or dug from pits and rivers. The material dug from many pits and river beds consists of a mixture of sand and particles of gravel and is called 'ballast' or 'all-in aggregate'. The name ballast derives from the use of this material to load empty ships and barges. The term 'all-in aggregate' is used to describe the natural mixture of fine grains of sand and larger coarse particles of gravel.

All-in aggregate

Ballast

All-in aggregate (ballast) is one of the cheapest materials that can be used for making concrete and is used for mass concrete work, such as large open foundations. The proportion of fine to coarse particles in an all-in aggregate cannot be varied and the proportion may vary from batch to batch so that it is not possible to control the mix and therefore the strength of concrete made with all-in aggregate. Accepted practice today is to make concrete for building from a separate mix of fine and coarse aggregate which is produced from ballast by washing, sieving and separating the fine from the coarse aggregate.

Fine aggregate and coarse aggregate

Fine aggregate is natural sand which has been washed and sieved to remove particles larger than 5 mm and coarse aggregate is gravel which has been crushed, washed and sieved so that the particles vary from 5 up to 50 mm in size. The fine and coarse aggregate are delivered separately. Because they have to be sieved, a prepared mixture of fine and coarse aggregate is more expensive than natural all-in aggregate. The reason for using a mixture of fine and coarse aggregate is that by combining them in the correct proportions, a concrete with very few voids or spaces in it can be made and this reduces the quantity of comparatively expensive cement required to produce a strong concrete.

Cement

The cement most used is ordinary Portland cement. It is manufactured by heating a mixture of finely powdered clay and limestone with water to a temperature of about 1200°C, at which the lime and clay fuse to form a clinker. This clinker is ground with the addition of a little gypsum to a fine powder of cement. Cement powder reacts with water and its composition gradually changes and the particles of cement bind together and adhere strongly to materials with which they are mixed. Cement hardens gradually after it is mixed with water.

Some thirty minutes to an hour after mixing with water the cement is no longer plastic and it is said that the initial set has occurred. About 10 hours after mixing with water, the cement has solidified and it increasingly hardens until some 7 days after mixing with water when it is a dense solid mass.

Water–cement ratio

The materials used for making concrete are mixed with water for two reasons. Firstly to cause the reaction between cement and water which results in the cement acting as a binding agent and secondly to make the materials of concrete sufficiently plastic to be placed in position. The ratio of water to cement used in concrete affects its ultimate strength, and a certain water–cement ratio produces the best concrete. If too little water is used the concrete is so stiff that it cannot be compacted and if too much water is used the concrete does not develop full strength.

The amount of water required to make concrete sufficiently plastic depends on the position in which the concrete is to be placed. The extreme examples of this are concrete for large foundations, which can be mixed with comparatively little water and yet be consolidated, and concrete to be placed inside formwork for narrow reinforced concrete beams where the concrete has to be comparatively wet to be placed. In the first example, as little water is used, the proportion of cement to aggregate can be as low as say 1 part of cement to 9 of aggregate and in the second, as more water has to be used, the proportion of cement to aggregate has to be as high as say 1 part of cement to 4 of aggregate. As cement is expensive compared with aggregate it is usual to use as little water and therefore cement as the necessary plasticity of the concrete will allow.

Proportioning materials

The materials used for mass concrete for foundations were often measured out by volume, the amount of sand and coarse aggregate being measured in wooden boxes constructed for the purpose. This is a crude method of measuring the materials because it is laborious to have to fill boxes and then empty them into mixers and no account is taken of the amount of water in the aggregate. The amount of water in aggregate affects the finished concrete in two ways: (a) if the aggregate is very wet the mix of concrete may be too weak, have an incorrect ratio of water to cement and not develop full strength and, (b) damp sand occupies a greater volume than dry. This increase in volume of wet sand is termed bulking.

The more accurate method of proportioning the materials for concrete is to measure them by weight. The materials used in reinforced concrete are commonly weighed and mixed in large concrete mixers. It is not economical for builders to employ expensive concrete mixing machinery for small buildings and the concrete for foundations, floors and lintels is usually delivered to site ready mixed, except for small batches that are mixed by hand or in a portable petrol driven mixer. The materials are measured out by volume and providing the concrete is thoroughly mixed, is not too wet and is properly consolidated the finished concrete is quite satisfactory.

Concrete mixes

British Standard 5328: Specifying concrete, including ready-mixed concrete, gives a range of mixes. One range of concrete mixes in the Standard, ordinary prescribed mixes, is suited to general building work such as foundations and floors. These prescribed mixes should be used in place of the traditional nominal volume mixes such as 1:3:6 cement, fine and coarse aggregate by volume, that have been used in the past. The prescribed mixes, specified by dry weight of aggregate, used with 100 kg of cement, provide a more accurate method of measuring the proportion of cement to aggregate and as they are measured against the dry weight of aggregate, allow for close control of the water content and therefore the strength of the concrete.

The prescribed mixes are designated by letters and numbers as C7.5P, C10P, C15P, C20P, C25P and C30P. The letter C stands for 'compressive', the letter P for 'prescribed' and the number indicates the 28-day characteristic cube crushing strength in newtons per square millimetre (N/mm^2) which the concrete is expected to attain. The prescribed mix specifies the proportions of the mix to give an indication of the strength of the concrete sufficient for most building purposes, other than designed reinforced concrete work.

Table 1 equates the old nominal volumetric mixes of cement and aggregate with the prescribed mixes and indicates uses for these mixes.

Table 1 Concrete mixes.

Nominal volume mix	BS 5328 Standard mixes	Uses
1:8 all-in 1:3:6	ST1	Foundations
1:3:6 1:2:4	ST2 ST3	Site concrete
$1:1\frac{1}{2}:3$	ST4	Site concrete reinforced

Ready-mixed concrete

The very many ready-mixed concrete plants in the United Kingdom are able to supply to all but the most isolated building sites. These plants prepare carefully controlled concrete mixes which are delivered to site by lorries on which the concrete is churned to delay setting.

Because of the convenience and the close control of these mixes, much of the concrete used in building today is provided by ready-mixed suppliers. To order ready-mixed concrete it is only necessary to specify the prescribed mix, for example C10P, the cement, type and size of aggregate and workability, that is medium or high workability, depending on the ease with which the concrete can be placed and compacted.

Soluble sulphates

There are water soluble sulphates in some soils, such as plastic clay, which react with ordinary cement and in time will weaken concrete. It is usual practice, therefore, to use one of the sulphate-resistant cements for concrete in contact with sulphate bearing soils.

Portland blast-furnace cement

This cement is more resistant to the destructive action of sulphates than ordinary Portland cement and is often used for concrete foundations in plastic clay subsoils. This cement is made by grinding a mixture of ordinary Portland cement with blast-furnace slag. Alternatively another type of cement known as 'sulphate resisting cement' is often used.

Sulphate resisting Portland cement

This cement has a reduced content of aluminates that combine with soluble sulphates in some soils and is used for concrete in contact with those soils.

OVERSITE CONCRETE (CONCRETE OVERSITE)

On firm non-cohesive subsoils and rocks such as sand, gravel and sound rock beds which are near the surface, under vegetable top soil and are well drained or dry it is satisfactory to lay the concrete oversite directly on a bed of hardcore or broken rock rubble as there is little likelihood of any appreciable amount of moisture rising and being absorbed by the concrete. The concrete is laid within the confines of the external walls and load bearing internal walls and consolidated and levelled to a thickness of 100 mm ready for solid floor finishes or a raised ground floor.

On much of the low lying land that is most suitable for building, the subsoil such as clay retains moisture which will tend to rise through a hardcore bed to concrete oversite. The damp concrete will be cold underfoot and require additional energy from heating systems to maintain an equable indoor temperature. It is practice today to form a continuous layer of some material that is impervious to water under, in or over the concrete oversite as a damp-proof membrane on the site of all inhabited buildings where there is a likelihood of moisture rising to the concrete.

DAMP-PROOF MEMBRANE

Concrete is spread oversite as a solid base for floors and as a barrier to moisture rising from the ground. Concrete is to some degree permeable to water and will absorb moisture from the ground; a damp oversite concrete slab will be cold and draw appreciable heat from rooms.

A requirement of the Building Regulations is that floors shall adequately resist the passage of moisture to the inside of the building. As concrete is permeable to moisture, it is generally necessary to use a damp-proof membrane under, in or on top of ground supported floor slabs as an effective barrier to moisture rising from the ground. The membrane should be continuous with the damp-proof course in walls, as a barrier to moisture rising between the edges of the concrete slab and walls.

A damp-proof membrane should be impermeable to water in either liquid or vapour form and be tough enough to withstand possible damage during the laying of screeds, concrete or floor finishes. The damp-proof membrane may be on top, sandwiched in or under the concrete slab.

Being impermeable to water the membrane will delay the drying out of wet concrete to ground if it is under the concrete, and of screeds to concrete if it is on top of the concrete.

Damp-proof membrane below site concrete

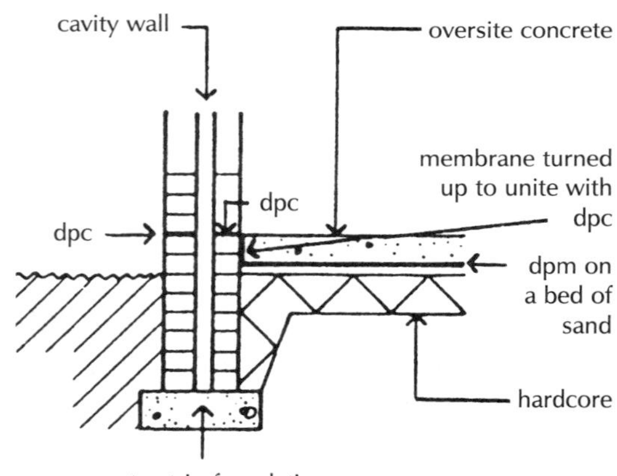

Fig. 26 Below concrete damp-proof membrane.

The obvious place to use a continuous damp-proof membrane is under the oversite concrete. The membrane is spread on a layer of comparatively dry concrete, clinker or ash which is spread and levelled over the hardcore as illustrated in Fig. 26. The edges of the membrane are turned up the face of external and internal walls ready for concrete laying so that it may unite and overlap the dpc in walls.

The membrane should be spread with some care to ensure that thin membranes are not punctured by sharp, upstanding particles in the blinding and that the edge upstands are kept in place as the concrete is laid.

The advantage of a damp-proof membrane under the site concrete is that it will be protected from damage during subsequent building operations. A disadvantage is that the membrane will delay the drying out of the oversite concrete that can only lose moisture by upwards evaporation to air.

Where underfloor heating is used the membrane should be under the concrete.

Surface damp-proof membrane

Floor finishes such as pitch mastic and mastic asphalt that are impermeable to water can serve as a combined damp-proof membrane and floor finish. These floor finishes should be laid to overlap

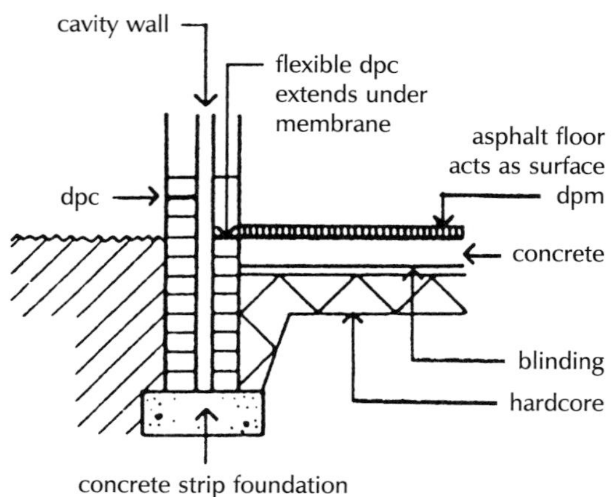

Fig. 27 Surface damp-proof membrane.

the damp-proof course in the wall as illustrated in Fig. 27 to seal the joint between the concrete and the wall.

Where hot soft bitumen or coal tar pitch are used as an adhesive for wood block floor finishes the continuous layer of the impervious adhesive can serve as a waterproof membrane.

The disadvantage of impervious floor finishes and impervious adhesives for floor finishes as a damp-proof membrane are that the concrete under the floor finish and the floor finish itself will be cold underfoot and make calls on the heating system and if the old floor finish is replaced with another there may be no damp-proof membrane.

Damp-proof membrane below a floor screed

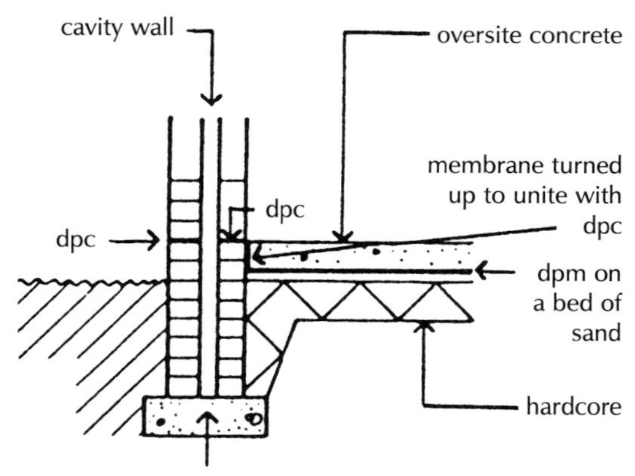

Fig. 28 Sandwich damp-proof membrane.

The oversite concrete is laid during the early stages of the erection of buildings. It is practice to lay floor finishes to solid ground floors after the roof is on and wet trades such as plastering are completed to avoid damage to floor finishes. By this time the site concrete will have thoroughly dried out. A layer of fine grained material such as sand and cement is usually spread and levelled over the surface of the dry concrete to provide a true level surface for a floor finish. As the wet finishing layer, called a screed, will not strongly adhere to dry concrete it is made at least 65 mm thick so that it does not dry too quickly and crack. Electric conduits and water service pipes are commonly run in the underside of the screed.

As an alternative to under concrete or surface damp-proof membranes a damp-proof membrane may be sandwiched between the site concrete and the floor screed, as illustrated in Fig. 28. At the junction of wall and floor the membrane overlaps the damp-proof course in the wall.

Materials for damp-proof membrane

The materials used as damp-proof membrane must be impermeable to water both in liquid and vapour form and sufficiently robust to withstand damage by later building operations.

Polythene and polyethylene sheet

Polythene or polyethylene sheet is commonly used as a damp-proof membrane with oversite concrete for all but severe conditions of dampness. It is recommended that the sheet should be at least 0.25 mm thick (1200 gauge). The sheet is supplied in rolls 4 m wide by 25 m long. When used under concrete oversite the sheet should be laid on a blinding layer of sand or compacted fuel ash spread over the hardcore.

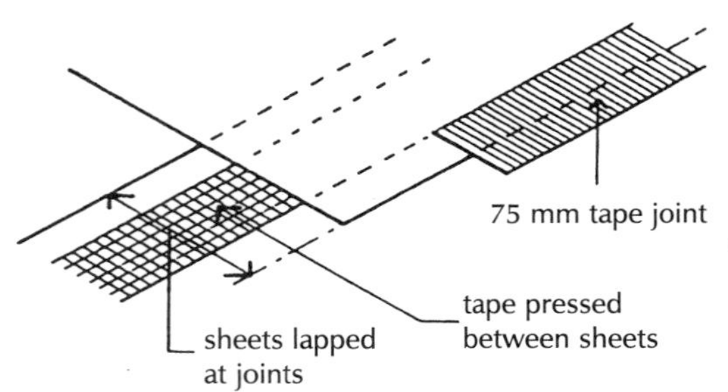

Fig. 29 Jointing laps in polythene sheet.

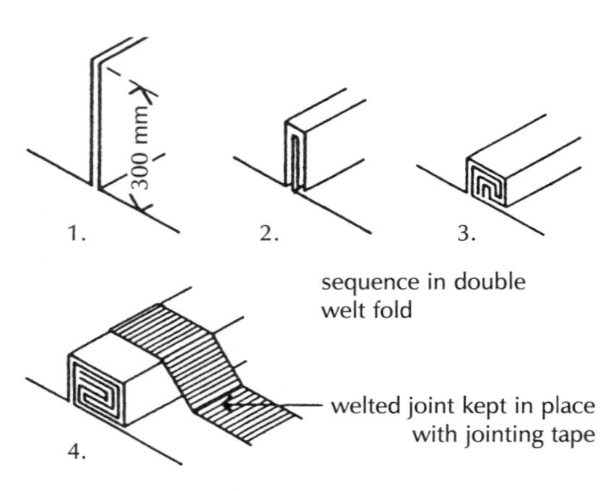

Fig. 30 Double welted fold joint in polythene sheet.

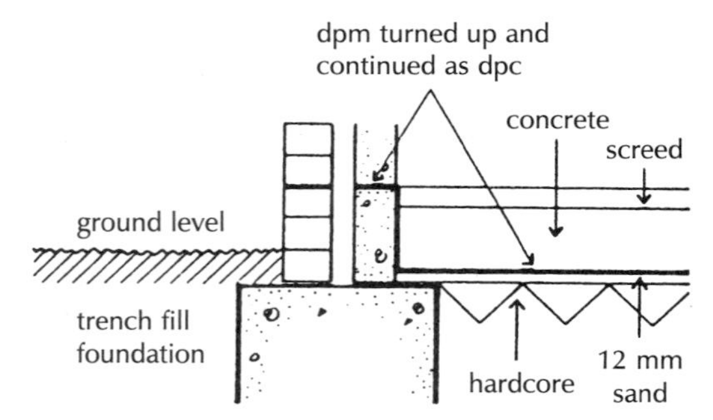

Fig. 31 Damp-proof membrane turn up.

The sheets are spread over the blinding and lapped 150 mm at joints and continued across surrounding walls, under the dpc for the thickness of the wall.

Where site conditions are reasonably dry and clean, the overlap joints between the sheets are sealed with mastic or mastic tape between the overlapping sheets and the joint completed with a polythene jointing tape as illustrated in Fig. 29.

For this lapped joint to be successful the sheets must be dry and clean else the jointing tape will not adhere to the surface of the sheets and the joint will depend on the weight of the concrete or screed pressing the joint sufficiently heavily to make a watertight joint. As clean and dry conditions on a building site are rare, this type of joint should be only used where there is unlikely to be heavy absorption of ground moisture.

Where site conditions are too wet to use mastic and tape, the joint is made by welting the overlapping sheets with a double welted fold as illustrated in Fig. 30, and this fold is kept in place by weighing it down with bricks or securing it with tape until the screed or concrete has been placed. The double welt is formed by folding the edges of sheets together and then making a welt which is flattened.

The plastic sheet is effectively impossible to fold and so stiff and elastic that it will always tend to unfold so that it requires a deal of patience to fold, hold in place and then contrive to fold along the joint. By using the maximum size of sheet available it is possible to minimise the number of joints.

The sheet should be used so that there are only joints one way as it is impractical to form a welt at junctions of joints.

Where the level of the damp-proof membrane is below that of the dpc in walls it is necessary to turn it up against walls so that it can overlap the dpc or be turned over as dpc as illustrated in Fig. 31. To keep the sheet in place as an upstand to walls it is necessary to keep it in place with bricks or blocks laid on the sheet against walls until the concrete has been placed and the bricks or blocks removed as the concrete is run up the wall.

At the internal angle of walls a cut is made in the upstand sheet to facilitate making an overlap of sheet at corners. These sheets which are commonly used as a damp-proof membrane will serve as an effective barrier to rising damp, providing they are not punctured or displaced during subsequent building operations.

Hot pitch or bitumen

A continuous layer of hot applied coal-tar pitch or soft bitumen is poured on the surface and spread to a thickness of not less than 3 mm. In dry weather a concrete blinding layer is ready for the membrane 3 days after placing. The surface of the concrete should be brushed to remove dust and primed with a solution of coal-tar pitch or bitumen solution or emulsion. The pitch is heated to 35°C to 45°C and bitumen to 50°C to 55°C.

Properly applied pitch or bitumen layers serve as an effective damp-proof membrane both horizontally and spread up inside wall faces to unite with dpcs in walls and require less patient application than plastic sheet materials.

Bitumen solution, bitumen/rubber emulsion or tar/rubber emulsion

These cold applied solutions are brushed on to the surface of concrete in three coats to a finished thickness of not less than 2.5 mm, allowing each coat to harden before the next is applied.

Bitumen sheet

Sheets of bitumen with hessian, fibre or mineral fibre base are spread on the concrete oversite or on a blinding of stiff concrete below the concrete, in a single layer with the joints between adjacent sheets lapped 75 mm. The joints are then sealed with a gas torch which melts the bitumen in the overlap of the sheets sufficient to bond them together. Alternatively the lap is made with hot bitumen spread between the overlap of the sheets which are then pressed together to make a damp-proof joint. The bonded sheets may be carried across adjacent walls as a dpc, or up against the walls and then across as dpc where the membrane and dpc are at different levels.

The polythene or polyester film and self-adhesive rubber/bitumen compound sheets, described in Volume 4 under 'Tanking', can also be used as damp-proof membranes, with the purpose cut, shaped cloaks and gussets for upstand edges and angles. This type of membrane is particularly useful where the membrane is below the level of the dpc in walls.

Bitumen sheets, which may be damaged on building sites, should be covered for protection as soon as possible by the screed or site concrete.

Mastic asphalt or pitch mastic

These materials are spread hot and finished to a thickness of at least 12.5 mm. This expensive damp-proof membrane is used where there is appreciable water pressure under the floor and as 'tanking' to basements as described in Volume 4.

RESISTANCE TO THE PASSAGE OF HEAT

The requirements of the Building Regulations and practical advice in Approved Document L include provision for insulation to some ground floors. The requirement is that ground floors should have a maximum insulation value (U value) of 0.45 W/m^2K. Some ground floor slabs that are larger than 10 m in both length and breadth may not need the addition of an insulating layer to provide the U value of 0.45.

Of the heat that is transferred through a solid, ground supported floor a significant part of the transfer occurs around the perimeter of the floor to the ground below, foundation walls and ground around the edges of the floor, so that the cost of insulating the whole floor is seldom justified. Insulation around or under the edges of a solid floor will significantly reduce heat losses to the extent that overall insulation is unnecessary.

In the CIBS guide to the thermal properties of building structures, the U value of an uninsulated solid floor 20 × 20 m on plan, with four edges exposed, is given as 0.36 W/m^2K and one 10 × 10 m as 0.6 W/m^2K. The 20 × 20 floor has a U value below that in the requirement of the Building Regulations and will not require insulation. The U value of a 10 m^2 floor can be reduced by the use of edge insulation. With edge insulation of a metre deep all around and under the floor, the U value can be reduced to 0.48 W/m^2K which is somewhat higher than the U value in the requirement of the Building Regulations and may necessitate some small overall insulation. This is the basis for the assumption that floor slabs that are larger than 10 m in both length and width may not need an overall insulation layer.

To reduce heat losses through thermal bridges around the edges of solid floors that do not need overall insulation, and so minimise problems of condensation and mould growth, it may be wise to build in edge insulation, particularly where the wall insulation is not carried down below the ground floor slab. Edge insulation is formed either as a vertical strip between the edge of the slab and the wall or under the slab around the edges of the floor as illustrated in Fig. 32. The depth or width of the strips of insulation vary from 0.25 m to 1 m and the thickness of the insulation will be similar to that needed for overall insulation.

The only practical way of improving the insulation of a solid ground floor to the required U value is to add a layer of some material with a high insulation value to the floor. The layer of insulation may be laid below a chipboard or plywood panel floor finish or below a timber boarded finish or below the screed finish to a floor or under the concrete floor slab. With insulation under the screed or slab it is important that the density of the insulation board is sufficient to support the load of the floor itself and imposed loads on the floor. A density of at least 16 kg/m^3 is recommended for domestic buildings.

The advantage of laying the insulation below the floor slab is that the high density slab, which warms and cools slowly (slow thermal response) in response to changes in temperature of the constant low output heating systems, will not lose heat to the ground. The damp-proof membrane may be laid under or over the insulation layer or under the floor screed. The damp-proof membrane should be under insulation that absorbs water and may be over insulation with low water absorption and high resistance to ground contaminants.

With the insulation layer and the dpm below the concrete floor slab

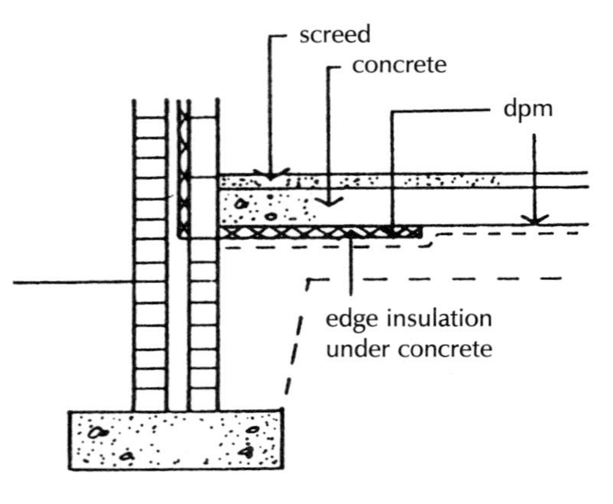

Edge insulation under concrete

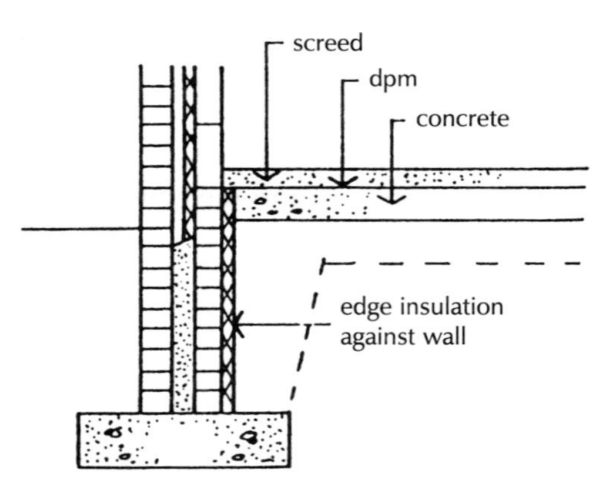

Edge insulation vertical

Fig. 32 Perimeter insulation to ground slab.

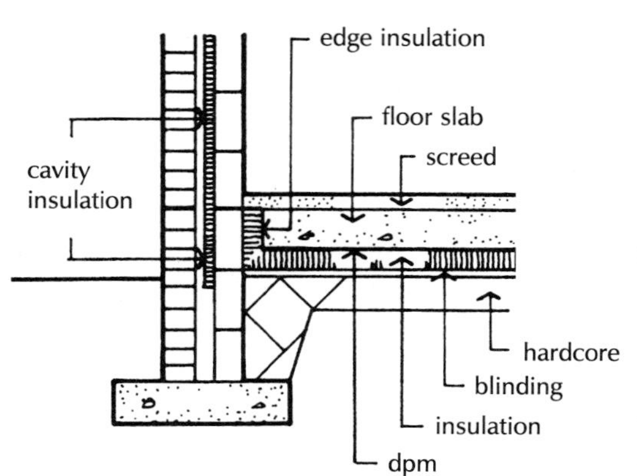

Fig. 33 dpm over insulation under floor slab.

it is necessary to continue the dpm and insulation up vertically around the edges of the slab to unite with the dpc in walls as illustrated in Fig. 33.

One method of determining the required thickness of insulation is to use a thickness of insulation related to the U value of the chosen insulation material, as for example thicknesses of 25 mm for a U value of 0.02 W/m^2K, 37 mm for 0.03 W/m^2K, 49 mm for 0.04 W/m^2K and 60 mm for 0.05 W/m^2K, ignoring the inherent resistance of the floor.

Another more exacting method is to calculate the required thickness related to the actual size of the floor and its uninsulated U value, taken from a table in the CIBS guide to the thermal properties of building structures. For example, from the CIBS table the U value of a solid floor 10×6 m, with four edges exposed is 0.74 W/m^2K.

$$\text{The inherent resistance of the floor} = \frac{1}{0.74} = 1.36\,m^2K/W$$

$$\text{Thermal resistance required} = \frac{1}{0.45} = 2.22\,m^2K/W$$

$$\text{Additional resistance required} = 2.22 - 1.36 = 0.86\,m^2K/W$$

Assume an insulant with U value of 0.04 W/m^2K

$$\text{Thickness of insulation layer required} = 0.86 \times 0.04 \times 1000 = 34.4\,\text{mm (49)}$$

or with an insulation value of 0.03 W/m^2K = 25.8 mm (37)

These thicknesses are appreciably less than those given by the first method, shown in brackets.

Where the wall insulation is in the cavity or on the inside face of the wall it is necessary to avoid a cold bridge across the foundation wall and the edges of the slab, by fitting insulation around the edges of the slab or by continuing the insulation down inside the cavity, as illustrated in Fig. 34.

An advantage of fitting the dpm above the insulation is that it can be used to secure the upstand edge insulation in place while concrete is being placed.

The disadvantage of the dpm being below the concrete floor slab is that it will prevent the wet concrete drying out below and so lengthen the time required for it to adequately dry out, to up to 6 months. A concrete floor slab that has not been sufficiently dried out may adversely affect water sensitive floor finishes such as wood.

The advantage of laying the insulation layer under the screed is that it can be laid inside a sheltered building on a dried slab after the roof is

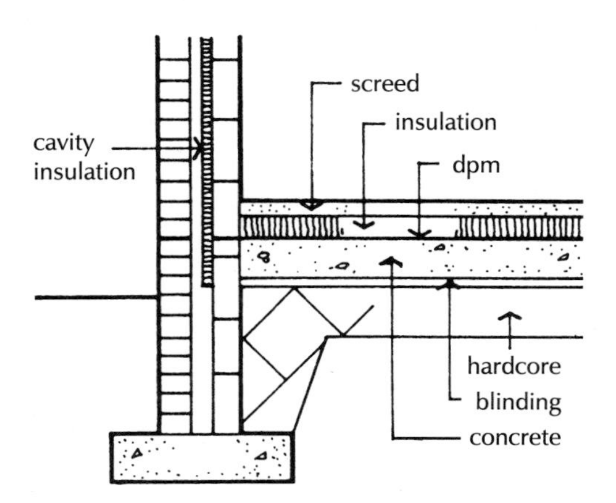

Fig. 34 dpm under insulation and screed.

finished and that the dpm, whether over or under the insulation layer, can more readily be joined to the dpc in walls, as illustrated in Fig. 34. Where the wall insulation is in the cavity it should be continued down below the floor slab to minimise the cold bridge across the wall to the screed as illustrated in Fig. 34.

If the dpm is laid below the insulation it is necessary to spread a separating layer over the insulation to prevent wet screed running into the joints between the insulation boards. The separating layer should be building paper or 500 gauge polythene sheet.

To avoid damage to the insulation layer and the dpm it is necessary to take care in tipping, spreading and compacting wet concrete or screed. Scaffold boards should be used for barrowing and tipping concrete and screed and a light mesh of chicken wire can be used over separating layers or dpms over insulation under screeds as added protection.

Materials for underfloor insulation

Any material used as an insulation layer to a solid, ground supported floor must be sufficiently strong and rigid to support the weight of the floor or the weight of the screed and floor loads without undue compression and deformation. To meet this requirement one of the rigid board or slab insulants is used. The thickness of the insulation is determined by the nature of the material from which it is made and the construction of the floor, to provide the required U value.

Some insulants absorb moisture more readily than others and some insulants may be affected by ground contaminants. Where the insulation layer is below the concrete floor slab, with the dpm above the insulation one of the insulants with low moisture absorption characteristics should be used.

The materials commonly used for floor insulation are rockwool slabs, extruded polystyrene, cellular glass and rigid polyurethane foam boards.

DAMP-PROOF COURSES

The function of a dpc is to act as a barrier to the passage of moisture or water between the parts separated by the dpc. The movement of moisture or water may be upwards in the foundation of walls and ground floors, downwards in parapets and chimneys or horizontal where a cavity wall is closed at the jambs of openings.

One of the functional requirements of walls (see Chapter 2) is resistance to moisture. A requirement of the Building Regulations is that walls shall adequately resist the passage of moisture to the inside of the building. To meet this requirement it is necessary to form a barrier to moisture rising from the ground in walls. This barrier is the horizontal, above ground, dpc.

Damp-proof courses above ground

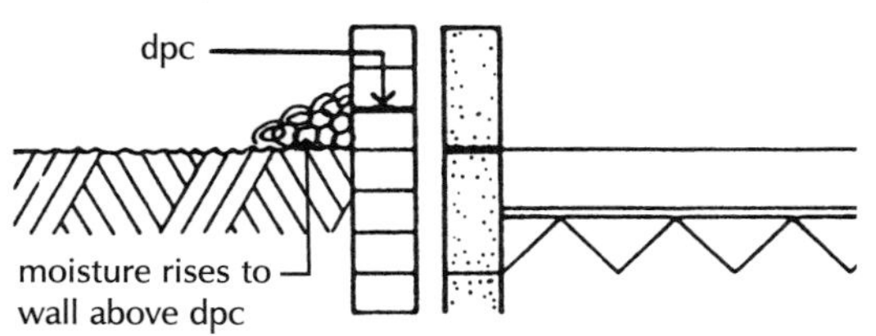

Fig. 35

There should be a continuous horizontal dpc above ground in walls whose foundations are in contact with the ground, to prevent moisture from the ground rising through the foundation to the wall above ground, which otherwise would make wall surfaces damp and damage wall finishes. The dpc above ground should be continuous for the whole length and thickness of the wall and be at least 150 mm above finished ground level to avoid the possibility of a build up of material against the wall acting as a bridge for moisture from the ground as illustrated in Fig. 35.

Material for damp-proof courses above ground

It is convenient to group the materials used for dpcs as flexible, semi-rigid and rigid. Flexible materials such as metal, bitumen and polythene sheet can accommodate moderate settlement movement in a building which may fracture the semi-rigid material mastic asphalt and will probably fracture the rigid materials brick and slate.

Flexible dpcs

Lead

Lead for use as a dpc should weigh not less than 19.5 kg/m^2 (Code No 4, 1.8 mm thick). Lead is an effective barrier to moisture and water. It is liable to corrosion in contact with freshly laid lime or cement mortar and should be protected by a coating of bitumen or bitumen paint applied to the mortar surface and both surfaces of the lead. Lead is durable and flexible and can suffer distortion due to moderate settlement in walls without damage. Lead is an expensive material and is little used today other than for ashlar stonework or as a shaped dpc in chimneys. Lead should be laid in rolls the full thickness of the wall or leaf of cavity walls and be lapped at joints along the length of the wall and at intersections at least 100 mm or the width of the dpc.

Copper

Copper as a dpc should be annealed, at least 0.25 mm thick and have a nominal weight of 2.28 kg/m^2. Copper is an effective barrier to moisture and water, it is flexible, has high tensile strength and can suffer distortion due to moderate settlement in a wall without damage. It is an expensive material and is little used today as a dpc above ground. When used as a dpc, it may cause staining of wall surfaces due to the oxide that forms. It is spread on an even bed of mortar and lapped at least 100 mm or the width of the dpc at running joints and intersections.

Bitumen damp-proof course

There are four types of bitumen dpc, as follows:

(1) bitumen dpc with hessian base
(2) bituman dpc with fibre base

(3) bituman dpc with hessian base and lead
(4) bitumen dpc with fibre base and lead.

Bitumen dpcs are reasonably flexible and can withstand distortion due to moderate settlement in walls without damage. They may extrude under heavy loads without affecting their efficiency as a barrier to moisture. Bitumen dpcs, which are made in rolls to suit the thickness of walls, are bedded on a level bed of mortar and lapped at least 100 mm or the width of the dpc at running joints and intersections.

Bitumen is much used for dpcs because it is at once economical, flexible, reasonably durable and convenient to lay. There is little to choose between hessian or fibre as a base for a bitumen dpc above ground. The fibre base is cheaper but less tough than hessian.

The lead cored dpc, with a lead strip weighing not less than 1.20 kg/m^2, joined with soldered joints, is more expensive and more effective than the bitumen alone types. It is generally used as the horizontal dpc for houses.

The combination of a mortar bed, bitumen dpc and the mortar bed over the dpc for brickwork makes a comparatively deep mortar joint that may look unsightly.

Polythene sheet

Polythene sheet for use as a dpc should be black, low density polythene sheet of single thickness not less than 0.46 mm, weighing approximately 0.48 kg/m^2. Polythene sheet is flexible, can withstand distortion due to moderate settlement in a wall without damage and is an effective barrier against moisture. It is laid on an even bed of mortar and lapped at least the width of the dpc at running joints and intersections. Being a thin sheet material, polythene makes a thinner mortar joint than bitumen dpc, and is sometimes preferred for that reason.

Its disadvantage as a dpc is that it is fairly readily damaged by sharp particles in mortar or the coarse edges of brick.

Polymer-based sheets

Polymer-based sheets are thinner than bitumen sheets and are used where the thicker bitumen dpc mortar joint would be unsightly. This dpc material, which has its laps sealed with adhesive, may be punctured by sharp particles and edges.

Semi-rigid damp-proof courses

Mastic asphalt

Mastic asphalt, spread hot in one coat to a thickness of 13 mm, is a semi-rigid dpc, impervious to moisture and water. Moderate settlement in a wall may well cause a crack in the asphalt through which moisture or water may penetrate. It is an expensive form of dpc, which shows on the face of walls as a thick joint, and it is rarely used as a dpc.

Rigid damp-proof courses

Up to the twentieth century, damp-proof courses in walls were not common. The inevitability of some moisture rising in walling on damp soils was accepted. Infrequently a few courses of dense bricks might be used at the base of walls as a solid bearing for walls and to act as a dpc to an extent. With the extensive building, both commercial and domestic, that occurred after the Industrial Revolution it became more common to use one of the rigid systems of dpc in the form of bricks in lowland areas and slates where the natural material was quarried and was comparatively cheap. With the introduction of bitumen felts, and later the synthetic sheet materials, bricks and slates were largely abandoned as dpcs.

Slate damp-proof courses

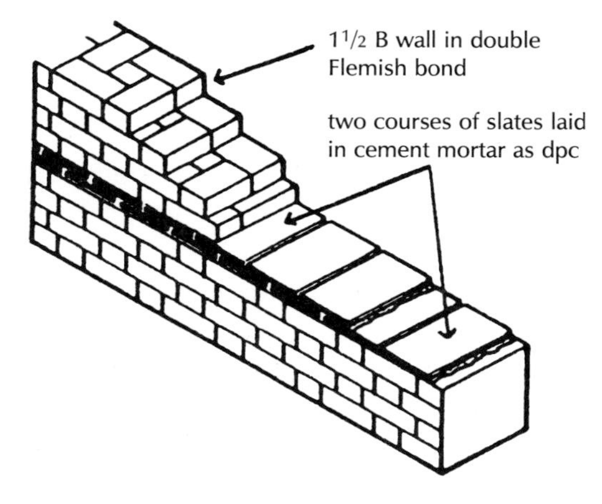

Fig. 36 Slate damp-proof course.

Beds of natural slate were quarried and the heavily compressed, dense material that was formed in layers was split to thin slates that were sufficiently impermeable to water to serve as an effective dpc.

Two courses of dense Welsh slates were laid at first in lime or hydraulic lime and sand, and later in cement and sand. The slates were laid on a bed of mortar in two courses, breaking joint as illustrated in Fig. 36. Because of the small units of slate and the joints being staggered this dpc could remain reasonably effective where moderate settlement occurred.

To be effective the edges of the slates should be exposed on a wall face and not be covered, which made a deep, somewhat ugly joint.

The majority of external brick walls are built as a cavity today and it would be laborious, wasteful and therefore expensive to use a separate slate dpc in each leaf of the wall.

Brick damp-proof courses

Two or three courses of dense, semi-engineering or engineering bricks were laid in hydraulic lime and later cement mortar. There is little likelihood of these dense bricks fracturing under moderate settlement.

Because of the dissimilar colour and texture of these bricks to that of facing bricks and the cost of the material this form of dpc is little used.

Damp-proof courses in cavity walls

A cavity wall is built as two leaves separated by a cavity. The purpose of the cavity is to act as a barrier to the penetration of rainwater to the inside of buildings. It is practice to build a cavity wall directly off the foundation so that the cavity extends below ground. A requirement of the Building Regulations is that the cavity should be carried down at least 150 mm below the level of the lowest dpc.

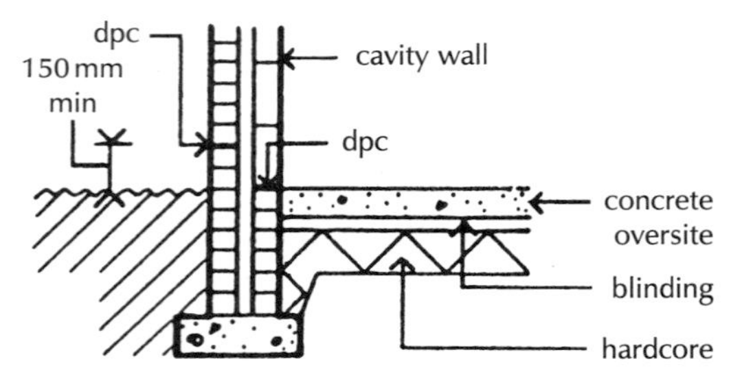

Fig. 37 Damp-proof course at different levels.

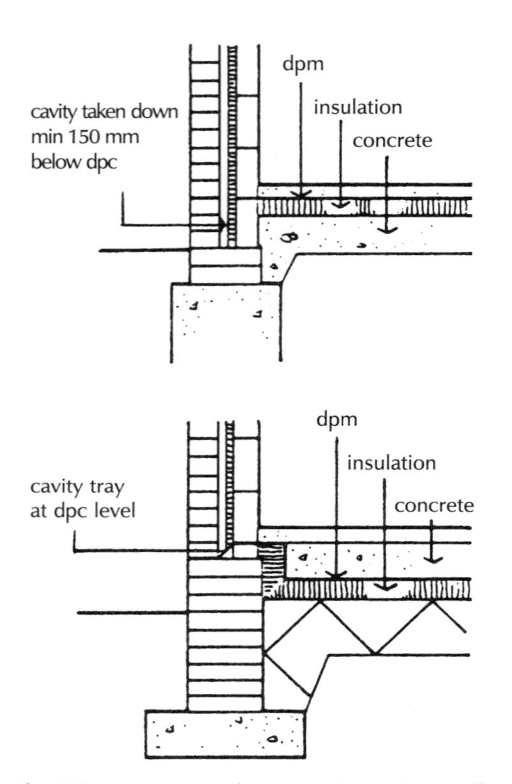

Fig. 38 Damp-proof courses in cavity walls.

A dpc in external walls should ideally be at the same level as the dpm in the concrete oversite for the convenience of overlapping the two materials to make a damp-proof joint.

Where the dpcs in both leaves of a cavity wall are at least 150 mm above outside ground level and the floor level is at, or just above, ground level, it is necessary to dress the dpm up the wall and into the level of the dpc. This is a laborious operation which makes it difficult to make a moisture tight joint at angles and intersections.

The solution is to lay the dpc in the inner leaf of the cavity wall, level with the dpm in the floor, as illustrated in Fig. 37.

Where the level of the foundation is near the surface, as with trench fill systems, it may be convenient to build two courses of solid brickwork up to ground level on which the cavity wall is raised, as illustrated in Fig. 38. As little vegetable top soil has been removed the floor level finishes some way above ground and the dpm in the floor can be united with the dpc at the same level.

The cavity insulation is taken down to the base of the cavity to continue wall insulation down to serve in part as edge insulation to the floor construction.

It is accepted practice to finish the cavity in external walling at the level of the dpc, at least 150 mm above ground, where the wall is built as a solid wall up to the dpc, as illustrated in Fig. 38. This form of construction may be used where the inner leaf of the cavity wall was built with light weight concrete blocks, used for their insulating property. These blocks fairly readily absorb moisture, expand when wet and might be affected by frost and deteriorate, whereas solid brickwork below ground will provide a stable base.

With this arrangement the requirements of the Building Regulations recommend the use of a cavity tray at the bottom of the cavity. This tray takes the form of a sheet of a flexible, impermeable material such as one of the flexible dpc materials which is laid across the cavity from a level higher in the inner leaf so that it falls towards the outer leaf to catch and drain any snow or moisture that might enter the cavity. The cavity thus acts as both tray and dpc to the cavity wall leaves.

In this detail of construction the under concrete insulation is below the lowest level of the cavity and should be turned up against the outer walls as edge insulation.

SUPPORT FOR FOUNDATION TRENCHES

The trenches which have to be dug for the foundations of walls may be excavated by hand for single small buildings but where, for example, several houses are being built at the same time it is often economical to use mechanical trench diggers.

If the trenches are of any depth it may be necessary to fix temporary timber supports to stop the sides of the trench from falling in. The

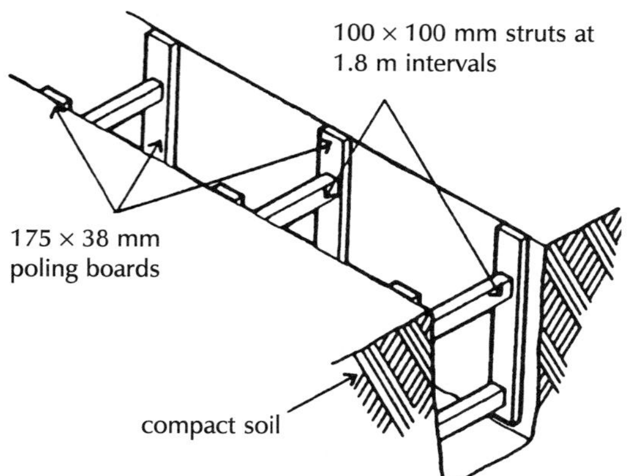

Fig. 39 Struts and poling boards.

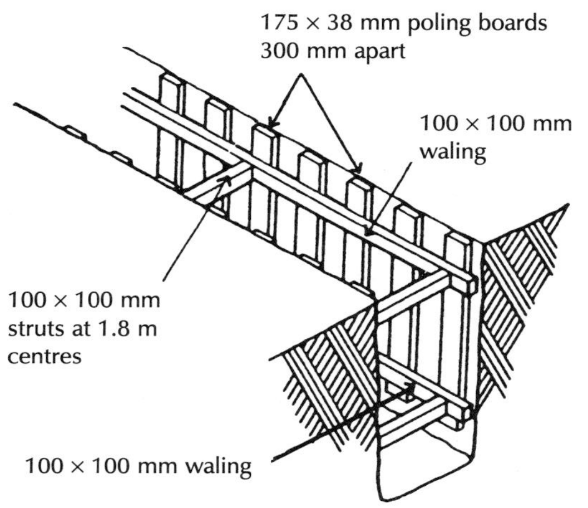

Fig. 40 Struts, waling and poling boards.

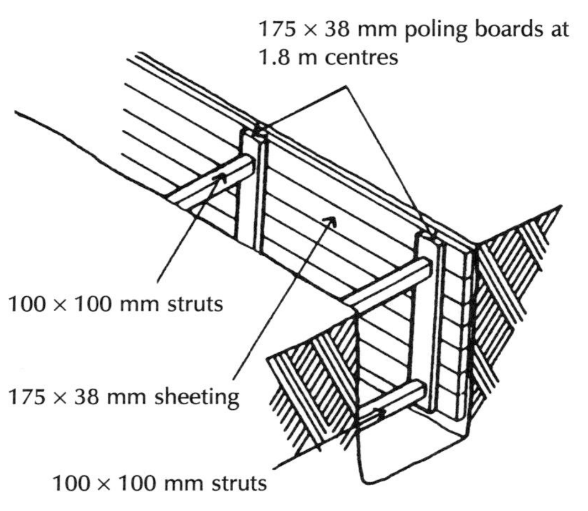

Fig. 41 Struts, poling boards and sheeting.

nature of the soil being excavated mainly determines the depth of trench for which timber supports to the sides should be used.

Soft granular soils readily crumble and the sides of trenches in such soil may have to be supported for the full depth of the trench. The sides of trenches in clay soil do not usually require support for some depth, say up to 1.5 m, particularly in dry weather. In rainy weather, if the bottom of the trench in clay soil gets filled with water, the water may wash out the clay from the sides at the bottom of the trench and the whole of the sides above may cave in.

The purpose of temporary timbering supports to trenches is to uphold the sides of the excavation as necessary to avoid collapse of the sides, which may endanger the lives of those working in the trench, and to avoid the wasteful labour of constantly clearing falling earth from the trench bottoms.

The material most used for temporary support for the sides of excavations for strip foundations is rough sawn timber. The timbers used are square section struts, across the width of the trench, supporting open poling boards, close poling boards and walings or poling boards and sheeting.

Whichever system of timbering is used there should be as few struts, that is horizontal members, fixed across the width of the trench as possible as these obstruct ease of working in the trench. Struts should be cut to fit tightly between poling or waling boards and secured in position so that they are not easily knocked out of place.

For excavations more than 1.5 m deep in compact clay soils it is generally sufficient to use a comparatively open timbering system as the sides of clay will not readily fall in unless very wet or supporting heavy nearby loads. A system of struts between poling boards spaced at about 1.8 m intervals as illustrated in Fig. 39 will usually suffice.

Where the soil is soft, such as soft clay or sand, it will be necessary to use more closely spaced poling boards to prevent the sides of the trench between the struts from falling in. To support the poling boards horizontal walings are strutted across the trench, as illustrated in Fig. 40.

For trenches in dry granular soil it may be necessary to use sheeting to the whole of the sides of trenches. Rough timber sheeting boards are fixed along the length and up the sides of the trench to which poling boards are strutted, as illustrated in Fig. 41.

The three basic arrangements of timber supports for trenches are indicative of some common system used and the sizes given are those that might be used.

2: Walls

A wall is a continuous, usually vertical structure, thin in proportion to its length and height, built to provide shelter as an external wall or divide buildings into rooms or compartments as an internal wall.

Prime function

The prime function of an external wall is to provide shelter against wind, rain and the daily and seasonal variations of outside temperature normal to its location, for reasonable indoor comfort. The basic function of shelter may be served by crude systems of interlaced branches of trees covered with dried mud, the more permanent protection of a brick wall or a screen of sheets of glass fixed to or hung from a structural frame.

Strength and stability

To provide adequate shelter a wall should have sufficient strength and stability to be self-supporting and to support roofs and upper floors. To differentiate the structural requirements of those walls that carry the loads from roofs and upper floors in addition to their own weight from those that are freestanding and only carry their own weight, the terms loadbearing and non-loadbearing are used. In practice non-loadbearing, internal walls are often described as partitions.

Thermal resistance

For reasonable indoor comfort a wall should provide resistance to excessive transfer of heat both from inside to outside and from outside to inside during periods of cold or hot, seasonal, outside temperatures.

The materials that are most effective in resisting heat transfer are of a fibrous or cellular nature in which very many small pockets of air are trapped to act as insulation against the transfer of heat.

Because of their lightweight nature these materials do not have sufficient strength to serve as part of the structure of a wall by themselves. Lightweight insulating materials are either sandwiched between materials that have strength or behind those that resist penetration of wind and rain, or serve as internal wall finishes.

The majority of walls for traditional small buildings, such as houses, are constructed with solid blocks such as brick or are framed from small sections of timber. Which one of the two types of wall is used will generally depend on the availability of materials, such as clay for making bricks, stone for making blocks or timber for frames.

Walls may be classified as solid or framed. A solid wall (sometimes called a masonry wall) is constructed of either brick, or blocks of stone, or concrete laid in mortar with the blocks laid to overlap in

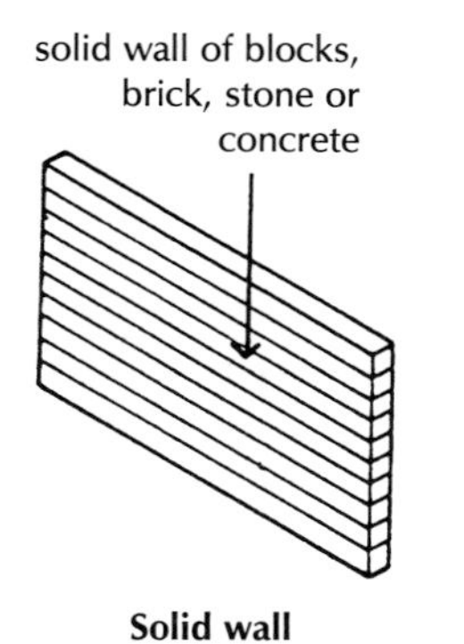

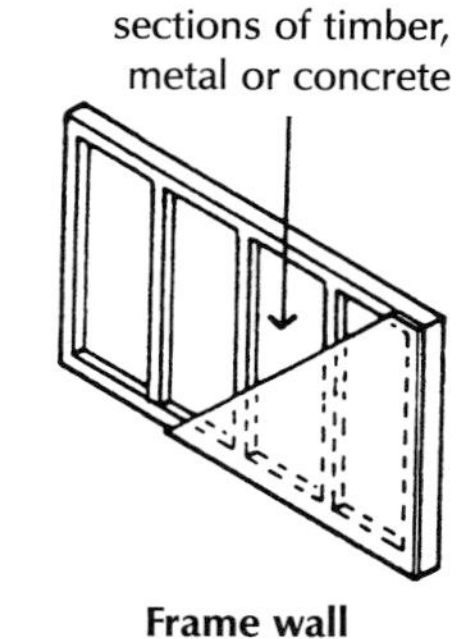

Fig. 42 Types of wall.

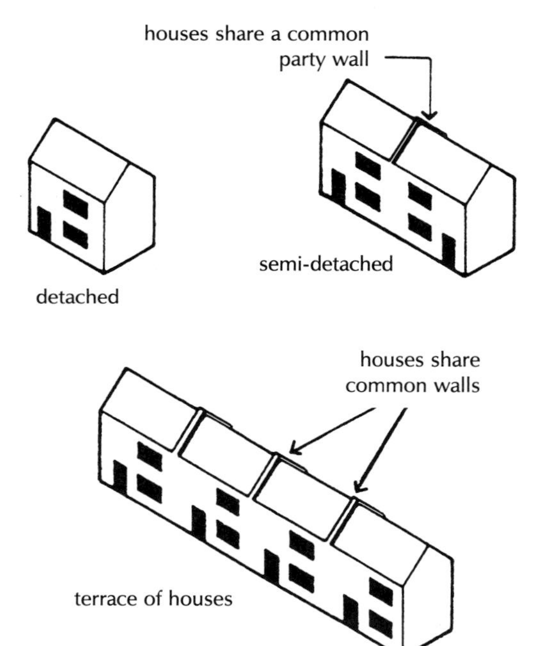

Fig. 43 Types of house.

some form of what is called bonding or as a monolith, that is, one solid uninterrupted material such as concrete which is poured wet and hardens into a solid monolith (one piece of stone). A solid wall of bricks or blocks may be termed a block (or masonry) wall, and a continuous solid wall of concrete, a monolithic wall.

A frame wall is constructed from a frame of small sections of timber, concrete or metal joined together to provide strength and rigidity, over both faces of which, or between the members of the frame, are fixed thin panels of some material to fulfil the functional requirements of the particular wall. Figure 42 is a diagram of the two types of wall.

Each of the two types of wall may serve as internal or external wall and as a loadbearing or non-loadbearing wall. Each of the two types of wall has different characteristics in fulfilling the functional requirements of a wall so that one type may have good resistance to fire but be a poor insulator against transfer of heat. There is no one material or type of wall that will fulfil all the functional requirements of a wall with maximum efficiency.

Traditional small buildings, such as houses, are commonly built as a square or rectangular box of enclosing walls as the most economical means of enclosing space. The walls of a single detached building are exposed on all sides to wind, rain and the variations of outside temperature.

Two buildings constructed on each side of a common separating wall, usually described as semi-detached, enjoy the advantage of a shared internal dividing wall and only three external walls exposed to wind and rain, as illustrated in Fig. 43. A disadvantage of the shared dividing wall is that it may not serve as an effective sound barrier.

A continuous terrace of houses enjoys the benefit of shared, common dividing walls, reduction in exposure to wind and rain and the likely disadvantage of the poor sound insulation through two common dividing walls.

FUNCTIONAL REQUIREMENTS

The function of a wall is to enclose and protect a building or to divide space within a building.

To provide a check that a particular wall construction satisfies a range of functional requirements it is convenient to adopt a list of specific requirements. The commonly accepted requirements of a wall are:

Strength and stability
Resistance to weather and ground moisture
Durability and freedom from maintenance
Fire safety

Resistance to the passage of heat
Resistance to airborne and impact sound
Security

Strength and stability

Strength

The strength of the materials used in wall construction is determined by the strength of a material in resisting compressive and tensile stress and the way in which the materials are put together. The usual method of determining the compressive and tensile strength of a material is to subject samples of the material to tests to assess the ultimate compressive and tensile stress at which the material fails in compression and in tension. From these tests the safe working strengths of materials in compression and in tension are set. The safe working strength of a material is considerably less than the ultimate strength, to provide a safety factor against variations in the strength of materials and their behaviour under stress. The characteristic working strengths of materials, to an extent, determine their use in the construction of buildings.

The traditional building materials timber, brick and stone have been in use since man first built permanent settlements, because of the ready availability of these natural materials and their particular strength characteristics. The moderate compressive and tensile strength of timber members has long been used to construct a frame of walls, floors and roofs for houses.

The compressive strength of well burned brick combined with the durability, fire resistance and appearance of the material commends it as a walling material for the more permanent buildings.

The sense of solidity and permanence and compressive strength of sound building stone made it the traditional walling material for many larger buildings.

Steel and concrete, which have been used in building since the Industrial Revolution, are used principally for their very considerable strength as the structural frame members of large buildings where the compressive strength of concrete, separately or in combination with steel, is used for both columns and beams.

In the majority of small buildings, such as houses, the compressive strength of brick and stone is rarely fully utilised because the functional requirements of stability and exclusion of weather dictate a thickness of wall in excess of that required for strength alone. To support the very modest loads on the walls of small buildings the thinnest brick or stone wall would be quite adequate.

Stability

The stability of a wall may be affected by foundation movement (see Chapter 1), eccentric loads, lateral forces (wind) and expansion due to temperature and moisture changes.

Eccentric loads, that is those not acting on the centre of the

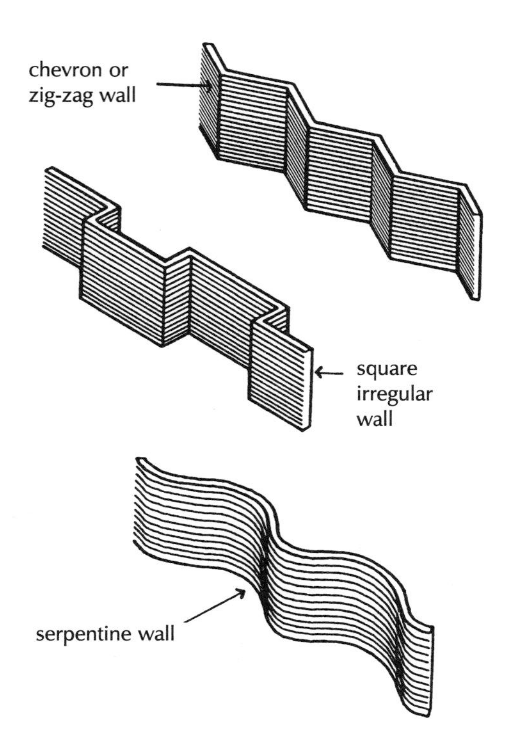

Fig. 44 Irregular profile walls.

thickness of the wall, such as from floors and roofs, and lateral forces, such as wind, tend to deform and overturn walls. The greater the eccentricity of the loads and the greater the lateral forces, the greater the tendency of a wall to deform, bow out of the vertical and lose stability. To prevent loss of stability, due to deformation under loads, building regulations and structural design calculations set limits to the height or thickness ratios (slenderness ratios) to provide reasonable stiffness against loss of stability due to deformation under load.

To provide stiffness against deformation under load, lateral, that is horizontal, restraint is provided by walls and roofs tied to the wall for stiffening up the height of the wall and by intersecting walls and piers that are bonded or tied to the wall as stiffening against deformation along the length of walls.

Irregular profile walls have greater stiffness against deformation than straight walls because of the buttressing effect of the angle of the zigzag, chevron, offset or serpentine profile of the walls, illustrated in Fig. 44. The more pronounced the chevron, zigzag, offset or serpentine of the wall, the stiffer it will be.

Similarly the diaphragm and fin walls, described in Volume 3, are stiffened against overturning and loss of stability by the cross ribs or diaphragms built across the wide cavity to diaphragm walls and the fins or piers that are built and bonded to straight walls in the fin wall construction.

Resistance to weather and ground moisture

A requirement of the Building Regulations is that walls should adequately resist the passage of moisture to the inside of the building. Moisture includes water vapour and liquid water. Moisture may penetrate a wall by absorption of water from the ground that is in contact with foundation walls or through rain falling on the wall.

To prevent water, which is absorbed from the ground by foundation walls, rising in a wall to a level where it might affect the inside of a building it is necessary to form a continuous, horizontal layer of some impermeable material in the wall. This impermeable layer, the damp-proof course, is built in, some 150 mm above ground level, to all foundation walls in contact with the ground and is joined to the damp-proof membrane in solid ground floors as described and illustrated in Chapter 1.

The ability of a wall to resist the passage of water to its inside face depends on its exposure to wind driven rain and the construction of the wall. The exposure of a wall is determined by its location and the extent to which it is protected by surrounding higher ground, or sheltered by surrounding buildings or trees, from rain driven by the prevailing winds. In Great Britain the prevailing, warm westerly winds from the Atlantic Ocean cause more severe exposure to driving rain along the west coast of the country than do the cooler easterly winds on the east coast.

British Standard 5628: Part 3 defines five categories of exposure as: very severe; moderate/severe; sheltered/moderate; sheltered; and very sheltered. A map of Great Britain, published by the Building Research Establishment, shows contours of the variations of exposure across the country. The contour lines, indicating the areas of the categories of exposure, are determined from an analysis of the most severe likely spells of wind driven rain, occurring on average every 3 years, plotted on a 10 km grid. The analysis is based on the 'worst case' for each geographical area, where a wall faces open country and the prevailing wind, such as a gable end wall on the edge of a suburban site facing the prevailing wind or a wall of a tall building on an urban site rising above the surrounding buildings and facing the prevailing wind.

Where a wall is sheltered from the prevailing winds by adjacent high ground or surrounding buildings or trees the exposure can be reduced by one category in sheltered areas of the country and two in very severe exposure areas of the country. The small-scale and large-scale maps showing categories of exposure to driving rain provide an overall picture of the likely severity of exposure over the country. To estimate the likely severity of exposure to driving rain, of the walls of a building on a particular site, it is wise to take account of the categories of exposure shown on the maps, make due allowance for the overlap of categories around contour lines and obtain local knowledge of conditions from adjacent buildings and make allowance for shelter from high ground, trees and surrounding buildings.

The behaviour of a wall in excluding wind and rain will depend on the nature of the materials used in the construction of the wall and how they are put together. A wall of facing bricks laid in mortar will absorb an appreciable amount of the rain driven on to it so that the wall must be designed so that the rain is not absorbed to the inside face of the wall. This may be effected by making the wall of sufficient thickness, by applying an external facing of say rendering or slate hanging, or by building the wall as a cavity wall of two skins or leaves with a separating cavity.

A curtain wall of glass (see Volume 4) on the other hand will not absorb water through the impermeable sheets of glass so that driving rain will pour down the face of the glass and penetrate the joints between the sheets of glass and the supporting frame of metal or wood, so that close attention has to be made to the design of these joints that at once have to be sufficiently resilient to accommodate thermal movement and at the same time compact enough to exclude wind and rain.

It is generally accepted practice today to construct walls of brick, stone or blocks as a cavity wall with an outer and inner leaf or skin separated by a cavity of at least 50 mm. The outer leaf will either be sufficiently thick to exclude rain or be protected by an outer skin of rendering or cladding of slate or tile and the inner leaf will be constructed of brick or block to support the weight of floors and roofs with either the inner leaf providing insulation against transfer of heat or the cavity filled with some thermal insulating material.

Durability and freedom from maintenance

The durability of a wall is indicated by the frequency and extent of the work necessary to maintain minimum functional requirements and an acceptable appearance. Where there are agreed minimum functional requirements such as exclusion of rain and thermal properties, the durability of walls may be compared through the cost of maintenance over a number of years. Standards of acceptable appearance may vary widely from person to person, particularly with unfamiliar wall surface materials such as glass and plastic coated sheeting, so that it is difficult to establish even broadly-based comparative standards of acceptable appearance. With the traditional wall materials there is a generally accepted view that a wall built of sound, well burned bricks or wisely chosen stone 'looks good' so that there is to a considerable extent a consensus of acceptable appearance for the traditional walling materials.

A wall built with sound, well burned bricks laid in a mortar of roughly the same density as the bricks and designed with due regard to the exposure of the wall to driving rain, and with sensible provisions of dpcs to walls around openings and to parapets and chimneys, should be durable for the anticipated life of the majority of buildings and require little if any maintenance and repair. In time, these materials exposed to wind and rain will slowly change colour. This imperceptible change will take place over many years and is described as weathering, that is a change of colour due to exposure to weather. It is generally accepted that this change enhances the appearance of brick and stone walls.

Walls built of brick laid in lime mortar may in time need repointing, to protect the mortar joints and maintain resistance to rain penetration and to improve the appearance of the wall.

Fire safety

Fires in buildings generally start from a small source of ignition, the 'outbreak of fire', which leads to the 'spread of fire' followed by a steady state during which all combustible material burns steadily up to the final 'decay stage'. It is in the early stages of a fire that there is most danger to the occupants of buildings from smoke and noxious fumes. The Building Regulations set standards for means of escape, limitation of spread of fire and containment of fire.

Fire safety regulations are concerned to assure a reasonable standard of safety in case of fire. The application of the regulations, as set out in the practical guidance given in Approved Document B, is directed to the safe escape of people from buildings in case of fire rather than the protection of the building and its contents.

The requirements of Part B of Schedule 1 to the Building Regulations are concerned to:

(1) provide adequate means of escape
(2) limit internal fire spread (linings)
(3) limit internal fire spread (structure)
(4) limit external fire spread
(5) provide access and facilities for the fire services

Means of escape

The requirements for means of escape from one and two storey houses are that each habitable room either opens directly on to a hallway or stair leading to the entrance, or that it has a window or door through which escape could be made and that means are provided for giving early warning in the case of fire. With increased height and size, where floors are more than 4.5 m above ground, it is necessary to protect internal stairways or provide alternative means of escape. Where windows and doors may be used as a means of escape their minimum size and the minimum and maximum height of window cills are defined.

Smoke alarms

To ensure the minimum level of safety it is recommended that all new houses should be fitted with self-contained smoke alarms permanently wired to a separately fused circuit at the distribution board. Battery-operated alarms are not acceptable. Where more than one smoke alarm is fitted they should be interconnected so that the detection of smoke by any one unit operates in all of them.

Internal fire spread (linings)

Fire may spread within a building over the surface of materials that encourage spread of flame across their surfaces, when subject to intense radiant heat, and those which give off appreciable heat when burning.

In Approved Document B is a classification of the performance of linings relative to surface spread of flame over wall and ceiling linings and limitations in the use of thermoplastic materials used in rooflights and lighting diffusers.

Internal fire spread (structure)

The premature failure of the structural stability of a building during fires is restricted by specifying a minimum period of fire resistance for the elements of the structure. An element of structure is defined as part of a structural frame, a loadbearing wall and a floor.

The requirements are that the elements should resist collapse for a minimum period of time in which the occupants may escape in the event of fire. Periods of fire resistance vary from 30 minutes for dwelling houses with a top floor not more than 5 m above ground, to 120 minutes for an industrial building, without sprinklers, whose top floor is not more than 30 m above ground.

For information on the requirements for buildings other than dwellings, concerning purpose groups, compartments, concealed spaces, external fire spread and access for the Fire Services, see Volume 4.

Resistance to the passage of heat

The traditional method of heating buildings was by burning wood or coal in open fireplaces in England and in freestanding stoves in much of northern Europe. The ready availability of wood and coal was adequate to the then modest demands for heating of the comparatively small population of those times. The highly inefficient open fire had the advantage of being a cosy focus for social life and the disadvantage of generating draughts of cold air necessary for combustion. The more efficient freestanding stove, which lacked the obvious cheery blaze of the open fire, was more suited to burning wood, the fuel most readily available in many parts of Europe.

The considerable increase in population that followed the Industrial Revolution and the accelerating move from country to town and city increased demand for the dwindling supplies of wood for burning. Coal became the principal fuel for open fires and freestanding stoves.

During the eighteenth century town gas became the principal source for lighting and by the nineteenth century had largely replaced solid fuels as the heat source for cooking. From about the middle of the twentieth century oil was used as the heat source for heating. Following the steep increase in the price of oil, town gas and later on natural gas was adopted as the fuel most used for heating.

Before the advent of oil and then gas as fuels for heating, it was possible to heat individual rooms by means of solid fuel burning open fires or stoves and people accepted the need for comparatively thick clothing for warmth indoors in winter.

With the adoption of oil and gas as fuels for heating it was possible to dispense with the considerable labour of keeping open fires and stoves alight and the considerable area required to store an adequate supply of solid fuels. With the adoption of oil and gas as fuel for heating it was practical to heat whole buildings and there was no longer the inconvenience of cold corridors, toilets and bathrooms and the draughts of cold air associated with open fireplaces. The population increasingly worked in heated buildings, many in sedentary occupations, so that tolerance of cold diminished and the expectation of thermal comfort increased.

For a description of the history of the development of heating appliances over the centuries and the increased use of thermal insulation, see Volume 2, Fires and Stoves.

Of recent years the expectation of improved thermal comfort in buildings, the need to conserve natural resources and the increasing cost of fuels have led to the necessity for improved insulation against transfer of heat. To maintain reasonable and economical conditions of thermal comfort in buildings, walls should provide adequate insulation against excessive loss or gain of heat, have adequate thermal storage capacity and the internal face of walls should be at a reasonable temperature.

For insulation against loss of heat, lightweight materials with low conductivity are more effective than dense materials with high conductivity, whereas dense materials have better thermal storage capacity than lightweight materials.

Where a building is continuously heated it is of advantage to use the thermal storage capacity of a dense material on the inside face of the wall with the insulating properties of a lightweight material behind it. Here the combination of a brick or dense block inner leaf, a cavity filled with some lightweight insulating material and an outer leaf of brick against penetration of rain is of advantage.

Where buildings are intermittently heated it is important that inside faces of walls warm rapidly, otherwise if the inside face were to remain cold, the radiation of heat from the body to the cold wall face would make people feel cold. The rate of heating of smooth wall surfaces is improved by the use of low density, lightweight materials on or immediately behind the inside face of walls.

The interior of buildings is heated by the transfer of heat from heaters and radiators to air (conduction), the circulation of heated air (convection) and the radiation of energy from heaters and radiators to surrounding colder surfaces (radiation). This internal heat is transferred through colder enclosing walls, roofs and floors by conduction, convection and radiation to colder outside air.

Conduction

The rate at which heat is conducted through a material depends mainly on the density of the material. Dense metals conduct heat more rapidly than less dense gases. Metals have high conductivity and gases low conductivity. Conductivity is the amount of heat per unit area, conducted in unit time through a material of unit thickness, per degree of temperature difference. Conductivity is expressed in watts per metre of thickness of material per degree kelvin (W/mK) and usually denoted by the Greek letter λ (lambda).

Convection

The density of air that is heated falls, the heated air rises and is replaced by cooler air. This in turn is heated and rises so that there is a continuing movement of air as heated air loses heat to surrounding cooler air and cooler surfaces of ceilings, walls and floors. Because the rate of transfer of heat to cooler surfaces varies from rapid transfer through thin sheet glass in windows to an appreciably slower rate of transfer through insulated walls, and because of the variability of the rate of exchange of cold outside air with warm inside air by ventilation, it is not possible to quantify heat transfer by convection. Usual practice is to make an assumption of likely total air changes per hour or volume (litres) per second and then calculate the heat required to raise the temperature of the incoming cooler air introduced by ventilation.

Radiation

Radiant energy from a body, radiating equally in all directions, is partly reflected and partly absorbed by another body and converted to heat. The rate of emission and absorption of radiant energy depends on the temperature and the nature of the surface of the radiating and receiving bodies. The heat transfer by low temperature radiation from heaters and radiators is small, whereas the very considerable radiant energy from the sun that may penetrate glass and that from high levels of artificial illumination is converted to appreciable heat inside buildings. An estimate of the solar heat gain and heat gain from artificial illumination may be assumed as part of the heat input to buildings.

Transmission of heat

Because of the complexity of the combined modes of heat transfer through the fabric of buildings it is convenient to use a coefficient of heat transmission as a comparative measure of transfer through the external fabric of buildings. This air-to-air heat transmittance coefficient, the U value, takes account of the transfer of heat by conduction through the solid materials and gases, convection of air in cavities and across inside and outside surfaces, and radiation to and from surfaces. The U value, which is expressed as W/m^2K, is the rate of heat transfer in watts through one square metre of a material or structure when the combined radiant and air temperatures on each side of the material or structure differ by 1 degree kelvin (1°C). A high rate of heat transfer is indicated by a high U value, such as that for single glazing of 5.3 (W/m^2K), and a low rate of heat transfer by a low U value, such as that for PIR insulation of 0.022 W/m^2K.

The U value may be used as a measure of the rate of transfer of heat through single materials or through a combination of materials such as those used in cavity wall construction.

Conservation of fuel and power

Standard assessment procedure (SAP) rating

The requirement in the Building Regulations for the conservation of fuel and power for dwellings alone is that the person carrying out building work of creating a new dwelling shall calculate the energy rating of the dwelling by a standard assessment procedure (SAP) and give notice of that rating to the local authority.

The SAP rating is based on an energy cost factor on a scale of 1 to 100, 1 being a maximum and 100 a minimum energy use to maintain a comfortable internal temperature and use of energy in water heating. While there is no obligation to achieve a particular rating, a rating of 60 or less indicates that there is inadequate insulation or inefficient heating systems or both, and the dwelling does not comply with the regulations.

Details of the notification of the SAP rating for new dwellings are held by the local authority. A prospective purchaser of the dwelling may well be put off where the rating is 60 or less and the local authority has not issued a Certificate of Compliance with the Regulations, whereas the purchaser will be encouraged by a rating of say 85, which shows compliance with the Regulations.

The SAP rating is calculated by the completion of a four page worksheet by reference to 14 tables. The sequential completion of up to 99 entries by reference to the 14 tables is so tedious and difficult to follow as to confound all but those initiated in their use, and is hardly calculated to inform householders in a way that is simple and easy to understand as claimed by the authors of Approval Document L.

Calculation methods

Three methods of calculating the figures necessary for the SAP for dwellings are proposed in Approved Document L. They are:

(1) an elemental method
(2) a target U value method
(3) an energy rating method.

In the elemental method standard U values for the exposed elements of the fabric of buildings are shown under two headings: (a) for dwellings with SAP ratings of 60 or less and (b) for those with SAP ratings over 60. The standard U values are 0.2 and 0.25 W/m^2K for roofs, 0.45 W/m^2K for exposed walls, 0.35 and 0.45 W/m^2K for exposed floors and ground floors, 0.6 W/m^2K for semi-exposed walls and floors and 3.0 and 3.3 W/m^2K for windows, doors and rooflights, the two values being for headings (a) and (b), respectively. The basic allowance for the area of windows, doors and rooflights together is 22.5% of the total floor area. The area of windows, doors and rooflights, larger than those indicated by the percentage value, may be used providing there is a compensating improvement in the average U value by the use of glazing with a lower U value.

As it is unlikely that the SAP rating of the majority of new dwellings, complying with standard U values, will fall below 60, the over 60 rating values are the relevant ones.

The target U value method for dwellings is used to meet the requirement for conservation of fuel and power by relating a calculated average U value to a target U value, which it should not exceed. The average U value is the ratio of:

$$\frac{\text{total rate of heat loss per degree}}{\text{total external surface area}}$$

The target U value is:

$$\frac{\text{total floor area} \times 0.57}{\text{total area of exposed elements}} + 0.36 \text{ for dwellings with SAP ratings 60 or less, and}$$

$$\frac{\text{total floor area} \times 0.64}{\text{total area of exposed elements}} + 0.40 \text{ for dwellings with SAP ratings of more than 60}$$

The total area of exposed floors, windows, doors, walls and roof and the standard U values in the elemental method are used to calculate the heat loss per degree. Where the calculated average U value exceeds the target U value it is necessary to improve the thermal resistance of walls, windows or roof either separately or together so that the average U value does not exceed the target U value. As an option, account may be taken of solar heat gains other than those allowed for in the equation on which the method is based. This method is based on the assumption of a boiler with an efficiency of at least 72%. Where a boiler with an efficiency of 85% is used the target U value may be increased by 10%. The use of the elemental or target U value methods of showing compliance does not give exemption from the requirement to give notice of an SAP rating.

The energy rating method is a calculation based on SAP which allows the use of any valid conservation measures. The calculation takes account of ventilation rate, fabric losses, water heating requirements and internal and solar heat gains.

The requirement for conservation of fuel and power will be met if the SAP energy rating for the dwelling, or each dwelling in a block of flats or converted building, is related to the floor area of the dwelling and ranges from 80 for dwellings with a floor area of 80 m^2 or less to 85 for dwellings with a floor area of more than 120 m^2.

As there is a requirement to complete the SAP worksheet to determine an SAP rating, which has to be notified to the local authority, whichever method of showing compliance is used the most practical and economic method of approach is to use the standard U values for SAP ratings over 60 set out in the elemental method in the initial stages of design, and then to complete the SAP worksheet at a later stage and make adjustments to the envelope insulation, windows and boiler efficiency as is thought sensible to achieve a high SAP rating.

For a description of the requirements for conservation of fuel and power for all buildings other than dwellings see Volume 4.

Ventilation

The sensation of comfort is highly subjective and depends on the age, activity and to a large extent on the expectations of the subject. The young 'feel' cold less than the old and someone engaged in heavy manual work has less need of heating than another engaged in sedentary work. It is possible to provide conditions of thermal comfort that suit the general expectations of those living or working in a building. None the less, some may 'feel' cold and others 'feel' hot.

For comfort and good health in buildings it is necessary to provide means of ventilation through air changes through windows or ventilators, that can be controlled, depending on wind speed and direction and outside air temperature, to avoid the sensation of 'stuffiness' or cold associated with too infrequent or too frequent air changes respectively. As with heating, the sensation of stuffiness is highly subjective.

Air changes

For general guidance a number of air changes per hour is recommended, depending on the activity common to rooms or spaces. One air change each hour for dwellings and three for kitchens and sanitary accommodation is recommended. The more frequent air changes for kitchen and sanitary accommodation is recommended to minimise condensation of moisture-laden, warm air on cold internal surfaces in those rooms.

Condensation

Condensation is the effect of moisture from air collecting on a surface colder than the air, for example in a bathroom or kitchen where water from warm moisture-laden air condenses on to the cold surfaces of walls and glass. To minimise condensation, ventilation of the room to exchange moisture-laden air with drier outside air and good insulation of the inner face of the wall are required.

A consequence of the need for internal air change in buildings is that the heat source must be capable of warming the incoming air to maintain conditions of thermal comfort, and the more frequent the air change the greater the heat input needed. The major source of heat loss through walls is by window glass which is highly conductive to heat transfer. This heat loss can be reduced to some small extent by the use of double glazing. Most of the suppliers of double glazed windows provide one of the very effective air seals around all of the opening parts of their windows. These air seals are very effective in excluding the draughts of cold air that otherwise would penetrate the necessary gaps around opening windows and so serve to a large extent to reduce the heat loss associated with opening windows to an extent that they may reduce air changes to an uncomfortable level.

There is a fine balance between the need for air change and the expectations of thermal comfort that receives too little consideration in the design of windows.

Resistance to the passage of sound

Sound is transmitted as airborne sound and impact sound. Airborne sound is generated as cyclical disturbances of air from, for example, a radio, that radiate from the source of the sound with diminishing intensity. The vibrations in the air caused by the sound source will set up vibrations in enclosing walls and floors which will cause vibrations of air on the opposite side of walls and floors.

Impact sound is caused by contact with a surface, as for example the slamming of a door or footsteps on a floor which set up vibrations in walls and floors that in turn cause vibrations of air around them that are heard as sound.

The most effective insulation against airborne sound is a dense barrier such as a solid wall which absorbs the energy of the airborne sound waves. The heavier and more dense the material of the wall the more effective it is in reducing sound. The Building Regulations require walls and floors to provide reasonable resistance to airborne sound between dwellings and between machine rooms, tank rooms, refuse chutes and habitable rooms. A solid wall, one brick thick, or a solid cavity wall plastered on both sides is generally considered to provide reasonable sound reduction between dwellings at a reasonable cost. The small reduction in sound transmission obtained by doubling the thickness of a wall is considered prohibitive in relation to cost.

For reasonable reduction of airborne sound between dwellings one above the other, a concrete floor is advisable.

The more dense the material the more readily it will transmit impact sound. A knock on a part of a rigid concrete frame may be heard some considerable distance away. Insulation against impact sound will therefore consist of some absorbent material that will act to cushion the impact, such as a carpet on a floor, or serve to interrupt the path of the sound, as for example the absorbent pads under a floating floor.

Noise generated in a room may be reflected from the walls and ceilings and build up to an uncomfortable intensity inside the room, particularly where the wall and ceiling surfaces are hard and smooth. To prevent the build-up of reflected sound some absorbent material should be applied to walls and ceilings, such as acoustic tiles or curtains, to absorb the energy of the sound waves.

BRICK AND BLOCK WALLS

The majority of the walls of small buildings in this country are built of brick or block. The external walls of heated buildings, such as houses, are built as a cavity wall with an outer leaf of brick, a cavity and an inner leaf of concrete blocks. Internal walls and partitions are built, in the main, of concrete blocks.

BRICKS

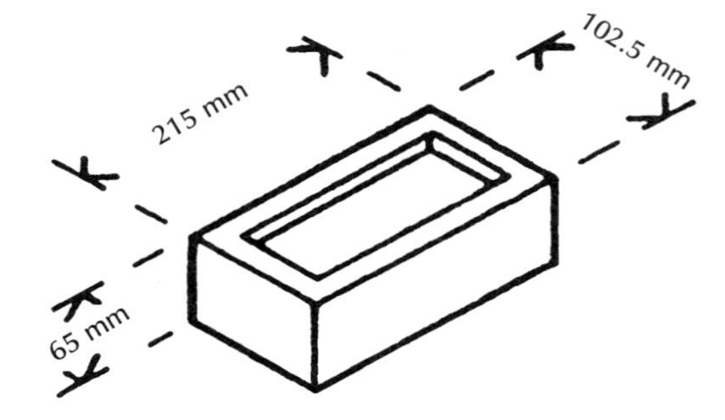

Fig. 45 Standard brick.

The word brick is used to describe a small block of burned clay of such size that it can be conveniently held in one hand and is slightly longer than twice its width. Blocks made from sand and lime or concrete are manufactured in clay brick size and these are also called bricks. The great majority of bricks in use today are of clay.

The standard brick is 215 × 102.5 × 65 mm, as illustrated in Fig. 45, which with a 10 mm mortar joint becomes 225 × 112.5 × 75 mm.

Materials from which bricks are made

Clay

In this country there are very extensive areas of clay soil suitable for brickmaking. Clay differs quite widely in composition from place to place and the clay dug from one part of a field may well be quite different from that dug from another part of the same field. Clay is ground in mills, mixed with water to make it plastic and moulded, either by hand or machine, to the shape and size of a brick.

Bricks that are shaped and pressed by hand in a sanded wood mould and then dried and fired have a sandy texture, are irregular in shape and colour and are used as facing bricks due to the variety of their shape, colour and texture.

Machine made bricks are either hydraulically pressed in steel moulds or extruded as a continuous band of clay. The continuous band of clay, the section of which is the length and width of a brick, is cut into bricks by a wire frame. Bricks made this way are called ‘wire cuts’.

Wire cuts

Press moulded bricks generally have a frog or indent and wire cuts have none. The moulded brick is baked to dry out the water and burned at a high temperature so that part of the clay fuses the whole mass of the brick into a hard durable unit. If the moulded brick is burned at too high a temperature part of the clay fuses into a solid glass-like mass and if it is burned at too low a temperature no part of the clay fuses and the brick is soft. Neither overburned nor underburned brick is satisfactory for building purposes.

A brick wall has very good fire resistance, is a poor insulator against transference of heat, does not, if well built, deteriorate structurally and requires very little maintenance over a long period of time. Bricks are cheap because there is an abundance of the natural material from which they are made, that is clay. The clay can easily be dug out of the ground, it can readily be made plastic for moulding into brick shapes and it can be burned into a hard, durable mass at a temperature which can be achieved with quite primitive equipment.

Because there is wide variation in the composition of the clays suitable for brick making and because it is possible to burn bricks over quite a wide range of temperatures sufficient to fuse the material into a durable mass, a large variety of bricks are produced in this country. The bricks produced which are suitable for building vary in colour from almost dead white to practically black and in texture from almost as smooth as glass to open coarse grained. Some are quite light in weight and others dense and heavy and there is a wide selection of colours, textures and densities between the extremes noted.

It is not possible to classify bricks simply as good and bad as some are good for one purpose and not for another. Bricks may be classified in accordance with their uses as commons, facing and engineering bricks or by their quality as internal quality, ordinary quality and special quality. The use and quality classifications roughly coincide, as commons are much used for internal walls, facing or ordinary quality for external walls and engineering or special quality bricks for their density and durability in positions of extreme exposure. In cost, commons are cheaper than facings and facings cheaper than engineering bricks.

Types of brick

Commons

These are bricks which are sufficiently hard to safely carry the loads normally supported by brickwork, but because they have a dull texture or poor colour they are not in demand for use as facing bricks which show on the outside when built and affect the appearance of buildings. These 'common' bricks are used for internal walls and for rear walls which are not usually exposed to view. Any brick which is sufficiently hard and of reasonably good shape and of moderate price may be used as a 'common' brick. The type of brick most used as a common brick is the Fletton brick.

Facings

This is by far the widest range of bricks as it includes any brick which is sufficiently hard burned to carry normal loads, is capable of withstanding the effects of rain, wind, soot and frost without breaking up and which is thought to have a pleasant appearance. As there are as many different ideas of what is a pleasant looking brick as there are bricks produced, this is a somewhat vague classification.

Engineering bricks

These are bricks which have been made from selected clay, which have been carefully prepared by crushing, have been very heavily moulded and carefully burned so that the finished brick is very solid and hard and is capable of safely carrying much heavier loads than other types of brick. These bricks are mainly used for walls carrying exceptionally heavy loads, for brick piers and general engineering works. The two best known engineering bricks are the red Southwater brick and the blue Staffordshire brick. Both are very hard, dense and do not readily absorb water. The ultimate crushing resistance of engineering bricks is greater than $50\,N/mm^2$.

Semi-engineering bricks

These are bricks which, whilst harder than most ordinary bricks, are not so hard as engineering bricks. It is a very vague classification without much meaning, more particularly as a so-called semi-engineering brick is not necessarily half the price of an engineering brick.

Composition of clay

Clays suitable for brick making are composed mainly of silica in the form of grains of sand and alumina, which is the soft plastic part of clay which readily absorbs water and makes the clay plastic and which melts when burned. Present in all clays are materials other than the two mentioned above such as lime, iron, manganese, sulphur and phosphates. The proportions of these materials vary widely and the following is a description of the composition, nature and uses of some of the most commonly used bricks classified according to the types of clay from which they are produced.

Flettons

There are extensive areas of what is known as Oxford clay. The clay is composed of just under half silica, or sand, about one-sixth alumina, one-tenth lime and small measures of other materials such as iron, potash and sulphur. The clay lies in thick beds which are economical to excavate. In the clay, in its natural state, is a small amount of mineral oil which, when the bricks are burned, ignites and assists in the burning.

Because there are extensive thick beds of the clay, which are economical to excavate, and because it contains some oil, the cheapest of all clay bricks can be produced from it. The name Fletton given to these bricks derives from the name of a suburb of Peterborough around which the clay is extensively dug for brickmaking. Flettons are cheap and many hundreds of millions of them are used in building every year. The bricks are machine moulded and burned and the finished brick is uniform in shape with sharp square edges or arises. The bricks are dense and hard and have moderately good strength; the average pressure at which these bricks fail, that is crumble, is around 21 N/mm^2

The bricks are light creamy pink to dull red in colour and because of the smooth face of the brick what are known as 'kiss marks' are quite distinct on the long faces. These 'kiss marks' take the form of three different colours, as illustrated in Fig. 46.

The surface is quite hard and smooth and if the brick is to be used for wall surfaces to be plastered, two faces are usually indented with grooves to give the surface a better grip or key for plaster. The bricks are then described as 'keyed Flettons'. Figure 47 is an illustration of a keyed Fletton.

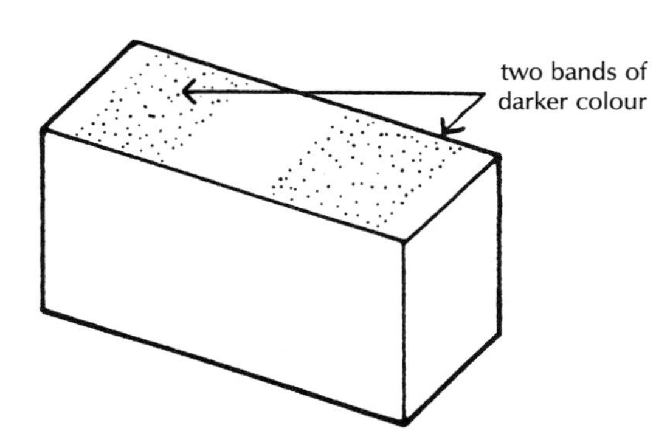

Fig. 46 Fletton brick showing kiss marks.

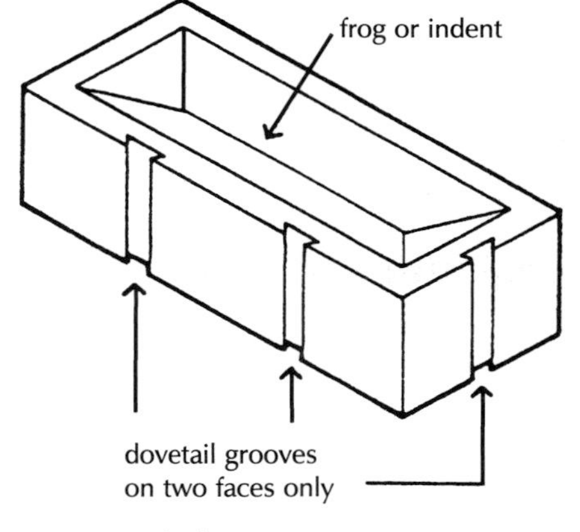

Fig 47 Keyed Fletton.

Stocks

The term 'stock brick' is generally used in the south-east counties of England to describe the London stock brick. This is a brick manufactured in Essex and Kent from clay composed of sand and alumina to which some chalk is added. Some combustible material is added to the clay to assist burning. The London stock is usually predominantly yellow after burning with shades of brown and purple. The manufacturers grade the bricks as 1st Hard, 2nd Hard and Mild, depending on how burned they are. The bricks are usually irregular in shape and have a fine sandy texture. Because of their colour they are sometimes called 'yellow stocks'. 1st Hard and 2nd Hard London stocks were much used in and around London as facings as they weather well and were of reasonable price. In other parts of England the term stock bricks describes the stock output of any given brick field.

Marls

By origin the word marl denotes a clay containing a high proportion of lime (calcium carbonate), but by usage the word marl is taken to denote any sandy clay. This derives from the use of sandy clays, containing some lime, as a top dressing to some soils to increase fertility. In most of the counties of England there are sandy clays, known today as marls, which are suitable for brick making. Most of the marl clays used for brick making contain little or no lime. Many of the popular facing bricks produced in the Midlands are made from this type of clay and they have a good shape, a rough sandy finish and vary in colour from a very light pink to dark mottled red.

Gaults

The gault clay does in fact contain a high proportion of lime and the burned brick is usually white or pale pink in colour. These bricks are of good shape and texture and make good facing bricks, and are more than averagely strong. The gault clay beds are not extensive in this country and lie around limestone and chalk hills in Sussex and Hampshire.

Clay shale bricks

Some clay beds have been so heavily compressed over the centuries by the weight of earth above them that the clay in its natural state is quite firm and has a compressed flaky nature. In the coal mining districts of this country a considerable quantity of clay shale has to be dug out to reach coal seams and in those districts the extracted shale is used extensively for brick making. The bricks produced from this shale are usually uniform in shape with smooth faces and the bricks are hard and durable. The colour of the bricks is usually dull buff, grey, brown or red. These bricks are used as facings, commons and semi-engineering, depending on their quality.

Calcium silicate bricks

Calcium silicate bricks are generally known as sand-lime bricks. The output of these bricks has increased over the past few years, principally because the output of Fletton bricks could not keep pace with the demand for a cheap common brick and sand-lime bricks have been mainly used as commons. The bricks are made from a carefully controlled mixture of clean sand and hydrated lime which is mixed together with water, heavily moulded to brick shape and then the moulded brick is hardened in a steam oven. The resulting bricks are very uniform in shape and colour and are normally a dull white. Coloured sand-lime bricks are made by adding a colouring matter during manufacture. These bricks are somewhat more expensive than Flettons and because of their uniformity in shape and colour they are not generally thought of as being a good facing brick. The advantage of them however is that the material from which they are made can be carefully selected and accurately proportioned to ensure a uniform hardness, shape and durability quite impossible with the clay used for most bricks.

Flint-lime bricks

Flint-lime bricks are manufactured from hydrated lime and crushed flint and are moulded and hardened as are sand-lime bricks. They are identical with sand-lime bricks in all respects.

Cellular and perforated bricks, illustrated in Fig. 48, are machine press moulded from plastic clays, either pressed from wire cuts or separately formed. The purpose of forming the hollows and perforations is to reduce the volume of moulded, wet clay, the better to control shrinkage and deformation during drying and burning to produce more uniformly shaped bricks.

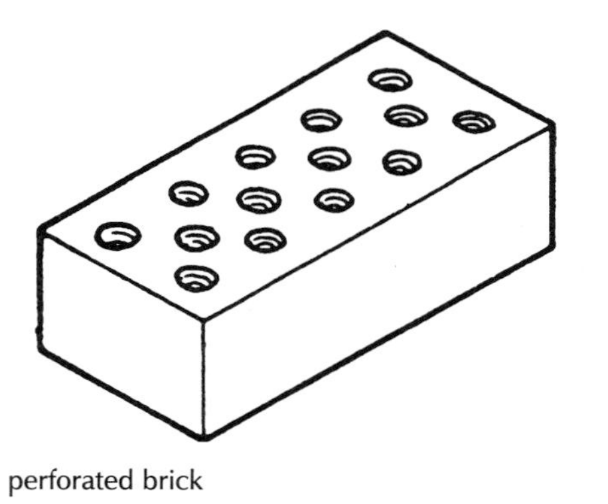

perforated brick

cellular pressed brick

Fig. 48 Cellular and perforated bricks.

These bricks, which have good crushing strength and are of semi-engineering quality, are used extensively in the brick enclosures to inspection pits and chambers for underground cable, for foundations and basements where the uniform shape and the density of the brick is an advantage.

They may be used for external walls as the perforations and hollows do not affect the weathering properties of the wall and may provide some little increase in insulation.

Special bricks

A range of special bricks is made for specific uses in fairface brickwork. These bricks are made from fine clays to control and reduce shrinkage deformation during firing. The finished bricks are dense, to

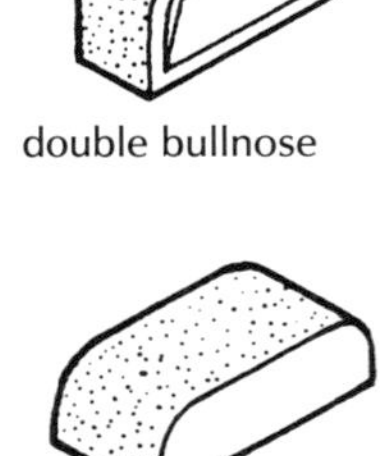

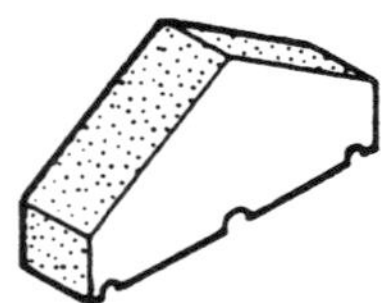

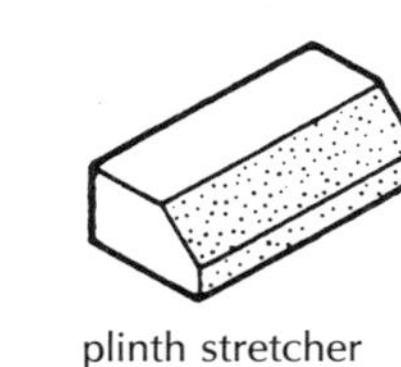

Fig. 49 Special bricks.

resist damage by exposure to rain and cold in the exposed positions in which they are most used. Figure 49 is an illustration of some typical specials.

The two coping bricks, the half round and the saddleback, are for use as coping to a 1B thick parapet wall. The bricks are some 50 mm or more wider than the thickness of a 1B wall so that when laid they overhang the wall each side to shed rainwater and the grooves on the underside of the hangover form a drip edge.

The two bullnose specials are made as a capping or coping for 1B walls. They are of the same length as the thickness of the wall on which they are laid, to provide protection and a flush finish to the wall.

The plinth bricks are used to provide and cap a thickening to the base of walls, a plinth, for the sake of appearance. A plinth at the base of a wall gives some definition to the base of a wall as compared to the wall being built flush out of the ground. A plinth may be formed by the junction of a solid $1\frac{1}{2}$B wall built from the foundation up to a cavity wall.

Properties of bricks

Hardness

This is a somewhat vague term commonly used in the description of bricks. By general agreement it is recognised that a brick which is to have a moderately good compressive strength, reasonable resistance to saturation by rainwater and sufficient resistance to the disruptive action of frost should be hard burned. Without some experience in the handling, and of the behaviour, of bricks in general it is very difficult to determine whether or not a particular brick is hard burned.

A method of testing for hardness is to hold the brick in one hand and give it a light tap with a hammer. The sound caused by the blow should be a dull ringing tone and not a dull thud. Obviously different types of brick will, when tapped, give off different sorts of sound and a brick that gives off a dull sound when struck may possibly be hard burned.

Compressive strength

This is a property of bricks which can be determined accurately. The compressive strength of bricks is found by crushing 12 of them individually until they fail or crumble. The pressure required to crush them is noted and the average compressive strength of the brick is stated as newtons per mm of surface area required to ultimately crush the brick. The crushing resistance varies from about 3.5 N/mm^2 for soft facing bricks up to 140 N/mm^2 for engineering bricks.

The required thickness of an external brick wall is determined primarily by its ability to absorb rainwater to the extent that water does not penetrate to the inside face of the wall. In positions of moderate exposure to wind driven rain a brick wall 215 mm thick may absorb so much water that it penetrates to the inside face.

The bearing strength of a brick wall 215 mm thick is very much greater than the loads a wall will usually carry. The current external wall to small buildings such as houses is built as a cavity wall with a 102.5 mm external leaf of brick, a cavity and an inner leaf of block. The external leaf is sufficiently thick, with the cavity, to prevent penetration of rain to the inside face and more than thick enough to support the loads it carries.

It is for heavily loaded brick piers and walls that the crushing strength of brick is a prime consideration.

The average compressive strength of some bricks commonly used is:

Mild (i.e. soft) stocks	3.5 N/mm^2
2nd Hard stocks	17.5 N/mm^2
Flettons	21 N/mm^2
Southwater A	70 N/mm^2

Absorption

Scientific work has been done to determine the amount of water absorbed by bricks and the rate of absorption, in an attempt to arrive at some scientific basis for grading bricks according to their resistance to the penetration of rain. This work has to date been of little use to those concerned with general building work. A wall built of very hard bricks which absorb little water may well be more readily penetrated by rainwater than one built of bricks which absorb a lot of water. This is because rain will more easily penetrate a small crack in the mortar between bricks if the bricks are dense than if the bricks around the mortar are absorptive.

Experimental soaking in water of bricks gives a far from reliable guide to the amount of water they can absorb as air in the pores and minute holes in the brick may prevent total absorption and to find total absorption the bricks have to be boiled in water or heated. The amount of water a brick will absorb is a guide to its density and therefore its strength in resisting crushing, but is not a reasonable guide to its ability to weather well in a wall. This term 'weather well' describes the ability of the bricks in a particular situation to suffer rain, frost and wind without losing strength, without crushing and to keep their colour and texture.

Frost resistance

A few failures of brickwork due to the disruptive action of frost have been reported during the last 30 years and scientific work has sought to determine a brick's resistance to frost failure. Most of the failures reported were in exposed parapet walls or chimney stacks where brickwork suffers most rain saturation and there is a likelihood of damage by frost. Few failures of ordinary brick walls below roof level have been reported. Providing sensible precautions are taken in the design of parapets and stacks above roof level and brick walls in general are protected from saturation by damaged rainwater gutters or blocked rainwater pipes there seems little likelihood of frost damage in this country.

Parapet walls, chimney stacks and garden walls should be built of sound, hard burned bricks protected with coping, cappings and damp-proof courses.

Efflorescence

Clay bricks contain soluble salts that migrate, in solution in water, to the surface of brickwork as water evaporates to outside air. These salts will collect on the face of brickwork as an efflorescence (flowering) of white crystals that appear in irregular, unsightly patches. This efflorescence of white salts is most pronounced in parapet walls, chimneys and below dpcs where brickwork is most liable to saturation. The concentration of salts depends on the soluble salt content of the bricks and the degree and persistence of saturation of brickwork.

The efflorescence of white salts on the surface is generally merely unsightly and causes no damage. In time these salts may be washed from surfaces by rain. Heavy concentration of salts can cause spalling and powdering of the surface of bricks, particularly those with smooth faces, such as Flettons. This effect is sometimes described as crypto efflorescence. The salts trapped behind the smooth face of bricks expand when wetted by rain and cause the face of the bricks to crumble and disintegrate.

Efflorescence may also be caused by absorption of soluble salts from a cement rich mortar or from the ground, that appear on the face of brickwork that might not otherwise be subject to efflorescence. Some impermeable coating between concrete and brick can prevent this (see Volume 4). There is no way of preventing the absorption of soluble salts from the ground by brickwork below the horizontal dpc level, although the effect can be reduced considerably by the use of dense bricks below the dpc.

Sulphate attack on mortars and renderings

When brickwork is persistently wet, as in foundations, retaining walls, parapets and chimneys, sulphates in bricks and mortar may in time crystallise and expand and cause mortar and renderings to disintegrate. To minimise this effect bricks with a low sulphate content should be used.

BONDING BRICKS

In building a wall it is usual to lay bricks in regular, horizontal courses so that each brick bears on two bricks below. The bricks are said to be bonded as they bind together by being laid across each other along the length of the wall, as illustrated in Fig. 50.

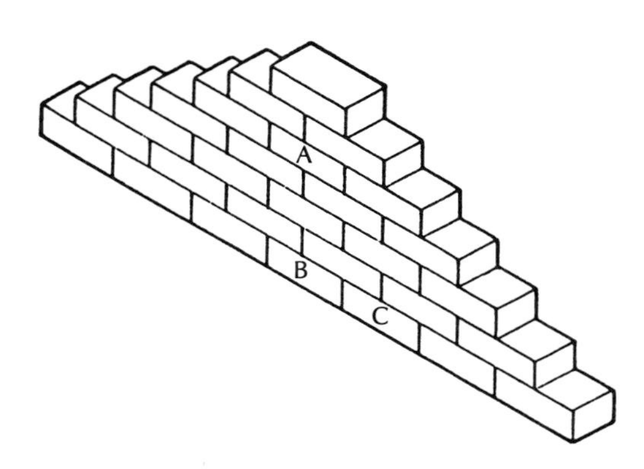

Fig. 50 Bricks stacked pyramid fashion.

The advantage of bonding is that the wall acts as a whole so that the load of a beam carried by the topmost brick in Fig. 50 is spread to the two bricks below it, then to the three below that and so on down to the base or foundation course of bricks.

The failure of one poor quality brick such as 'A in a wall and a slight settlement under part of the foundation such as 'B' and 'C' in Fig. 50 will not affect the strength and stability of the whole wall as the load carried by the weak brick and the two foundation bricks is transferred to the adjacent bricks.

Because of the bond, window and door openings may be formed in a wall, the load of the wall above the opening being transferred to the brickwork each side of the openings by an arch or lintel.

The effect of bonding is to stiffen a wall along its length and also to some small extent against lateral pressure, such as wind.

Stretcher bond

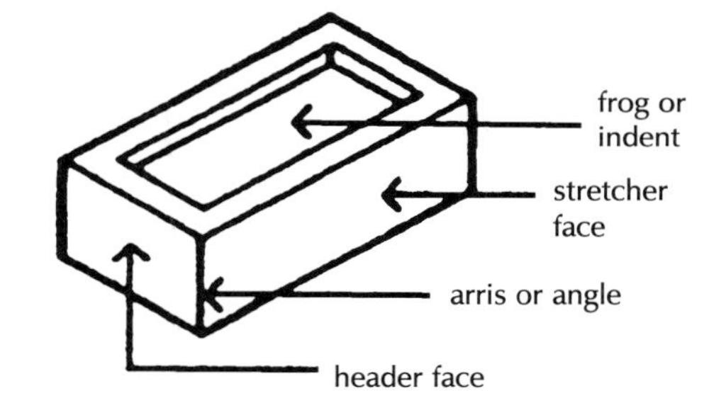

Fig. 51 Brick faces.

The four faces of a brick which may be exposed in fairface brickwork are the two, long, stretcher faces and the two header faces illustrated in Fig. 51. The face on which the brick is laid is the bed. Some bricks have an indent or frog formed in one of the bed faces. The purpose of the frog or indent is to assist in compressing the wet clay during moulding. The frog also serves as a reservoir of mortar on to which bricks in the course above may more easily be bedded.

The thickness of a wall is dictated primarily by the length of a brick. The length of bricks varies appreciably, especially those that are hand moulded and those made from plastic clays that will shrink differentially during firing.

It has been practice for some time to describe the thickness of a wall by reference to the length of a brick as a 1 B (brick) wall, a $1\frac{1}{2}$ B wall or a 2 B wall, rather than a precise dimension.

The external leaf of a cavity wall is often built of brick for the advantage of the appearance of brickwork. The most straightforward way of laying bricks in a thin outer leaf of a cavity wall is with the stretcher face of each brick showing externally. So that bricks are bonded along the length of the wall they are laid with the vertical joints between bricks lying directly under and over the centre of bricks in the courses under and over. This is described as stretcher bond as illustrated in Fig. 52. This wall is described as a $\frac{1}{2}$ B thick wall.

At the intersection of two half brick walls at corners or angles and at the jambs, sides of openings, the bricks are laid so that a header face shows in every other course to complete the bond, as illustrated in Fig. 52.

The appearance of a wall laid in stretcher bond may look somewhat monotonous because of the mass of stretcher faces showing. To

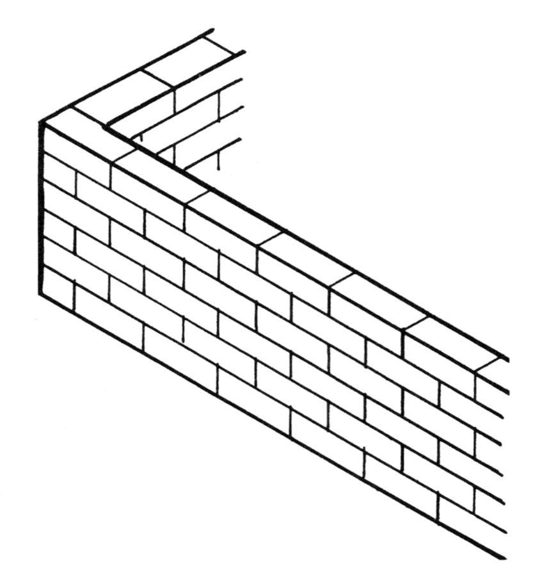

Fig. 52 Stretcher bond.

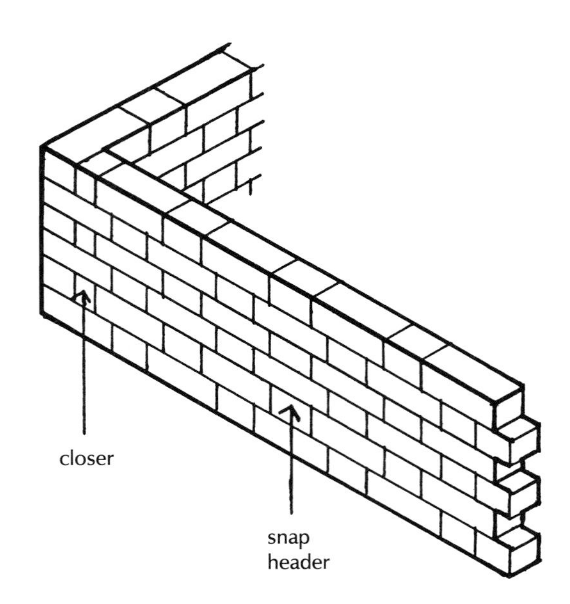

Fig. 53 Flemish bond with snap headers.

provide some variety the wall may be built with snap headers so that a stretcher face and a header face show alternately in each course with the centre of the header face lying directly under and over the centre of the stretcher faces in courses below and above, as illustrated in Fig. 53.

This form of fake Flemish bond is achieved by the use of half bricks, hence the name 'snap header'. The combination and variety in colour and shape can add appreciably to the appearance of a wall. Obviously the additional labour and likely wastage of bricks adds somewhat to cost.

English and Flemish bond

Because brick by itself does not provide adequate resistance to the transfer of heat, to meet the requirements of the Building Regulations for the conservation of fuel and power, it is used in combination with other materials in external cavity walling for most heated buildings. In consequence brick walling 1 B and thicker is less used than it was.

Solid brick walls may be used for heated and unheated buildings for arcades, screen walling and as boundary and earth retaining walling for the benefit of the appearance and durability of the material.

For the same reason that a $\frac{1}{2}$ B wall is bonded along its length a solid wall 1 B and thicker is bonded along its length and through its thickness.

The two basic ways in which a solid brick wall may be bonded are with every brick showing a header face with each header face lying directly over two header faces below or with header faces centrally over a stretcher face in the course below, as illustrated in Fig. 54.

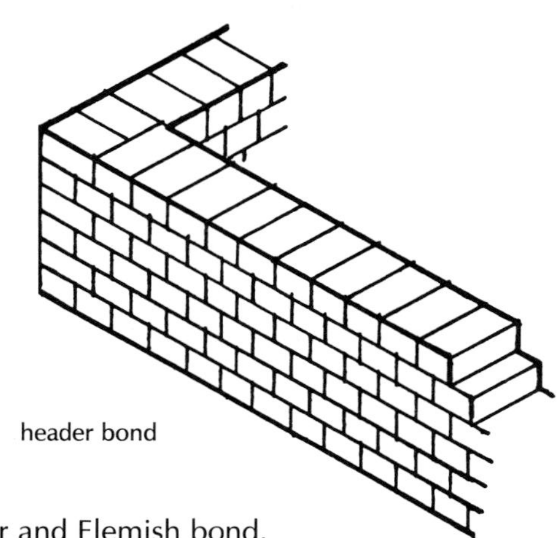

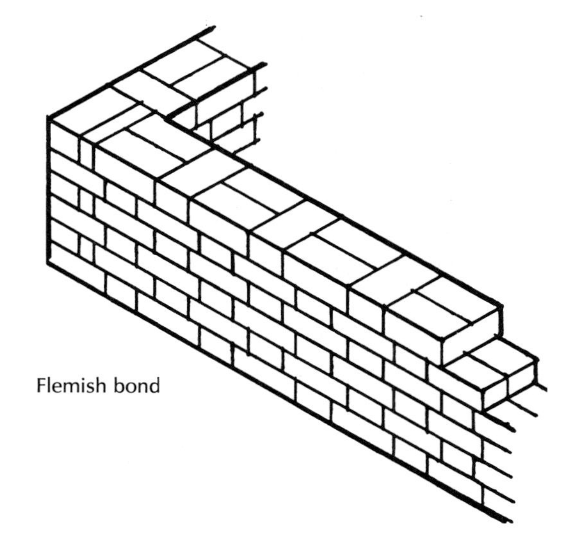

Fig. 54 Header and Flemish bond.

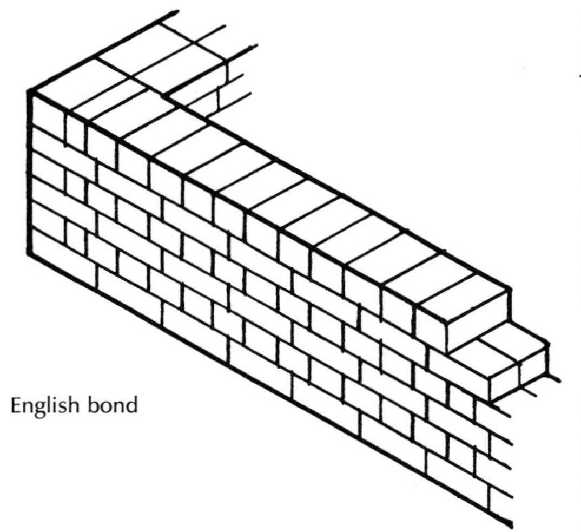

Fig. 55 English bond.

The bond in which header faces only show is termed 'heading' or 'header bond'. This bond is little used as the great number of vertical joints and header faces is generally considered unattractive.

The bond in which header faces lie directly above and below a stretcher face is termed Flemish bond. This bond is generally considered the most attractive bond for facing brickwork because of the variety of shades of colour between header and stretcher faces dispersed over the whole face of the walling. Figure 54 illustrates brickwork in Flemish bond.

English bond, illustrated in Fig. 55, avoids the repetition of header faces in each course by using alternate courses of header and stretcher faces with a header face lying directly over the centre of a stretcher face below. The colour of header faces, particularly in facing bricks, is often distinctly different from the colour of stretcher faces. In English bond this difference is shown in successive horizontal courses. In Flemish bond the different colours of header and stretcher faces are dispersed over the whole face of a wall, which by common consent is thought to be a more attractive arrangement.

Bonding at angles and jambs

At the end of a wall at a stop end, at an angle or quoin and at jambs of openings the bonding of bricks has to be finished up to a vertical angle. To complete the bond a brick $\frac{1}{4}$ B wide has to be used to close or complete the bond of the $\frac{1}{4}$ B overlap of face brickwork.

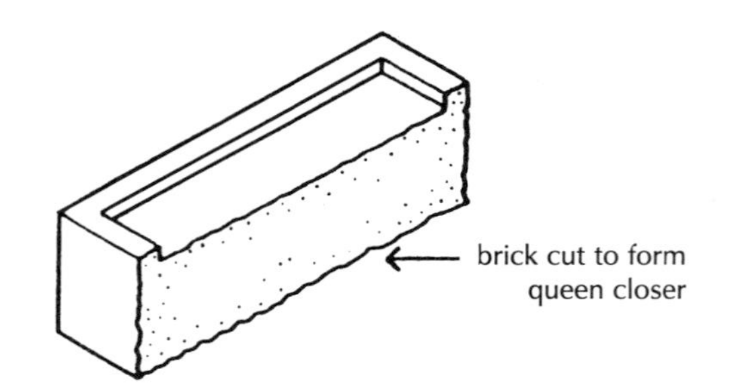

Fig. 56 Queen closer.

A brick, cut in half along its length, is used to close the bond at an angle. This cut brick is termed a 'queen closer', illustrated in Fig. 56. If the narrow width queen closer were laid at the angle, it might be displaced during bricklaying. To avoid this possibility the closer is laid next to a header, as illustrated in Fig. 57. The rule is that a closer is laid next to a quoin (corner) header.

There is often an appreciable difference in the length of facing bricks so that a solid wall 1 B thick may be difficult to finish as a wall fairface both sides. The word fairface describes a brick wall finished with a reasonably flat and level face for the sake of appearance. Where a 1 B wall is built with bricks of uneven length it may be necessary to select bricks of much the same length as headers and use longer bricks as stretchers. This additional care and labour will add appreciably to costs.

Walls $1\frac{1}{2}$B thick may be used for substantial walling for larger buildings, such as industrial, storage and civic, for the sake of the appearance of the brickwork and the durability and sense of solidity and permanence where the walling is finished fairface both sides.

To complete the bond of a solid wall $1\frac{1}{2}$B thick in double Flemish bond, that is Flemish bond on both faces, it is necessary to use cut half bricks in the thickness of the wall as illustrated in Fig. 57. At angles and stop ends of wall, queen closers are laid next to quoin headers and a three quarter length cut brick is used, as illustrated in Fig. 57.

Cutting the many half length bricks ($\frac{1}{2}$ bats) and three quarter length bricks and closers is time consuming and wasteful as it is not

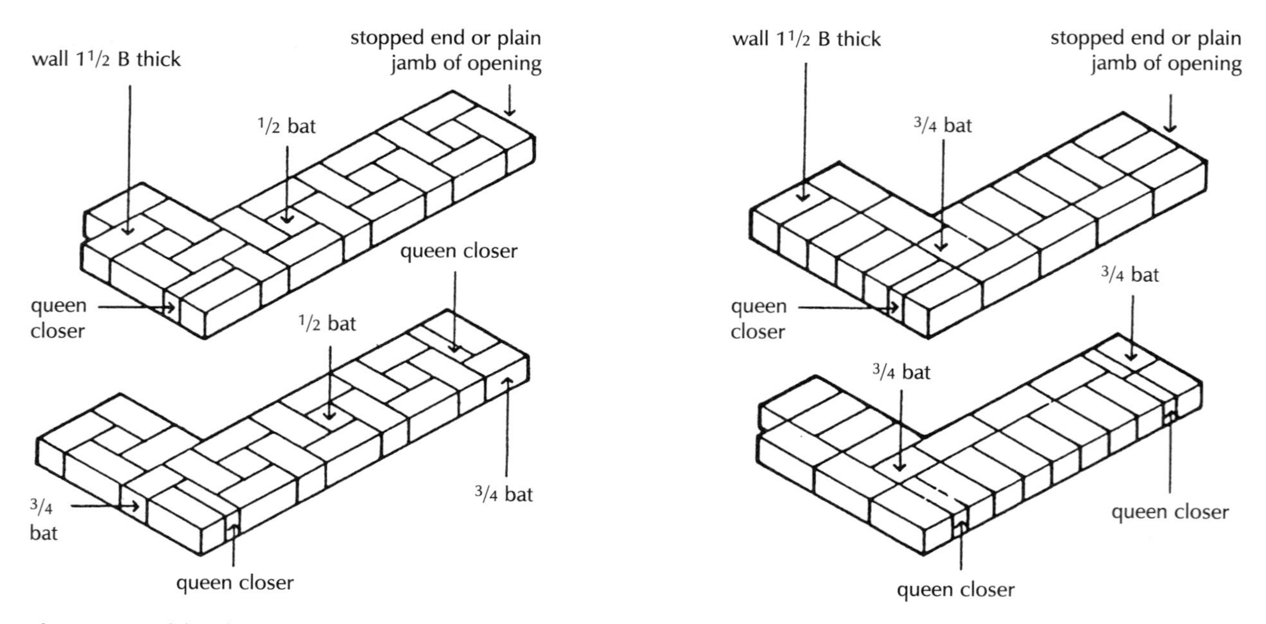

Fig. 57 Double Flemish bond.

Fig. 58 English bond.

always possible to cut a brick in half cleanly. This adds considerably to the cost of this walling, which is selected for appearance rather than economy.

A $1\frac{1}{2}$ B thick wall, finished fairface both sides and showing English bond both sides, requires considerably less cutting of bricks to complete the bond, as illustrated in Fig. 58. It is only necessary to cut closers and three quarter length bricks to complete the bond at angles and stop ends.

Walls $1\frac{1}{2}$ B thick that are to be finished fairface on one side only may be built with facing bricks for the fairface side and cheaper common bricks for the rest of the thickness of the wall, where the inside face is to be covered with plaster.

Garden wall bonds

Walls, such as garden walls, that are to be finished fairface both sides and built 1 B thick are often built in one of the garden wall bonds.

Because of the variations in size and shape of many facing bricks it is difficult to finish a 1 B wall fairface both sides because of the differences in length of bricks that are bonded through the thickness of the wall.

Garden wall bonds are designed specifically to reduce the number of through headers to minimise the labour in selecting bricks of roughly the same length for use as headers.

Usual garden wall bonds are three courses of stretchers to every one course of headers in English garden wall bond and one header to every three stretchers in Flemish garden wall bond, as illustrated in Fig. 59.

Fig. 59 Garden wall bonds.

The reduction in the number of through headers does to an extent weaken the through bond of the brickwork. This is of little consequence in a freestanding garden wall. Other combinations such as two or four stretchers to one header may be used.

The tops of garden walls are finished with one of the special coping bricks illustrated in Fig. 49 or one of the brick or stone cappings, coping, described for parapet walls in Chapter 4.

BUILDING BLOCKS

Building blocks are wall units, larger in size than a brick, that can be handled by one man. Building blocks are made of concrete or clay.

Concrete blocks

These are used extensively for both loadbearing and non-loadbearing walls, externally and internally. A concrete block wall can be laid in less time and may cost up to half as much as a similar brick wall. Lightweight aggregate concrete blocks have good insulating properties against transfer of heat and have been much used for the inner leaf of cavity walls with either a brick outer leaf or a concrete block outer leaf.

A disadvantage of some concrete blocks, particularly lightweight aggregate blocks, as a wall unit is that they may suffer moisture movement which causes cracking of applied finishes such as plaster. To minimise cracking due to shrinkage by loss of water, vertical movement joints should be built into long block walls, subject to moisture movement, at intervals of up to twice the height of the wall. These movement joints may be either a continuous vertical joint filled with mastic or they may be formed in the bonding of the blocks.

Because the block units are comparatively large, any settlement movement in a wall will show more pronounced cracking in mortar joints than is the case with the smaller brick wall unit.

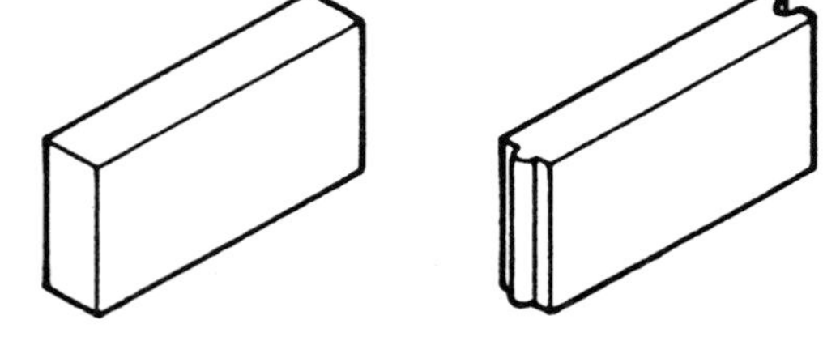

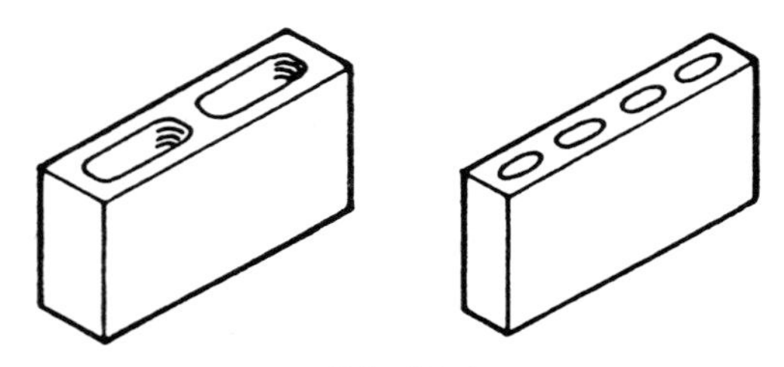

Fig. 60 Concrete blocks.

For some years it was fashionable to use concrete blocks as a fairface external wall finish. The blocks were accurately moulded to uniform sizes and made from aggregates to provide a variety of colours and textures. Blocks made to give an appearance of natural stone with plain or rugged exposed aggregate finish were used.

These special blocks are less used that they were, particularly because of the fairly rapid deterioration in the appearance of the blocks due to irregular weather staining of smooth faced blocks and the patchy dirt staining of coarse textured blocks.

Concrete blocks are manufactured from cement and either dense or lightweight aggregates as solid, cellular or hollow blocks as illustrated in Fig. 60. A cellular block has one or more holes or cavities that do not pass wholly through the block and a hollow block is one in which the holes pass through the block. The thicker blocks are made with cavities or holes to reduce weight and drying shrinkage.

The most commonly used size of both dense and lightweight concrete blocks is 440 mm long × 215 mm high. The height of the block is chosen to coincide with three courses of brick for the convenience of building in wall ties and also bonding to brickwork. The length of the block is chosen for laying in stretcher bond.

For the leaves of cavity walls and internal loadbearing walls 100 mm thick blocks are used. For non-loadbearing partition walls 60 or 75 mm thick lightweight aggregate blocks are used. Either 440 mm × 215 mm or 390 × 190 mm blocks may be used.

Concrete blocks may be specified by their minimum average compressive strength for:

(1) all blocks not less than 75 mm thick and
(2) a maximum average transverse strength for blocks less than 75 mm thick, which are used for non-loadbearing partitions.

The usual compressive strengths for blocks are 2.8, 3.5, 5.0, 7.0. 10.0, 15.0, 20.0 and 35.0 N/mm^2. The compressive strength of blocks used for the walls of small buildings of up to three storeys, recommended in Approved Document A to the Building Regulations, is 2.8 and 7 N/mm^2, depending on the loads carried.

Concrete blocks may also be classified in accordance with the aggregate used in making the block and some common uses.

Dense aggregate blocks for general use

The blocks are made of Portland cement, natural aggregate or blast-furnace slag. The usual mix is 1 part of cement to 6 or 8 of aggregate by volume. These blocks are as heavy per cubic metre as bricks, they are not good thermal insulators and their strength in resisting crushing is less than that of most well burned bricks. The colour and texture of these blocks is far from attractive and they are usually covered with plaster or a coat of rendering. These blocks are used for internal and external loadbearing walls, including walls below ground.

Lightweight aggregate concrete blocks for general use in building

The blocks are made of ordinary Portland cement and one of the following lightweight aggregates: granulated blast-furnace slag, foamed blast-furnace slag, expanded clay or shale, or well burned furnace clinker. The usual mix is 1 part cement to 6 or 8 of aggregate by volume.

Of the four lightweight aggregates noted, well burned furnace clinker produces the cheapest block which is about two-thirds the weight of a similar dense aggregate concrete block and is a considerably better thermal insulator. Blocks made from foamed blast-furnace slag are about twice the price of those made from furnace clinker, but they are only half the weight of a similar dense aggregate block and have good thermal insulating properties. The furnace clinker blocks are used extensively for walls of houses and the foamed blast-furnace slag blocks for walls of large framed buildings because of their lightness in weight.

Lightweight aggregate concrete blocks primarily for internal non-loadbearing walls

These thin blocks, usually 60 or 75 mm thick, are made with the same lightweight aggregate as those in Class 2. These blocks are more expensive than dense aggregate blocks and are used principally for non-loadbearing partitions. These blocks are manufactured as solid, hollow or cellular depending largely on the thickness of the block.

The thin blocks are solid and either square edged or with a tongue and groove in the short edges so that there is a mechanical bond between blocks to improve the stability of internal partitions. The poor structural stability may be improved by the use of storey height door linings which are secured at floor and ceiling level.

Thin block internal partitions afford negligible acoustic insulation and poor support for fittings, such as book shelves secured to them.

The thicker blocks are either hollow or cellular to reduce weight and drying shrinkage.

Moisture movement

As water dries out from precast concrete blocks shrinkage that occurs, particularly with lightweight blocks, may cause serious cracking of plaster and rendering applied to the surface of a wall built with them. Obviously the wetter the blocks the more they will shrink. It is essential that these blocks be protected on building sites from saturation by rain both when they are stacked on site before use and whilst walls are being built. Clay bricks are small and suffer very little drying shrinkage and therefore do not need to be protected from saturation by rain. Only the edges of these blocks should be wetted to increase their adhesion to mortar when the blocks are being laid.

Clay blocks

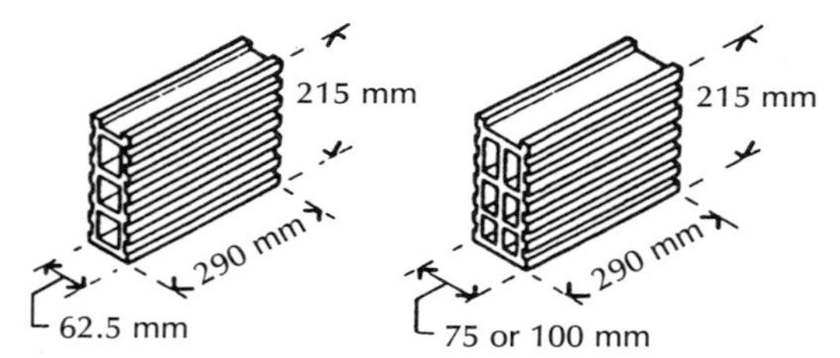

Fig. 61 Clay blocks.

Hollow clay building blocks are made for use as a wall unit. The blocks are made from selected brick clays that are press moulded and burnt. These hard, dense blocks are hollow to reduce shrinkage during firing and reduce their weight and they are grooved to provide a key for plaster, as illustrated in Fig. 61. The standard block is 290 long × 215 mm high and 62.5, 75, 100 and 150 mm thick.

Clay blocks are comparatively lightweight, do not suffer moisture movement, have good resistance to damage by fire and poor thermal insulating properties. These blocks are mainly used for non-load-bearing partitions in this country. They are extensively used in southern Europe as infill panel walls to framed buildings where the tradition is to render the external face of buildings on which the blocks provide a substantial mechanical key for rendering and do not suffer moisture movement that would otherwise cause shrinkage cracking.

Bonding blocks

Blocks are made in various thicknesses to suit most wall requirements and are laid in stretcher bond.

Thin blocks, used for non-loadbearing partitions, are laid in running stretcher bond with each block centred over and under blocks above and below. At return angles full blocks bond into the return wall in every other course, as illustrated in Fig. 62. So as not to disturb the full width bonding of blocks at angles, for the sake of stability, a short length of cut block is used as closer and infill block.

Thicker blocks are laid in off centre running bond with a three quarter length block at stop ends and sides of openings. The off centre bond is acceptable with thicker blocks as it avoids the use of cut blocks to complete the bond at angles, as illustrated in Fig. 62.

Thick blocks, whose length is twice their width, are laid in running (stretcher) bond as illustrated in Fig. 62, and cut blocks are only necessary to complete the bond at stop ends and sides of opening.

At the 'T' junctions of loadbearing concrete block walls it is sometimes considered good practice to butt the end face of the intersecting walls with a continuous vertical joint to accommodate shrinkage movements and to minimise cracking of plaster finishes.

Where one intersecting wall serves as a buttress to the other, the butt joint should be reinforced by building in split end wall ties at each horizontal joint across the butt joint to bond the walls. Similarly, non-loadbearing block walls should be butt jointed at intersections and the joint reinforced with strips of expanded metal bedded in horizontal joints across the butt joint.

Concrete block walls of specially produced blocks to be used as a fairface finish are bonded at angles to return walls with specially produced quoin blocks for the sake of appearance, as illustrated in Fig. 63. The 'L' shaped quoin blocks are made to continue the stretcher bond around the angle into the return walls.

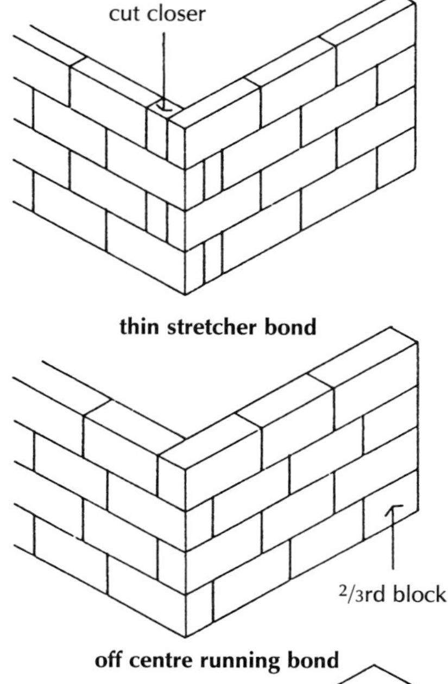

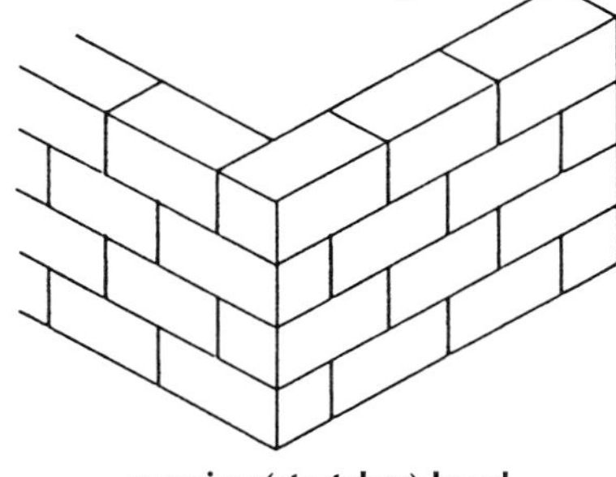

Fig. 62 Bonding building blocks.

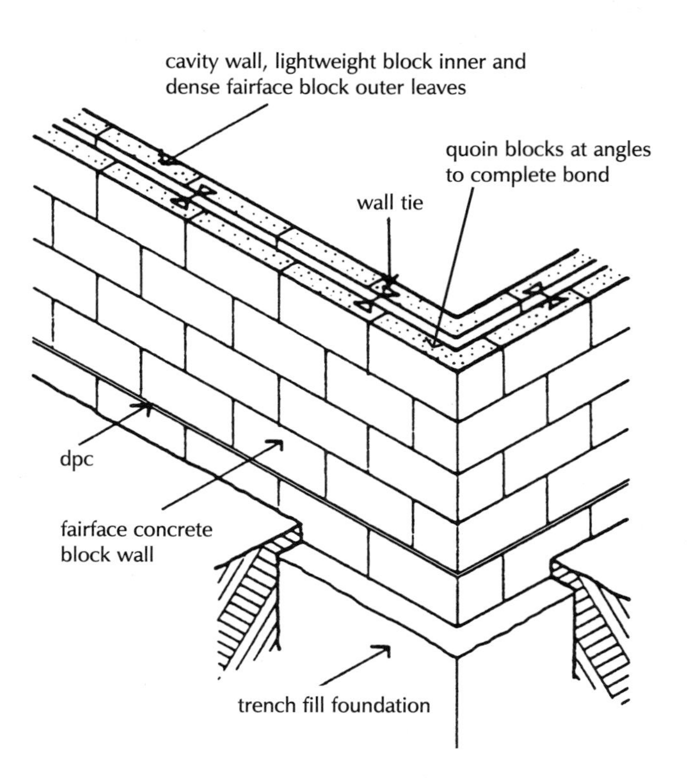

Fig. 63 Bonding block walls.

Quoin blocks are little used for other than fairface work as they are liable to damage in handling and use and add considerably to the cost of materials and labour.

MORTAR FOR BRICKWORK AND BLOCKWORK

Fig. 64 Badly shaped facing bricks laid without mortar.

Clay bricks are rarely exactly rectangular in shape and they vary in size. Some facing bricks are far from uniform in shape and size and if a wall were built of bricks laid without mortar and the bricks were bonded the result might be as shown, exaggerated, in Fig. 64.

Because of the variations in shape and size, the courses of bricks would not lie anywhere near horizontal. One of the functions of brickwork is to support floors and if a floor timber were to bear on the brick marked A it would tend to cause it to slide down the slope on which it would be resting. It is essential, therefore, that brickwork be laid in true horizontal courses, and the only way this can be done with bricks of differing shapes and sizes is to lay them on some material which is sufficiently plastic, while the bricks are being laid, to take up the difference in size, and which must be able to harden to such an extent that it can carry the weight normally carried by brickwork.

The material used is termed mortar. The basic requirements of a mortar are that it will harden to such an extent that it can carry the weight normally carried by bricks, without crushing, and that it be sufficiently plastic when laid to take the varying sizes of bricks. It must have a porosity similar to that of the bricks and it must not deteriorate due to the weathering action of rain or frost.

Sand is a natural material which is reasonably cheap and which, if mixed with water, can be made plastic, yet which has very good strength in resisting crushing. Its grains are also virtually impervious to the action of rain and frost. The material required to bind the grains of sand together into a solid mass is termed the matrix and the two materials used for this purpose are lime or cement.

Aggregate for mortar

Sand

The aggregate or main part of mortar is sand. The sand is dredged from pits or river beds and a good sand should consist of particles ranging up to 5 mm in size. In the ground, sand is usually found mixed with some clay earth which coats the particles of sand. If sand mixed with clay is used for mortar, the clay tends to prevent the cement or lime binding the sand particles together and in time the mortar crumbles. It is therefore important that the sand be thoroughly washed so that there is no more than 5% of clay in the sand delivered to the site.

Soft sand and sharp sand

Sand which is not washed and which contains a deal of clay in it feels soft and smooth when held in the hand, hence the term soft sand. Sand which is clean feels coarse in the hand, hence the term sharp. These are terms used by craftsmen. When soft sand is used, the mortar is very smooth and plastic and it is much easier to spread and to bed the bricks in than a mortar made of sharp or clean sand.

Naturally the bricklayer prefers to use a mortar made with soft or unwashed sand, often called 'builders' sand'. A good washed sand for mortar should, if clenched in the hand, leave no trace of yellow clay stains on the palm.

Matrix for mortar

The material that was used for many centuries before the advent of Portland cement as the matrix (binding agent) for mortar was lime. Lime, which mixes freely with water and sand, produces a material that is smooth, buttery and easily spread as mortar, into which the largely misshapen bricks in use at the time could be bedded.

The particular advantage of lime is that it is a cheap, readily available material that produces a plastic material ideal for bedding bricks. Its disadvantages are that it is a messy, laborious material to mix and as it is to an extent soluble in water it will lose its adhesive property in persistently damp situations. Protected from damp, a lime mortar will serve as an effective mortar for the life of most buildings.

Portland cement, which was first manufactured on a large scale in the latter part of the nineteenth century, as a matrix for mortar, produces a hard dense material that has more than adequate strength for use as mortar and is largely unaffected by damp conditions. A mixture of cement, sharp sand and water produces a coarse material that is not plastic and is difficult to spread. In use, cement has commonly been used with 'builders' sand' which is a natural mix of sand and clay. The clay content combines with water to make a reasonably plastic mortar at the expense of loss of strength and considerable drying shrinkage as the clay dries.

'Compo' mortar

During the last 50 years it has been considered good practice to use a mortar in which the advantages of lime and cement are combined. This combination or 'compo' mortar is somewhat messy to mix.

Mortar plasticiser

As an alternative to the use of lime it has become practice to use a mortar plasticiser with cement in the mix of cement mortars. A plasticiser is a liquid which, when combined with water, effervesces to produce minute bubbles of air that surround the coarse grains of sand and so render the mortar plastic, hence the name 'mortar plasticiser' mixes.

Ready mixed mortar

Of recent years ready mixed mortars have come into use particularly on sites where extensive areas of brickwork are laid. The wet material is delivered to site, ready mixed, to save the waste, labour and cost of mixing on site.

A wide range of lime and sand, lime cement and sand and cement and sand mixes is available. The sand may be selected to provide a chosen colour and texture for appearance sake or the mix may be pigmented for the same reason.

Lime mortar is delivered to site ready to use within the day of delivery. Cement mix and cement lime mortar is delivered to site ready mixed with a retarding admixture.

The retarding admixture is added to cement mix mortars to delay the initial set of cement. The initial set of ordinary Portland cement occurs some 30 minutes after the cement is mixed with water, so that an initial hardening occurs to assist in stiffening the material for use as rendering on vertical surfaces for example.

The advantages of ready mixed mortar are consistency of the mix, the wide range of mixes available and considerable saving in site labour costs and the inevitable waste of material common with site mixing.

Cement mortar

Cement is made by heating a finely ground mixture of clay and limestone, and water, to a temperature at which the clay and limestone fuse into a clinker. The clinker is ground to a fine powder called cement. The cement most commonly used is ordinary Portland cement which is delivered to site in 50 kg sacks. When the fine cement powder is mixed with water a chemical action between water and cement takes place and at the completion of this reaction the nature of the cement has so changed that it binds itself very firmly to most materials.

The cement is thoroughly mixed with sand and water, the reaction takes place and the excess water evaporates leaving the cement and sand to gradually harden into a solid mass. The hardening of the mortar becomes noticeable some few hours after mixing and is complete in a few days. The usual mix of cement and sand for mortar is from 1 part cement to 3 or 4 parts sand to 1 part of cement to 8 parts of sand by volume, mixed with just sufficient water to render the mixture plastic.

A mortar of cement and sand is very durable and is often used for brickwork below ground level and brickwork exposed to weather above roof level such as parapet walls and chimney stacks.

Cement mortar made with washed sand is not as plastic however as bricklayers would like it to be. Also when used with some types of bricks it can cause an unsightly effect known as efflorescence.

This word describes the appearance of an irregular white coating on the face of bricks, caused by minute crystals of water soluble salts in the brick. The salts go into solution in water inside the bricks and when the water evaporates in dry weather they are left on the face of bricks or plaster. Because cement mortar has greater compressive strength than required for most ordinary brickwork and because it is not very plastic by itself it is sometimes mixed with lime and sand.

Lime mortar

Lime is manufactured by burning limestone or chalk and the result of this burning is a dirty white, lumpy material known as quicklime. When this quicklime is mixed with water a chemical change occurs during which heat is generated in the lime and water, and the lime expands to about three times its former bulk. This change is gradual and takes some days to complete, and the quicklime afterwards is said to be slaked, that is it has no more thirst for water. More precisely the lime is said to be hydrated, which means much the same thing. Obviously the quicklime must be slaked before it is used in mortar otherwise the mortar would increase in bulk and squeeze out of the joints. Lime for building is delivered to site ready slaked and is termed 'hydrated lime'.

When mixed with water, lime combines chemically with carbon dioxide in the air and in undergoing this change it gradually hardens into a solid mass which firmly binds the sand.

A lime mortar is usually mixed with 1 part of lime to 3 parts of sand by volume. The mortar is plastic and easy to spread and hardens into a dense mass of good compressive strength. A lime mortar readily absorbs water and in time the effect is to reduce the adhesion of the lime to the sand and the mortar crumbles and falls out of the joints in the brickwork.

Mortar for general brickwork may be made from a mixture of cement, lime and sand in the proportions set out in Table 2. These mixtures combine the strength of cement with the plasticity of lime, have much the same porosity as most bricks and do not cause efflorescence on the face of the brickwork.

The mixes set out in Table 2 are tabulated from rich mixes (1) to weak mixes (2). A rich mix of mortar is one in which there is a high proportion of matrix, that is lime or cement or both, to sand as in the 1:3 mix and a weak mix is one in which there is a low proportion of lime or cement to sand as in the mix 1:3:12. The richer the mix of mortar the greater its compressive strength and the weaker the mix the greater the ability of the mortar to accommodate moisture or temperature movements.

Table 2 Mortar mixes.

Mortar designation	Air-entrained mixes		
	Cement:lime:sand	Masonry cement:sand	Cement:sand with plasticiser
1	1:0 to $\frac{1}{4}$:3		
2	1:$\frac{1}{2}$: 4 to $4\frac{1}{2}$	1:$2\frac{1}{2}$ to $3\frac{1}{2}$	1:3 to 4
3	1:1:5 to 6	1:4 to 5	1:5 to 6
4	1:2:8 to 9	1:$5\frac{1}{2}$ to $6\frac{1}{2}$	1:7 to 8
5	1:3:10 to 12	1:$6\frac{1}{2}$ to 7	1:8

Taken from BS 5628:Part 3:1985 (Table 15)

Proportions by volume

The general uses of the mortar mixes given in Table 2 are as mortar for brickwork or blockwork as follows:

Mix 1 For cills, copings and retaining walls
Mix 2 Parapets and chimneys
Mix 3 Walls below dpc
Mix 4 Walls above dpc
Mix 5 Internal walls and lightweight block inner leaf of cavity

Hydraulic lime

Hydraulic lime is made by burning a mixture of chalk or limestone that contains clay. Hydraulic lime is stronger than ordinary lime and will harden in wet conditions, hence the name. Ordinary Portland cement, made from similar materials and burnt at a higher temperature, has largely replaced hydraulic lime which is less used today.

Mortar plasticisers

Liquids known as mortar plasticisers are manufactured. When these liquids are added to water they effervesce, that is the mixture becomes bubbly like soda water. If very small quantities are added to mortar, when it is mixed, the millions of minute bubbles that form surround the hard sharp particles of sand and so make the mortar plastic and easy to spread. The particular application of these mortar plasticisers is that if they are used with cement mortar they increase its plasticity and there is no need to use lime. It seems that the plasticisers do not adversely affect the hardness and durability of the mortar and they are commonly and successfully used for mortars.

JOINTING AND POINTING

Jointing

Jointing is the word used to describe the finish of the mortar joints between bricks, to provide a neat joint in brickwork that is finished fairface. Fairface describes the finished face of brickwork that will not be subsequently covered with plaster, rendering or other finish.

Most fairface brickwork joints are finished, as the brickwork is raised, in the form of a flush or bucket handle joint. When the mortar

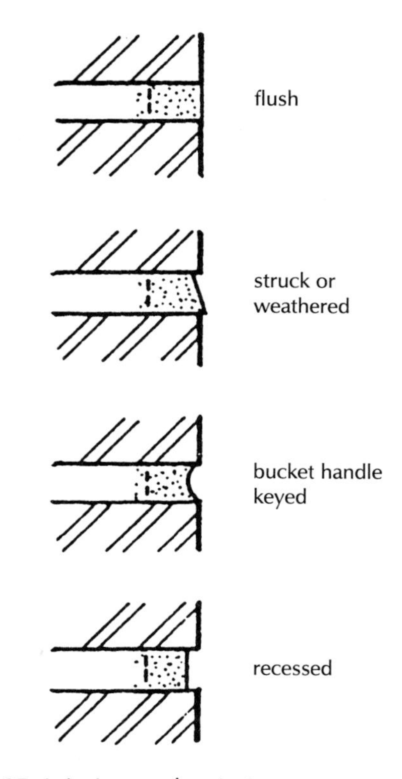

Fig. 65 Jointing and pointing.

has gone off, that is hardened sufficiently, the joint is made. Flush joints are generally made as a 'bagged' or a 'bagged in' joint. The joint is made by rubbing coarse sacking or a brush across the face of the brickwork to rub away all protruding mortar and leaving a flush joint. This type of joint, illustrated in Fig. 65, can most effectively be used on brickwork where the bricks are uniform in shape and comparatively smooth faced, where the mortar will not spread over the face of the brickwork.

A bucket handle joint is made by running the top face of a metal bucket handle or the handle of a spoon along the joint to form a concave, slightly recessed joint, illustrated in Fig. 65. The advantage of the bucket handle joint is that the operation compacts the mortar into the joint and improves weather resistance to some extent. A bucket handle joint may be formed by a jointing tool with or without a wheel attachment to facilitate running the tool along uniformly deep joints.

Flush and bucket handle joints are mainly used for jointing as the brickwork is raised.

The struck and recessed joints shown in Fig. 65 are more laborious to make and therefore considerably more expensive. The struck joint is made with a pointing trowel that is run along the joint either along the edges of uniformly shaped bricks or along a wood straight edge, where the bricks are irregular in shape or coarse textured, to form the splayed back joint. The recessed joint is similarly formed with a tool shaped for the purpose, with such filling of the joint as may be necessary to complete the joint.

Of the joints described the struck joint is mainly used for pointing the joints in old brickwork and the recessed joint to emphasise the profile, colour and textures of bricks for appearance sake to both new and old brickwork.

Pointing

The words jointing and pointing are commonly loosely used. Jointing is the operation of finishing off a mortar joint as the brickwork is raised, whereas pointing is the operation of filling the joint with a specially selected material for the sake of appearance or as weather protection to old lime mortar.

Pointing is the operation of filling mortar joints with a mortar selected for colour and texture to either new brickwork or to old brickwork. The mortar for pointing is a special mix of lime, cement and sand or stone dust chosen to produce a particular effect of colour and texture. The overall appearance of a fairface brick wall can be dramatically altered by the selection of mortar for pointing. The finished colour of the mortar can be affected through the selection of a particular sand or stone dust, the use of pigmented cement, the addition of a pigment and the proportion of the mix of materials.

The joints in new brickwork are raked out about 20 mm deep when the mortar has gone off sufficiently and before it has set hard and the joints are pointed as scaffolding is struck, that is taken down.

The mortar joints in old brickwork that was laid in lime mortar may in time crumble and be worn away by the action of wind and rain. To protect the lime mortar behind the face of the joints it is good practice to rake out the perished jointing or pointing and point or repoint all joints. The joints are raked out to a depth of about 20 mm and pointed with a mortar mix of cement, lime and sand that has roughly the same density as the brickwork. The operation of raking out joints is laborious and messy and the job of filling the joints with mortar for pointing is time consuming so that the cost of pointing old work is expensive.

Pointing or repointing old brickwork is carried out both as protection for the old lime mortar to improve weather resistance and also for appearance sake to improve the look of a wall.

Any one of the joints illustrated in Fig. 66 may be used for pointing.

WALLS OF BRICK AND BLOCK

Strength and stability

Up to the middle of the twentieth century the design and construction of small buildings, such as houses, was based on tried, traditional forms of construction. There were generally accepted rule of thumb methods for determining the necessary thickness for the walls of small buildings. By and large, the acceptance of tried and tested methods of construction, allied to the experience of local builders using traditional materials in traditional forms of construction, worked well. The advantage was that from a simple set of drawings an experienced builder could give a reasonable estimate of cost and build and complete small buildings, such as houses, without delay.

With the increasing use of unfamiliar materials, such as steel and concrete, in hitherto unused forms, it became necessary to make calculations to determine the least size of elements of structure for strength and stability in use. The practicability of constructing large multi-storey buildings provoked the need for standards of safety in case of fire and rising expectations of comfort and the need for the control of insulation, ventilation, daylight and hygiene.

During the last 50 years there has been a considerable increase in building control, that initially was the province of local authorities through building bylaws, later replaced by national building regulations. The Building Regulations 1985 set out functional requirements for buildings and health and safety requirements that may be met through the practical guidance given in 11 Approved Documents that in turn refer to British Standards and Codes of Practice.

In theory it is only necessary to satisfy the requirements of the Building Regulations, which are short and include no technical details of means of satisfying the requirements. The 11 Approved

Documents give practical guidance to meeting the requirements, but there is no obligation to adopt any particular solution in the documents if the requirements can be met in some other way.

The stated aim of the current Building Regulations is to allow freedom of choice of building form and construction so long as the stated requirements are satisfied. In practice the likelihood is that the practical guidance given in the Approved Documents will be accepted as if the guidance were statutory as the easier approach to building, rather than proposing some other form of building that would involve calculation and reference to a bewildering array of British Standards and Codes and Agrément Certificates.

In Approved Document A there is practical guidance to meeting the requirements of the Building Regulations for the walls of small buildings of the following three types:

(1) residential buildings of not more than three storeys
(2) small single storey non-residential buildings, and
(3) small buildings forming annexes to residential buildings (including garages and outbuildings).

Limitations as to the size of the building types included in the guidance are given in a disjointed and often confusing manner.

Height

The maximum height of residential buildings is given as 15 m from the lowest ground level to the highest point of any wall or roof, whereas the maximum allowable thickness of wall is limited to walls not more than 12 m. Height is separately defined, for example, as from the base of a gable and external wall to half the height of the gable. The height of single storey, non-residential buildings is given as 3 m from the ground to the top of the roof, which limits the guidance to very small buildings. The maximum height of an annexe is similarly given as 3 m, yet there is no definition of what is meant by annexe except that it includes garages and outbuildings.

Width

The least width of residential buildings is limited to not less than half the height. A diagram limits the dimensions of the wing of a residential building without defining the meaning of the term 'wing', which in the diagram looks more like an annexe than a wing. Whether the arms of a building which is 'L' or 'U' shaped on plan are wings or not is entirely a matter of conjecture. How the dimensions apply to semi-detached buildings or terraces of houses is open to speculation.

In seeking to give practical guidance to meeting functional requirements for strength and stability and at the same time impose limiting dimensions, the Approved Document has caused confusion.

One further limitation is that no floor enclosed by structural walls on all sides should exceed 70 m^2 and a floor without a structural wall on one side, 30 m^2. The floor referred to is presumably a suspended floor, though it does not say so. As the maximum allowable length of wall between buttressing walls, piers or chimneys is given as 12 m and the maximum span for floors as 6 m, the limitation is in effect a floor some 12 × 6 m on plan. It is difficult to understand the need for the limitation of floor area for certain 'small' buildings.

Strength

The guidance given in the Approved Document for walls of brick or block is based on compressive strengths of 5 N/mm^2 for bricks and 2.8 N/mm^2 for blocks for walls up to two storeys in height, where the storey height is not more than 2.7 m and 7 N/mm^2 for bricks and blocks of walls of three storey buildings where the storey height is greater than 2.7 m.

Stability

Thickness of walls

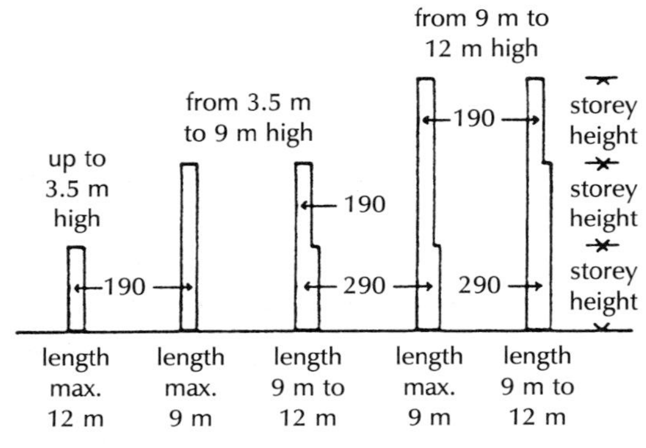

Fig. 66 Minimum thickness of walls.

The general limitation of wall thickness given for stability is that solid walls of brick or block should be at least as thick as one-sixteenth of the storey height. This is a limiting slenderness ratio relating thickness of wall to height, measured between floors and floor and roof that provide lateral support and give stability up the height of the wall. The minimum thickness of external, compartment and separating walls is given in a table in Approved Document A, relating thickness to height and length of wall as illustrated in Fig. 66. Compartment walls are those that are formed to limit the spread of fire and separating walls (party walls) those that separate adjoining buildings, such as the walls between terraced houses.

Cavity walls should have leaves at least 90 mm thick, cavity at least 50 mm wide and the combined thickness of the two leaves plus 20 mm, should be at least the thickness required for a solid wall of the same height and length.

Internal loadbearing walls, except compartment and separating walls, should be half the thickness of external walls illustrated in Fig. 66, minus 5 mm, except for the wall in the lowest storey of a three storey building which should be of the same thickness, or 140 mm, whichever is the greater.

Stability

Lateral support

For stability up the height of a wall lateral support is provided by floors and roofs as set out in Table 3.

Walls that provide support for timber floors are given lateral support by 30 × 5 mm galvanised iron or stainless steel 'L' straps fixed to the side of floor joists at not more than 2 m centres for houses up to three storeys and 1.25 m centres for all storeys in all other buildings. The straps are turned down 100 mm on the cavity face of the inner leaf of cavity walls and into solid wallings, as illustrated in Fig. 67.

Table 3 Lateral support for walls.

Wall type	Wall length	Lateral support required
Solid or cavity: external compartment separating	Any length	Roof lateral support by every roof forming a junction with the supported wall
	Greater than 3 m	Floor lateral support by every floor forming a junction with the supported wall
Internal loadbearing wall (not being a compartment or separating wall)	Any length	Roof or floor lateral support at the top of each storey

Taken from Approved Document A (Table 11)
The Building Regulations

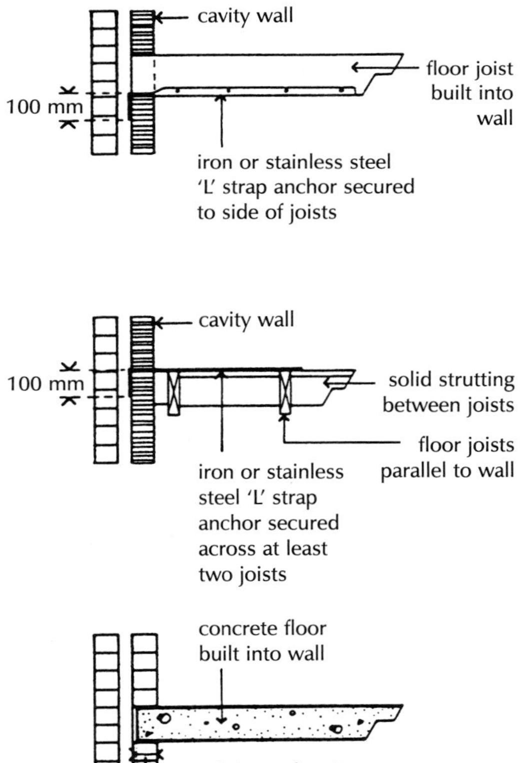

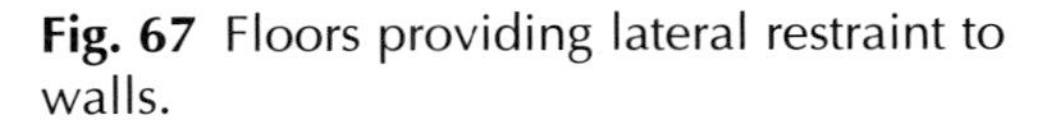

Fig. 67 Floors providing lateral restraint to walls.

Lateral support from timber floors, where the joists run parallel to the wall, is provided by 30×5 mm galvanised iron on stainless steel strap anchors secured across at least two joists at not more than 2 m centres for houses up to three storeys and 1.25 m for all storeys in all other buildings.

The 'L' straps are turned down a minimum of 100 mm on the cavity side of inner leaf of cavity walls and into solid walling. Solid timber strutting is fixed between joists under the straps as illustrated in Fig. 67.

Solid floors of concrete provide lateral support for walls where the floor bears for a minimum of 90 mm in both solid and cavity walls, as illustrated in Fig. 67.

To provide lateral support to gable end walls to roofs pitched at more than 15° a system of galvanised steel straps is used. Straps 30×5 mm are screwed to the underside of timber noggings fixed between three rafters, as illustrated in Fig. 68, with timber packing pieces between the rafter next to the gable and the wall.

The straps should be used at a maximum of 2 m centres and turned down against the cavity face of the inner leaf of a whole building block or down into a solid wall.

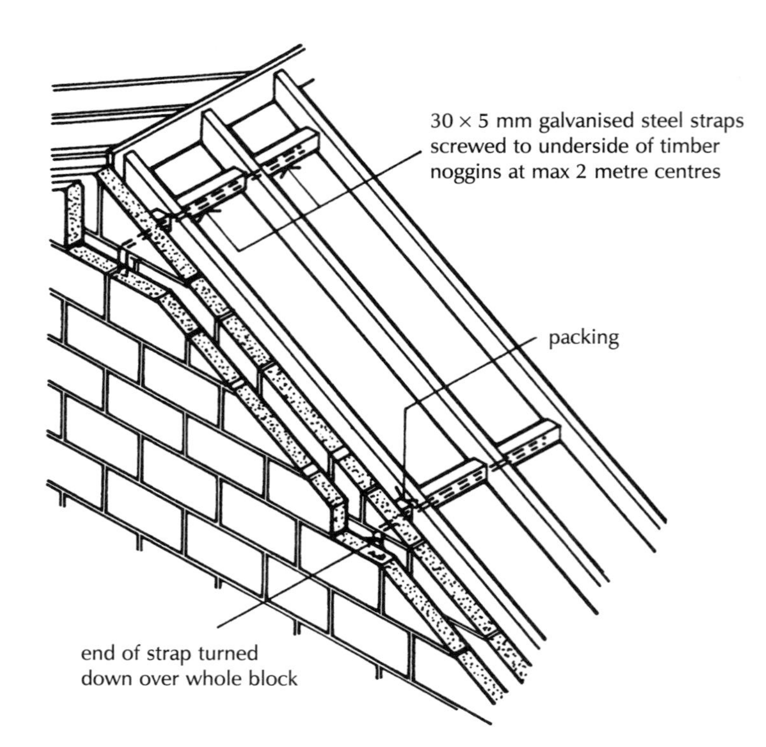

Fig. 68 Lateral support to gable ends.

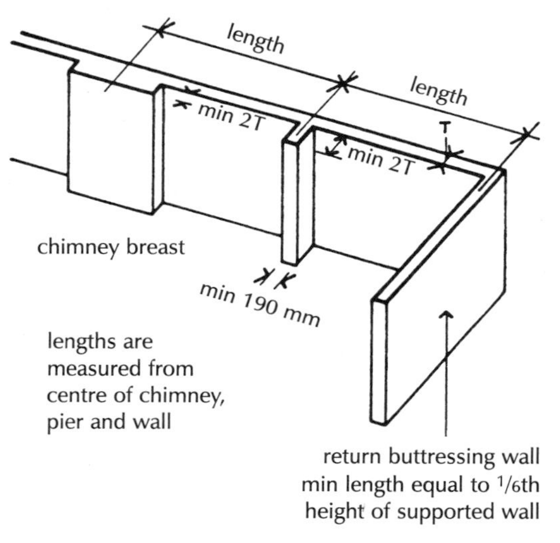

Fig. 69 Length of walls.

To provide stability along the length and at the ends of loadbearing walls there should be walls, piers or chimneys bonded to the wall at intervals of not more than 12 m, to buttress and stabilise the wall.

The maximum spacing of buttressing walls, piers and chimneys is measured from the centre line of the supports as illustrated in Fig. 69. The minimum length of a return buttressing wall should be equal to one-sixth of the height of the supported wall.

To be effective as buttresses to walls the return walls, piers and chimneys must be solidly bonded to the supported wall.

Chases in walls

To limit the effect of chases cut into walls in reducing strength or stability, vertical chases should not be deeper than one-third of the thickness of solid walls or a leaf of a cavity wall and horizontal chases not deeper than one-sixteenth. A chase is a recess, cut or built in a wall, inside which small service pipes are run and then covered with plaster or walling.

CAVITY WALLS

Resistance to weather

Between 1920 and 1940 it became more usual for external walls of small buildings to be constructed as cavity walls with an outer leaf of brick or block, an open cavity and an inner leaf of brick or block. The outer leaf and the cavity serve to resist the penetration of rain to the inside face and the inner leaf to support floors, provide a solid internal wall surface and to some extent act as insulation against transfer of heat.

The idea of forming a vertical cavity in brick walls was first proposed early in the nineteenth century and developed through the century. Various widths of cavity were proposed from the first 6 inch cavity, a later 2 inch cavity followed by proposals for 3, 4 or 5 inch wide cavities. The early cavity walls were first constructed with bonding bricks laid across the cavity at intervals, to tie the two leaves together. Either whole bricks with end closers or bricks specially made to size and shape for the purpose were used. Later on, during the middle of the century, iron ties were used instead of bond bricks and accepted as being adequate to tie the two leaves of cavity walls.

From the middle of the twentieth century it became common practice to construct the external walls of houses as a cavity wall with a 2 inch wide cavity and metal wall ties. It seems that the 2 inch width of cavity was adopted for the convenience of determining the cavity width, by placing a brick on edge inside the cavity so that the course height of a brick, about 65 mm, determined cavity width rather than any consideration of the width required to resist rain penetration. This was adapted to the 2 inch (50 mm) wide cavity for walls that became common until recent years.

In constructing the early open cavity walls it was considered good practice to suspend a batten of wood in the cavity to collect mortar droppings. The batten was removed from time to time, cleaned of mortar, and put back in the cavity as the work progressed. The practice, which was largely ignored by bricklayers as it impeded work, has since been abandoned in favour of care in workmanship to avoid mortar droppings becoming lodged inside the cavity.

With the increase in the price of fuels and expectations of thermal comfort, building regulations have of recent years made requirements for the thermal insulation of external walls that can best be met by the introduction of materials with high thermal resistance. The most convenient position for these lightweight materials in a cavity wall is inside the cavity, which is either fully or partially filled with insulation. A filled or partially filled cavity may well no longer be an efficient barrier to rain penetration so that, with the recent increase in requirement for the thermal insulation of walls it has now been accepted that the width of the cavity may be increased from the traditional 50 to 100 mm to accommodate increased thickness of insulation and still maintain a cavity against rain penetration.

Strength and stability

The practical guidance in Approved Document A to the Building Regulations accepts a cavity of from 50 to 100 mm for cavity walls having leaves at least 90 mm thick, built of coursed brickwork or blockwork with wall ties spaced at 450 mm vertically and from 900 to 750 mm horizontally for cavities of 50 to 100 mm wide respectively. As the limiting conditions for the thickness of walls related to height and length are the same for a solid bonded wall 190 mm thick as they are for a cavity wall of two leaves each 90 mm thick, it is accepted that the wall ties give the same strength and stability to two separate leaves of brickwork that the bond in solid walls does.

Wall ties

Iron ties which were used to tie the leaves of the early cavity walls were later replaced by mild steel ties that became standard for many years.

In contact with moisture, mild steel progressively corrodes by the formation of oxide of iron, called rust, which expands fiercely to the extent that brickwork around ties may become rust stained and disintegrate. Standard mild steel ties are coated with zinc to inhibit rust corrosion. The original zinc coating for ties, which was comparatively thin, has been increased in thickness in the current British Standard Specification, for improved resistance to corrosion. As added protection, the range of standard wall ties can be coated with plastic on a galvanised undercoating.

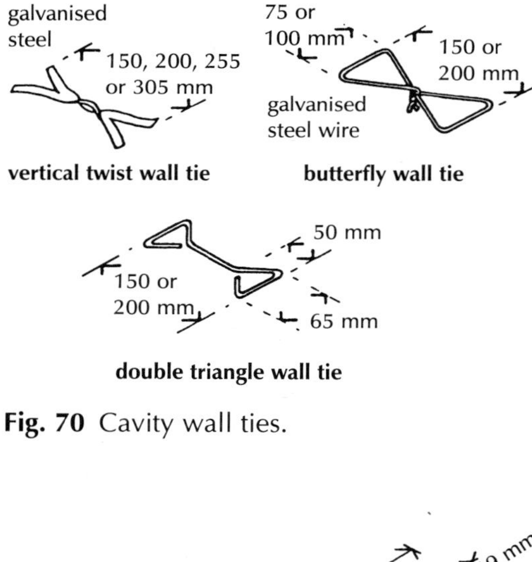

Fig. 70 Cavity wall ties.

On the majority of building sites wall ties are not commonly protected during delivery, storage, handling and use against the inevitable knocks that may perforate the toughest coating to mild steel and the consequent probability of rust occurring. There are, on the market, a range of standard and non-standard section wall ties made from stainless steel that will not suffer corrosion rusting during the useful life of buildings. It seems worthwhile to make the comparatively small additional expenditure on stainless steel ties as a precaution against staining and spalling of brickwork or blockwork around rusting mild steel ties.

The standard section wall ties, illustrated in Fig. 70, are the vertical twist strip, the butterfly and double triangle wire ties. As a check to moisture that may pass across the tie, the butterfly type is laid with the twisted wire ends hanging down into the cavity to act as a drip. The double triangle tie may have a bend in the middle of its length and the strip tie has a twist as a barrier to moisture passing across the tie. Of the three standard types the butterfly is more likely to collect mortar droppings than the others.

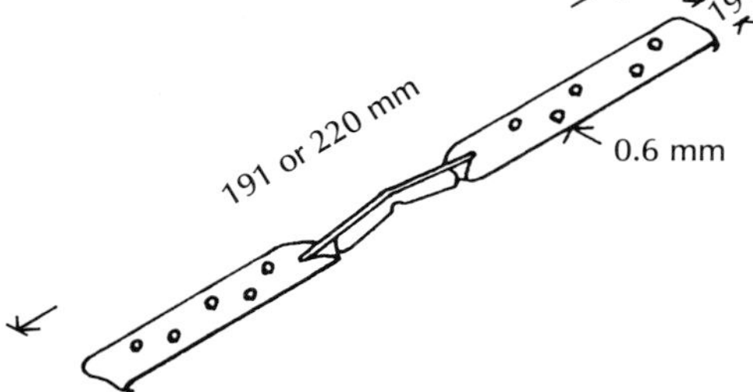

Fig. 71 Stainless steel wall tie.

The wall tie illustrated in Fig. 71 is made from corrosion resistant Austenitic stainless steel. The ridge at the centre of the length of the tie is designed for strength and to provide as small as possible a surface for the collection of mortar droppings. The perforations are to improve bond to mortar.

The length of wall ties varies to accommodate different widths of cavity and the thickness of the leaves of cavity walls. For a 50 mm cavity with brick leaves, a 191 mm or 200 mm long tie is made. For a 100 mm cavity with brick leaves, a 220 mm long tie is used.

Spacing of ties

The spacing of wall ties built across the cavity of a cavity wall is usually 900 mm horizontally and 450 mm vertically, or 2.47 ties per square metre, and staggered, as illustrated in Fig. 72, for the conventional 50 mm wide cavity, with the spacing reduced to 300 mm around the sides of openings. In Approved Document A to the Building Regulations, the practical guidance for the spacing of ties is given as 900 and 450 mm horizontally and vertically for 50 to 75 mm cavities, 750 and 450 mm horizontally and vertically for cavities from 76 to 100 mm wide and 300 mm vertically at unbonded jambs of all openings in cavity walls within 150 mm of openings to all widths of cavities.

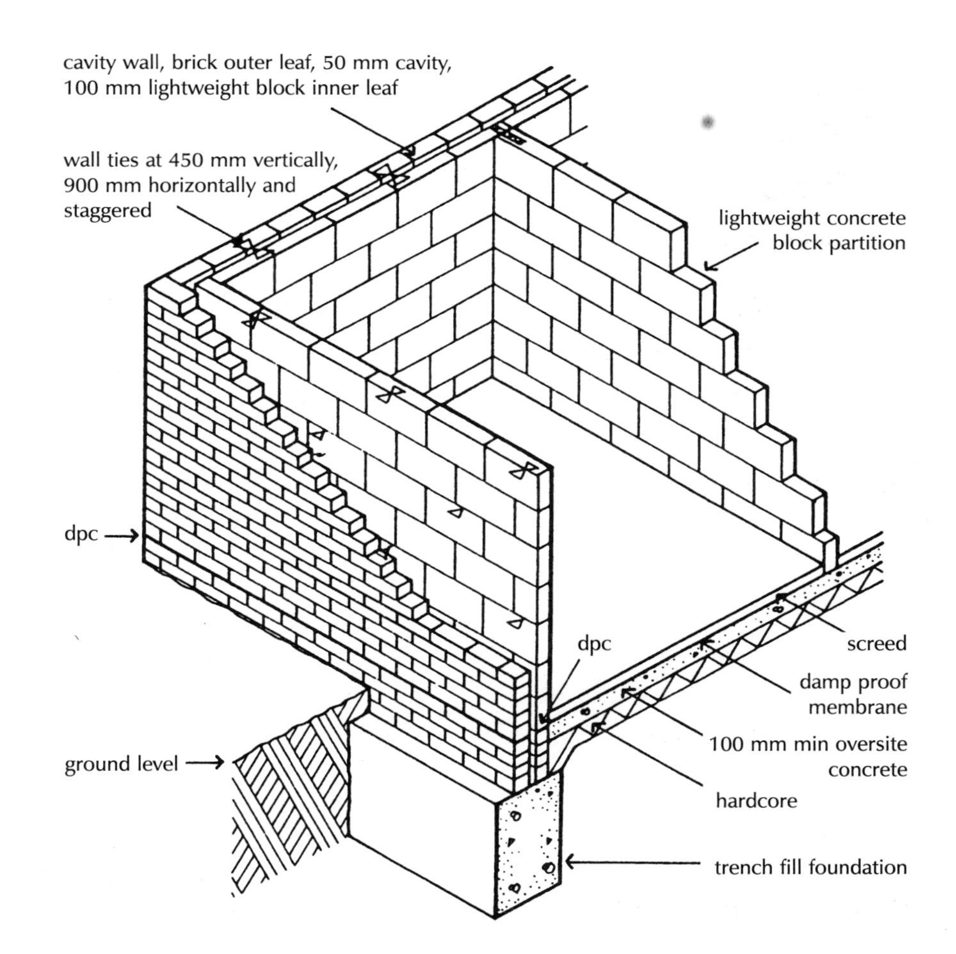

Fig. 72 Cavity wall.

Openings in walls

The practical guidance in Approved Document A in regard to openings in walls states that the number, size and position of openings should not impair the stability of a wall to the extent that the combined width of openings in walls between the centre line of buttressing walls or piers should not exceed two-thirds of the length of that wall together with more detailed requirements limiting the size of opening and recesses. There is a requirement that the bearing end of lintels with a clear span of 1200 mm or less may be 100 mm and above that span, 150 mm.

Figure 73 is an illustration of a window opening in a brick wall with the terms used to describe the parts noted.

For strength and stability the brickwork in the jambs of openings has to be strengthened with more closely spaced ties and the wall over the head of the opening supported by an arch, lintels or beams. The term jamb derives from the French word jambe, meaning leg. From Fig. 73, it will be seen that the brickwork on either side of the opening acts like legs which support brickwork over the head of the opening. The term jamb is not used to describe a particular width either side of openings and is merely a general term for the brickwork for full height of opening either side of the window. The word 'reveal' is used more definitely to describe the thickness of the wall revealed by cutting the opening and the reveal is a surface of brickwork as long as the height of the opening. The lower part of the opening is a cill for windows or a threshold for doors.

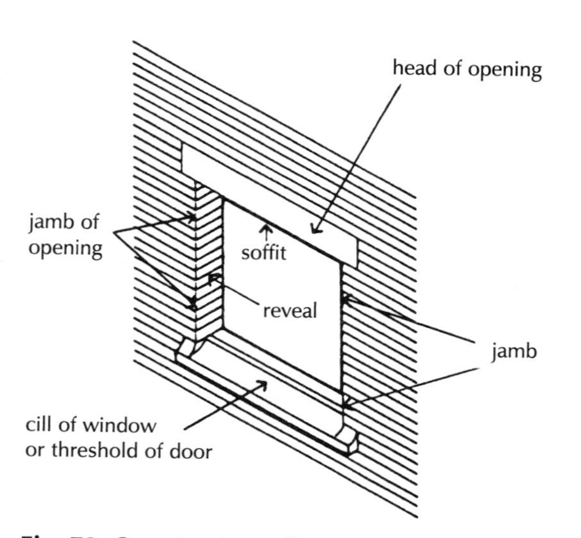

Fig. 73 Opening in wall.

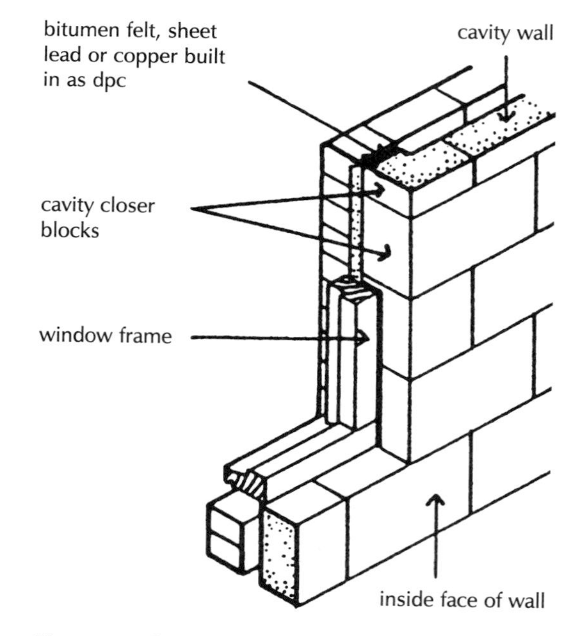

Fig. 74 Solid closing of cavity at jambs.

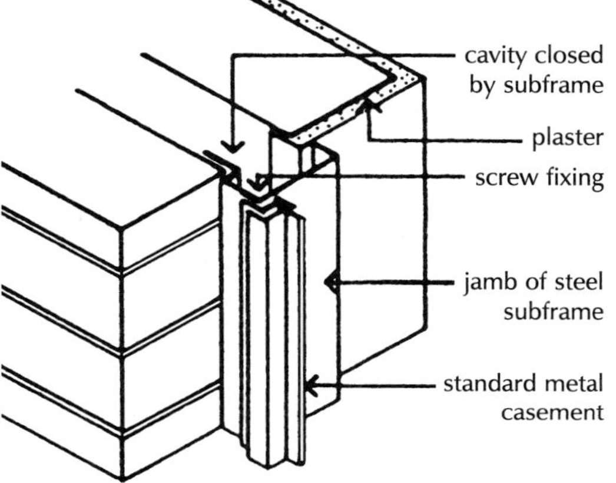

Fig. 75 Cavity closed with frame.

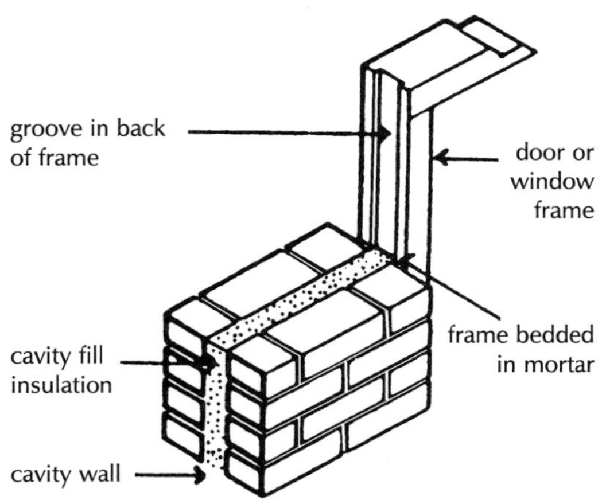

Fig. 76 Cavity fill insulation.

The jambs of openings may be plain or square into which the door or window frames are built or fixed or they may be rebated with a recess, behind which the door or window frame is built or fixed.

The cavity in a cavity wall serves to prevent penetration of water to the inner leaf. In the construction of the conventional cavity wall, before the adoption of cavity insulation, it was considered wise to close the cavity at the jambs of openings to maintain comparatively still air in the cavity as insulation. It was practice to build in cut bricks or blocks as cavity closers. To prevent penetration of water through the solid closing of cavity walls at jambs, a vertical dpc was built in as illustrated in Fig. 74. Strips of bitumen felt or lead were nailed to the back of wood frames and bedded between the solid filling and the outer leaf as shown.

As an alternative to solidly filling the cavity at jambs with cavity closers, window or door frames were used to cover and seal the cavity. Pressed metal subframes to windows were specifically designed for this purpose, as illustrated in Fig. 75. With mastic pointing between the metal subframe and the outer reveal, this is a satisfactory way of sealing cavities.

At the time when it first became common practice to fill the cavity solidly at jambs, there was no requirement for the insulation of walls. When insulation first became a requirement it was met by the use of lightweight, insulating concrete blocks as the inner leaf and the practice of solid filling of cavity at jambs, with a vertical dpc continued.

With the increasing requirement for insulation it has become practice to use cavity insulation as the most practical position for a layer of lightweight material. If the cavity insulation is to be effective for the whole of the wall it must be continued up to the back of window and door frames, as a solid filling of cavity at jambs would be a less effective insulator and act as a thermal or cold bridge.

With the revision of the requirement of the Building Regulations for enhanced insulation down to a standard U value of $0.45\,W/m^2K$ for the walls of dwellings it has become practice to use cavity insulation continued up to the frames of openings, as illustrated in Fig. 76, to avoid the cold bridge effect caused by solid filling. Door and window frames are set in position to overlap the outer leaf with a resilient mastic pointing as a barrier to rain penetration between the frame and the jamb. With a cavity 100 mm wide and cavity insulation as partial fill, it is necessary to cover that part of the cavity at jambs of openings, that is not covered by the frame. This can be effected by covering the cavity with plaster on metal lath or by the use of jamb linings of wood, as illustrated in Fig. 77.

With this form of construction at the jambs of openings there is no purpose in forming a vertical dpc at jambs.

The advantages of the wide cavity is that the benefit of the use of the cavity insulation can be combined with the cavity air space as resistance to the penetration of water to the inside face of the wall.

Cills and thresholds of openings

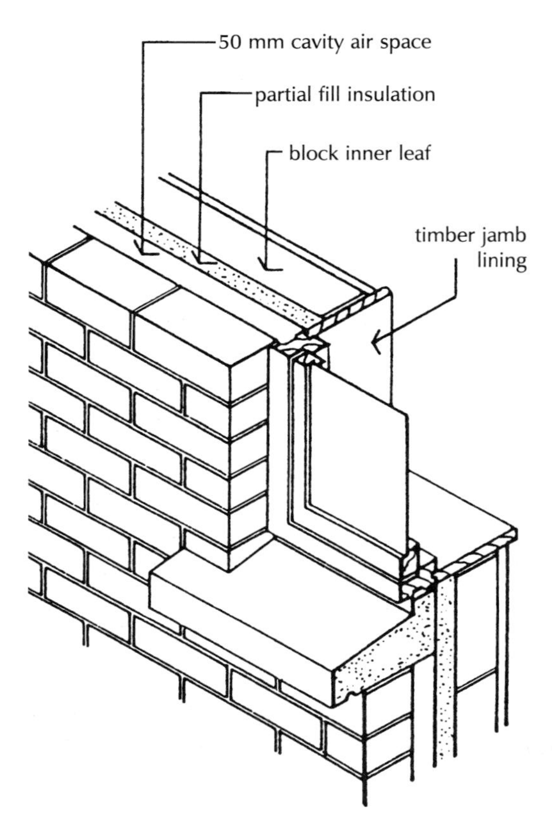

Fig. 77 Jamb lining to wide cavity.

A cill is the horizontal finish to the wall below the lower edge of a window opening on to which wind driven rain will run from the hard, smooth, impermeable surface of window glass. The function of a cill is to protect the wall below a window. Cills are formed below the edge of a window and shaped or formed to slope out and project beyond the external face of the wall, so that water runs off. The cill should project at least 45 mm beyond the face of the wall below and have a drip on the underside of the projection.

The cavity insulation shown in Fig. 77 is carried up behind the stone cill to avoid a cold bridge effect and a dpc is fixed behind the cill as a barrier to moisture penetration.

A variety of materials may be used as a cill such as natural stone, cast stone, concrete, tile, brick and non-ferrous metals. The choice of a particular material for a cill depends on cost, availability and to a large extent on appearance. Details of the materials used and the construction of cills are given in Volume 2.

As a barrier to the penetration of rain to the inside face of a cavity wall it is good practice to continue the cavity up and behind the cills as illustrated in Volume 2. Where cills of stone, cast stone and concrete are used the cill may extend across the cavity. As a barrier to rain penetration it has been practice to bed a dpc below these cills and extend it up behind the cill, as illustrated in Volume 2. Providing the cill has no joints in its length, its ends are built in at jambs and the material of the cill is sufficiently dense to cause most of the rainwater to run off, there seems little purpose in these under sill dpcs or trays.

The threshold to door openings serves as a finish to protect a wall or concrete floor slab below the door, as illustrated in Volume 2. Thresholds are commonly formed as part of a step up to external doors as part of the concrete floor slab with the top surface of the threshold sloping out. Alternatively, a natural stone or cast stone threshold may be formed.

Head of openings in cavity walls

The brickwork and blockwork over the head of openings in cavity walls has to be supported. Because of the bonding of brickwork and blockwork over the opening it is necessary to provide support for the weight of the brickwork or blockwork within a 45° isosceles triangle formed by the stretcher bond and the weight of floors and roofs carried by the wall over the opening.

The comparatively small loads over small openings are carried by a lintel or arch. With the adoption of cavity insulation as a principal method of enhancing the resistance of walls to the transfer of heat and the need to continue cavity insulation up to the back of window and door frames to minimise cold bridges, it is practice today to use lintels to support the inner and outer leaves over openings.

Steel lintels

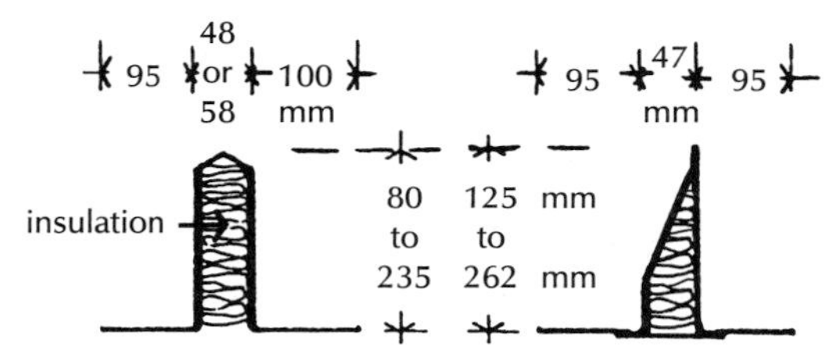

Fig. 78 Lintels for cavity walls.

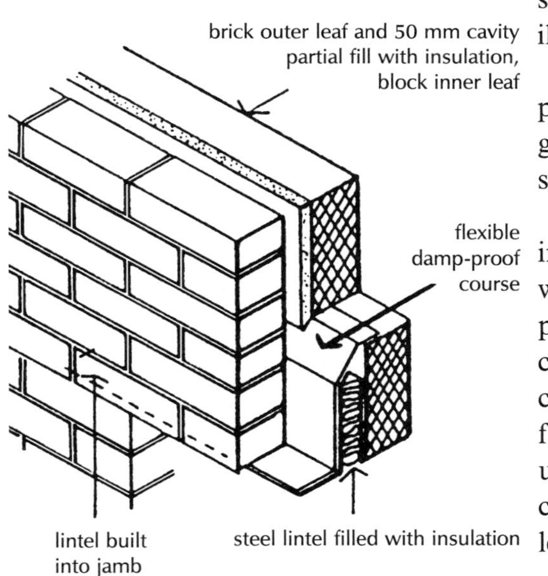

Fig. 79 Top hat lintel.

Most loadbearing brick or blockwork walls over openings, where the cavity insulation is continued down to the head of the window or door frame, are supported by steel section lintels. The advantage of these lintels is that they are comparatively lightweight and easy to handle, they provide adequate support for walling over openings in small buildings and once they are bedded in place the work can proceed without delay. Because of their ease of handling and use these lintels have largely replaced concrete lintels.

The lintels are formed either from mild steel strip that is pressed to shape, and galvanised with a zinc coating to inhibit rust, or from stainless steel. The lintels for use in cavity walling are formed with either a splay to act as an integral damp-proof tray or as a top hat section over which a damp-proof tray is dressed. Typical sections are illustrated in Fig. 78.

The splay section lintels are galvanised and coated with epoxy powder coating as corrosion protection and the top hat section with a galvanised coating. For insulation the splay section and top hat section lintels are filled with expanded polystyrene.

The top hat section steel lintel is built into the jambs of both the inner and outer leaf to provide support for both leaves of the cavity wall, as illustrated in Fig. 79. The two wings at the bottom of the lintel provide support for the brick outer and block inner leaves over the comparatively narrow openings for windows and doors. Where the cavity is partly filled with insulation it is usual to dress a flexible dpc from the block inner leaf down to a lower brick course or down to the underside of the brick outer leaf. The purpose of the damp-proof course or tray is to collect any water that might penetrate the outer leaf and direct it to weep holes in the wall.

The splay section lintel is built into the jambs of openings to provide support for the outer and inner leaf of the cavity wall over the

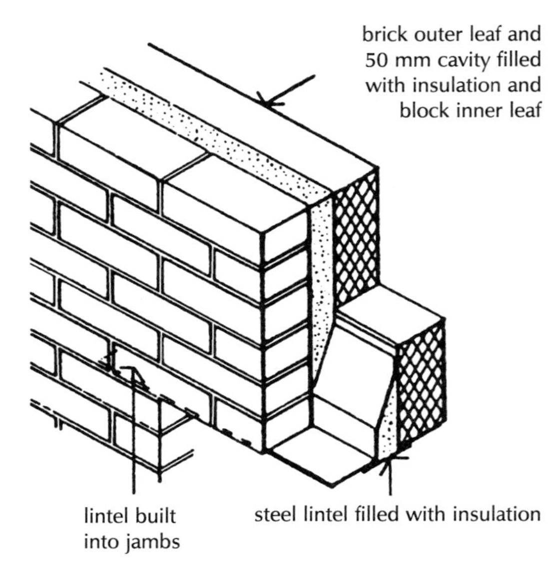

Fig. 80 Splay lintel.

openings, as illustrated in Fig. 80. Where the cavity is filled with insulation there is no need to build in a damp-proof course or tray. Any water that might penetrate the outer leaf will be directed towards the outside by the splay of the lintel.

Unless the window or door frame is built-in or fixed with its external face close to the outside face of the wall, the edge of the wing of the lintel will be exposed on the soffit of the opening. In handling and building in, there is a possibility that the edge of this wing might suffer damage to the protective galvanised coating. On the external face of a wall, water may penetrate the zinc coat and cause corrosion of the steel below. Rust very quickly spreads around the initial fracture of the protective coating. It is worthwhile making the comparatively small outlay on a stainless steel lintel as insurance against a possibly much larger expenditure on replacement of a corroded galvanised steel lintel.

Fairface brickwork supported by steel lintels may be laid as horizontal course brickwork or as a flat brick on edge or end lintel.

Concrete lintels

As an alternative to the use of steel lintels reinforced concrete lintels may be used to support the separate leaves over openings. This construction may be used where the appearance of a concrete lintel over openings in fairface brick is acceptable and where an outer leaf of brick or block is to be rendered to enhance protection against rain penetration or for appearance sake.

A range of precast reinforced concrete lintels is available to suit the widths of most standard door and window openings with adequate allowance for building in ends of lintels each side of openings. For use with fairface brickwork the lintel depth should match the depth of brick course heights to avoid untidy cutting of bricks around lintel ends.

These comparatively lightweight lintels are bedded on walling as support for outer and inner leaves.

The partial fill cavity insulation shown in Fig. 81 is carried down between the leaves of the wall to the head of the window or door frame.

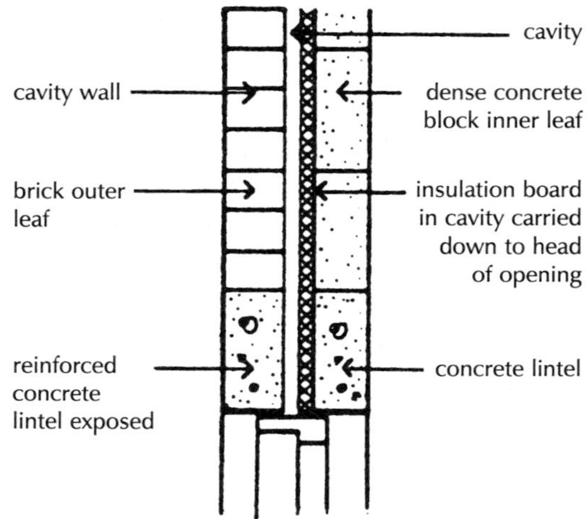

Fig. 81 Concrete lintels.

Cavity trays

In positions of severe exposure to wind driven rain, the outer leaf of a cavity wall may absorb water to the extent that rainwater penetrates to the cavity side of the outer leaf. It is unlikely, however, that water will enter the cavity unless there are faults in construction.

Where the mortar joints in the outer leaf of a cavity wall are inadequately flushed up with mortar, or the bricks in the outer leaf are grossly porous and where the wall is subject to severe or very severe exposure, there is a possibility that wind driven rain may penetrate through the outer leaf to the cavity.

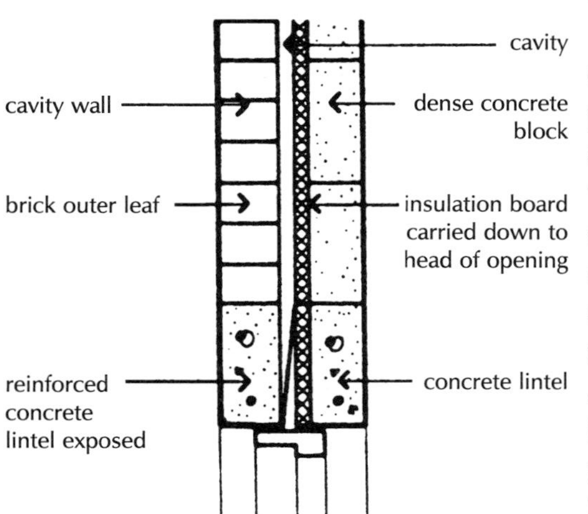

Fig. 82 Dpc and tray.

Where rain penetrates the outer leaf and enters the cavity, it is certain that either the construction of the wall is poorly executed, or the bricks have been unwisely chosen or the outer leaf is of inadequate thickness for the position of exposure in which it is built. The solution is to choose some alternative form of construction such as a thicker outer leaf, more suited to the position of exposure.

Some consider it good practice to use some form of impermeable cavity tray above all horizontal breaks in a cavity wall against the possibility of rainwater penetrating the outer leaf. The dubious argument is that freely flowing water may enter the cavity in sufficient volume and run down on to horizontal breaks in walls, such as openings, and cause damp staining. Having been persuaded that it is sense to accept this idea, it has become common practice to build in some form of damp-proof course or tray of flexible, impermeable material to direct freely flowing water out to the external face of walls. A strip of polymer based polythene, bitumen felt or sheet lead is used for the purpose. The dpc and tray shown in Fig. 82 is built in at the top of the inner lintel and dressed down to the underside of the outer lintel over the head of the window. As an alternative the dpc tray could be built in on top of the second block course and dressed down to the top of the outer lintel, with weep holes in the vertical brick joints.

If there were a real need for these trays it would be common to observe the evidence of water seeping through these weep holes to the outer face of cavity walls. No such evidence exists where cavity walls are sensibly designed and soundly constructed. A disadvantage of the weep holes is that cold air may enter and cool the air space and increase thermal transmission.

Where the cavity is solidly filled with insulation any water penetrating the outer leaf will be trapped between closed cell insulation and the wall or saturate open cell insulation.

Good sense dictates that the notion of the dpc trays in cavity walls be abandoned.

Resistance to the passage of heat

Because thin solid walls of brick or block may offer poor resistance to the penetration of wind driven rain, many loadbearing walls are built with two leaves of brick or block separated by a cavity, whose prime purpose is as a barrier to rain penetration.

Because the resistance to the passage of heat of a cavity wall by itself is poor, it is necessary to introduce a material with high resistance to heat transfer to the wall construction.

Most of the materials, thermal insulators, that afford high resistance to heat transfer are fibrous or cellular, lightweight, have comparatively poor mechanical strength and are not suitable by themselves for use as part of the wall structure. The logical position for such material in a cavity wall, therefore, is inside the cavity.

Cavity wall insulation

Partial fill

The purpose of the air space in a cavity wall is as a barrier to the penetration of rainwater to the inside face of the wall. If the clear air space is to be effective as a barrier to rain penetration it should not be bridged by anything other than cavity ties. If the cavity is then filled with some insulating material, no matter how impermeable to water the material is, there will inevitably be narrow capillary paths around wall ties and between edges of insulation boards or slabs across which water may penetrate. As a clear air space is considered necessary as a barrier to rain penetration there is good reason to fix insulation material inside a cavity so that it only partly fills the cavity and a cavity is maintained between the outer leaf and the insulating material. This construction, which is described as partial fill insulation of cavity, requires the use of some insulating material in the form of boards that are sufficiently rigid to be secured against the inner leaf of the cavity.

In theory a 25 mm wide air space between the outer leaf and the cavity insulation should be adequate to resist the penetration of rain providing the air space is clear of all mortar droppings and other building debris that might serve as a path for water. In practice, it is difficult to maintain a clear 25 mm wide air gap because of protrusion of mortar from joints in the outer leaf and the difficulty of keeping so narrow a space clear of mortar droppings. Good practice, therefore, is to use a 50 mm wide air space between the outer leaf and the partial fill insulation.

To meet insulation requirements and the use of a 100 mm cavity with partial fill insulation it may be economic to use a lightweight block inner leaf to augment the cavity insulation to bring the wall to the required U value.

Usual practice is to build the inner leaf of the cavity wall first, up to the first horizontal row of wall ties, then place the insulation boards in position against the inner leaf. Then as the outer leaf is built, a batten may be suspended in the cavity air space and raised to the level of the first row of wall ties and the batten is then withdrawn and cleared of droppings. Insulation retaining wall ties are then bedded across the cavity to tie the leaves and retain the insulation in position and the sequence of operations is repeated at each level of wall ties.

The suspension of a batten in the air space and its withdrawal and cleaning at each level of ties does considerably slow the process of brick and block laying.

Insulation retaining ties are usually standard galvanised steel or stainless steel wall ties to which a plastic disc is clipped to retain the edges of the insulation, as illustrated in Fig. 83. The ties may be set in line one over the other at the edges of boards, so that the retaining clips retain the corners of four insulation boards.

The materials used for partial fill insulation should be of boards, slabs or batts that are sufficiently rigid for ease of handling and to be retained in a vertical position against the inner leaf inside the cavity without sagging or losing shape, so that the edges of the boards remain close butted throughout the useful life of the building. For small dwellings the Building Regulations do not limit the use of combustible materials as partial fill insulation in a cavity in a cavity wall.

To provide a clear air space of 50 mm inside the cavity as a barrier to rain penetration and to provide sufficient space to keep the cavity clear during building, an insulant with a low U value is of advantage if a nominal 75 mm wide cavity is formed between the outer and inner leaves.

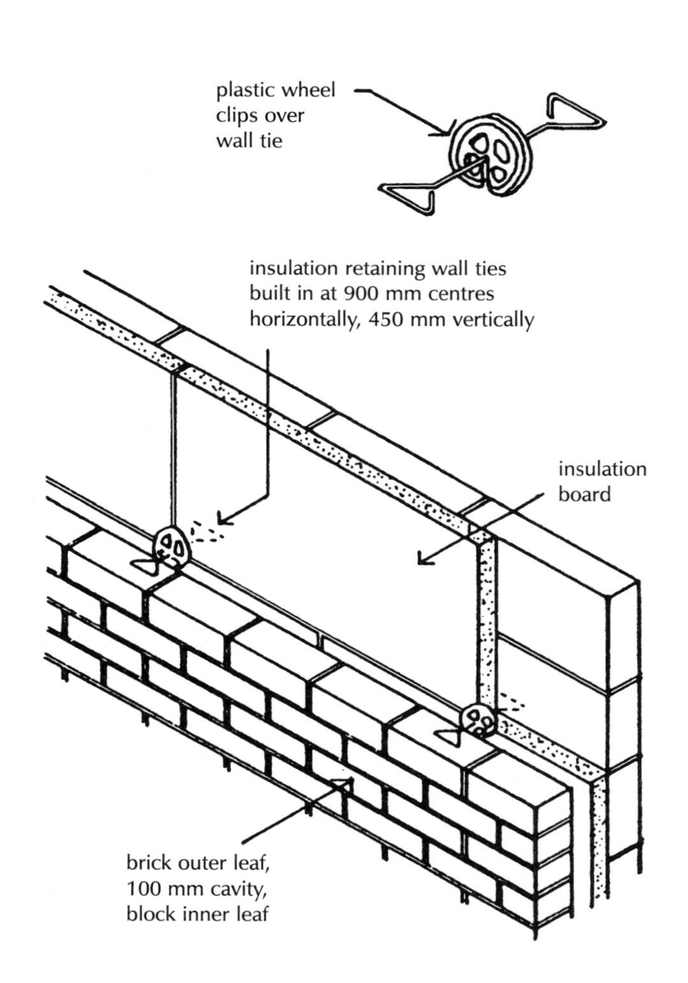

Fig. 83 Partial fill cavity insulation.

Insulation materials

The materials used as insulation for the fabric of buildings may be grouped as inorganic and organic insulants.

Inorganic insulants are made from naturally occurring materials that are formed into fibre, powder or cellular structures that have a high void content, as for example, glass fibre, mineral fibre (rockwool), cellular glass beads, vermiculite, calcium silicate and magnesia or as compressed cork.

Inorganic insulants are generally incombustible, do not support spread of flame, are rot and vermin proof and generally have a higher U value than organic insulants.

The inorganic insulants most used in the fabric of buildings are glass fibre and rockwool in the form of loose fibres, mats and rolls of felted fibres and semi-rigid and rigid boards, batts and slabs of compressed fibres, cellular glass beads fused together as rigid boards, compressed cork boards and vermiculite grains.

Organic insulants are based on hydrdocarbon polymers in the form of thermosetting or thermoplastic resins to form structures with a high void content, as for example polystyrene, polyurethane, isocyanurate and phenolic. Organic insulants generally have a lower U value than inorganic insulants, are combustible, support spread of flame more readily than inorganic insulants and have a comparatively low melting point.

The organic insulants most used for the fabric of buildings are expanded polystyrene in the form of beads or boards, extruded polystyrene in the form of boards and polyurethane, isocyanurate and phenolic foams in the form of preformed boards or spray coatings.

The materials that are cheapest, most readily available and used for cavity insulation are glass fibre, rockwool and EPS (expanded polystyrene), in the form of slabs or boards, in sizes to suit cavity tie spacing. With the recent increase in requirements for the insulation of walls it may well be advantageous to use one of the somewhat more expensive organic insulants such as XPS (extruded polystyrene), PIR (polyisocyanurate) or PUR (polyurethane) because of their lower U value, where a 50 mm clear air space is to be maintained in the cavity, without greatly increasing the overall width of the cavity.

Table 4 gives details of insulants made for use as partial fill to cavity walls.

Insulation thickness

A rough guide to determine the required thickness of insulation for a wall to achieve a U value of 0.45 W/m^{2}K is to assume the insulant provides the whole or a major part of the insulation by using 30 mm thickness with a U value of 0.02, 46 with 0.03, 61 with 0.04, 76 with 0.05 and 92 with 0.06 W/m^{2}K.

Table 4 Insulating materials.

Cavity wall partial fill	Thickness mm	U valve W/m²K
Glass fibre rigid slab 455 × 1200 mm	30, 35, 40, 45	0.033
Rockwool rigid slab 455 × 1200 mm	30, 40, 50	0.033
Cellular glass 450 × 600	40, 45, 50	0.042
EPS boards 450 × 1200 mm	25, 40, 50	0.037
XPS boards 450 × 1200 mm	25, 30, 50	0.028
PIR boards 450 × 1200 mm	20, 25, 30, 35, 50	0.022
PUR boards 450 × 1200 mm	20, 25, 30, 35, 40, 50	0.022

EPS expanded polystyrene
XPS extruded polystyrene
PIR rigid polyisocyanurate
PUR rigid polyurethane

A more exact method is by calculation as follows.

$$\text{Thermal resistance required} = \frac{1}{0.45} = 2.22\,\text{m}^2\text{K/W}$$

Thermal resistance of construction is:

external surface	0.06 m^2K/W
102 brick outer leaf	0.12
cavity at least 50	0.18
115 block inner leaf	1.05
13 plasterboard	0.03
inside surface	0.12
Total	1.56

$$\begin{aligned}\text{Additional resistance to be provided by insulation} &= 2.22 - 1.56\\ &= 0.66\,\text{m}^2\text{K/W}\end{aligned}$$

Assuming insulation with a U value of 0.03, then the thickness of insulation required = 0.66 × 0.03 × 1000 = 19.8 mm. This thickness is about one-third of that proposed by the first method that takes little account of the resistance of the rest of the wall construction.

Total fill

The thermal insulation of external walls by totally filling the cavity has been in use for many years. There have been remarkably few reported incidents of penetration of water through the total fill of cavities to the inside face of walls and the system of total fill has become an accepted method of insulating cavity walls.

The method of totally filling cavities with an insulant was developed after the steep increase in the price of oil and other fuels in the mid-1960s, as being the most practical way to improve the thermal insulation of existing cavity walls. Small particles of glass or rock wool fibre or foaming organic materials were blown through holes drilled in the outer leaf of existing walls to completely fill the cavity.

This system of totally filling the cavity of existing walls has been very extensively and successfully used. The few reported failures due to penetration of rainwater to the inside face were due to poor workmanship in the construction of the walls. Water penetrated across wall ties sloping down into the inside face of the wall, across mortar droppings bridging the cavity or from mortar protruding into the cavity from the outer leaf.

From the few failures due to rain penetration it would seem likely that the cavity in existing walls that have been totally filled was of little, if any, critical importance in resisting rain penetration in the position of exposure in which the walls were situated. None the less it is wise to provide a clear air space in a cavity wherever practical, against the possibility of rain penetration.

Where insulation is used to fill totally a nominal 50 mm wide cavity there is no need to use insulation retaining wall ties.

With a brick outer and block inner leaf it is preferable to raise the outer brick leaf first so that mortar protrusions from the joints, sometimes called snots, can be cleaned off before the insulation is placed in position and the inner block leaf, with its more widely spaced joints is built, to minimise the number of mortar snots that may stick into the cavity. This sequence of operations will require scaffolding on both sides of the wall and so add to the cost.

Insulation that is built in as the cavity walls are raised, to fill the cavity totally, will to an extent be held in position by the wall ties and the two leaves of the cavity wall. Rolls or mats of loosely felted glass fibre or rockwool are often used. There is some likelihood that these materials may sink inside the cavity and gaps may open up in the insulation and so form cold bridges across the wall. To maintain a continuous, vertical layer of insulation inside the cavity one of the mineral fibre semi-rigid batts or slabs should be used. Fibre glass and rockwool semi-rigid batts or slabs in sizes suited to cavity tie spacing are made specifically for this purpose.

As the materials are made in widths to suit vertical wall tie spacing there is no need to push them down into the cavity after the wall is built, as is often the procedure with loose fibre rolls and mats, and so displace freshly laid brick or blockwork. There is no advantage in using one of the more expensive organic insulants such as XPS, PIR or PUR that have a lower U value than mineral fibre materials for the total cavity fill, as the width of the cavity can be adjusted to suit the required thickness of insulation.

The most effective way of insulating an existing cavity wall is to fill the cavity with some insulating material that can be blown into the cavity through small holes drilled in the outer leaf of the wall. The injection of the cavity fill is a comparatively simple job. The complication arises in forming sleeves around air vents penetrating the wall and sealing gaps around openings.

When filling the cavity of existing walls became common practice, a foamed organic insulant, ureaformaldehyde, was extensively used. The advantage of this material was that it could be blown, under pressure, through small holes in the outer leaf and as the constituents mixed they foamed and filled the cavity with an effective insulant. This material was extensively used, often by operatives ill trained in the sensible use of the material. The consequence was that through careless mixing of the components of the insulant and careless workmanship, the material gave off irritant fumes when used and later, when it was in place, these entered buildings and caused considerable distress to the occupants. Approved Document D of the Building Regulations details provisions for the use of this material in relation to the construction of the wall and its suitability, the composition of the materials, and control of those carrying out the work. As a result of past failures this material is less used than it was.

Glass fibre, granulated rockwool of EPS beads are used for the injection of insulation for existing cavity walls. These materials can also be used for blowing into the cavity of newly built walls.

Table 5 gives details of insulants for total cavity fill.

The required thickness of insulation can be taken from the two methods suggested for partial fill. In a calculation for total fill, the thermal resistance of the cavity is omitted.

Thermal bridge

A thermal bridge, more commonly known as a cold bridge in cold climates, is caused by appreciably greater thermal conductivity through one part of a wall than the rest of the wall. Where the cavity in a wall is partially or totally filled with insulation and the cavity is bridged with solid filling at the head, jambs or cill of an opening, there will be considerably greater transfer of heat through the solid filling than through the rest of the wall. Because of the greater transfer of

Table 5 Insulating materials.

Cavity wall total fill	Thickness mm	U value W/m^2K
Glass fibre		
semi-rigid batt 455 × 1200 mm	50, 65, 75, 100	0.036
glass fibres for blown fill		0.039
Rockwool		
semi-rigid batt 455 × 900 mm	50, 65, 75, 100	0.036
granulated for blown fill		0.037
EPS		
beads for blown fill		0.04

EPS expanded polystyrene

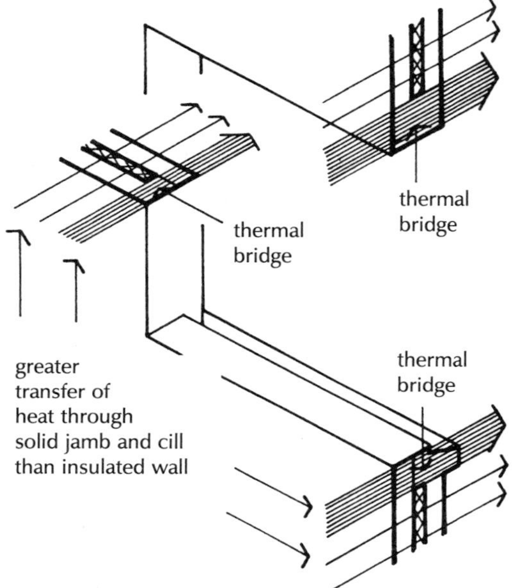

Fig. 84 Thermal bridge.

heat through the solid filling illustrated in Fig. 84, the inside face of the wall will be appreciably colder in winter than the rest of the wall and cause some loss of heat and encourage warm moist air to condense on the inside face of the wall on the inside of the cold bridge. This condensation water may cause unsightly stains around openings and encourage mould growth.

Thermal bridges around openings can be minimised by continuing cavity insulation to the head of windows and doors and to the sides and bottom of doors and windows.

Of late an inordinate fuss has been made about 'cold bridges' as though a cold bridge was some virulent disease or a heinous crime.

Solid filling of cavities around openings will allow greater transfer of heat than the surrounding insulated wall and so will window glass, both single and double, and window frames. To minimise heat transfer, cavity insulation should continue up to the back of window and door frames.

Where solid filling of cavities around openings is used the area of the solid filling should be included with that of the window and its frame for heat loss calculation.

SOLID WALLS

Up to the early part of the twentieth century walls were generally built as solid brickwork of adequate thickness to resist the penetration of rain to the inside face and to safely support the loads common to buildings both large and small.

At the time it was accepted that the interior of buildings would be cold during winter months when heating was provided by open fires and stoves, fired by coal or wood, to individual rooms. The people of northern Europe accepted the inevitability of a degree of indoor cold and dressed accordingly in thick clothing both during day and night time. There was an adequate supply of coal and wood to meet the expectations for some indoor heating for the majority.

The loss of heat through walls, windows and roofs was not a concern at the time. Thick curtains drawn across windows and external doors provided some appreciable degree of insulation against loss of heat.

From the middle of the twentieth century it became practical to heat the interior of whole buildings, with boilers fired by oil or gas. It is now considered a necessity to be able to heat the whole of the interior of dwellings so that the commonplace of icy cold bathrooms and corridors is an experience of the past.

In recent years an industry of scare stories has developed. Ill considered and unscientific claims by 'experts' that natural resources of fossil fuels such as oil and gas will soon be exhausted have been broadcast. These dire predictions have prompted the implementation of regulations to conserve fuel and power by introducing insulating materials to the envelope of all new buildings that are usually heated.

This 'bolting the stable door after the horse has gone' action will for very many years to come only affect new buildings, a minority of all buildings.

A consequence is that the cavity in external walls of buildings, originally proposed to exclude rain, has been converted to function as a prime position for lightweight insulating materials with exclusion of rain a largely ignored function of a cavity wall.

Resistance to weather

A solid wall of brick will resist the penetration of rain to its inside face by absorbing rainwater that subsequently, in dry periods, evaporates to outside air. The penetration of rainwater into the thickness of a solid wall depends on the exposure of the wall to driving rain and the permeability of the bricks and mortar to water.

The permeability of bricks to water varies widely and depends largely on the density of the brick. Dense engineering bricks absorb rainwater less readily than many of the less dense facing bricks. It would seem logical, therefore, to use dense bricks in the construction of walls to resist rain penetration.

In practice, a wall of facing bricks will generally resist the penetration of rainwater better than a wall of dense bricks. The reason for this is that a wall of dense bricks may absorb water through fine cracks between dense bricks and dense mortar, to a considerable depth of the thickness of a wall, and this water will not readily evaporate through the fine cracks to outside air in dry periods, whereas a wall of less dense bricks and mortar will absorb water to some depth of the thickness of the wall and this water will substantially evaporate to outside air. It is not unknown for a wall of dense bricks and mortar to show an outline of damp stains on its inside face through persistent wetting, corresponding to the mortar joints.

The general rule is that to resist the penetration of rain to its inside face a wall should be constructed of sound, well burned bricks of moderate density, laid in a mortar of similar density and of adequate thickness to prevent the penetration of rain to the inside face.

A solid 1 B thick wall may well be sufficiently thick to prevent the penetration of rainwater to its inside face in the sheltered positions common to urban settlements on low lying land. In positions of moderate exposure a solid wall $1\frac{1}{2}$ B thick will be effective in resisting the penetration of rainwater to its inside face.

In exposed positions such as high ground and near the coast a wall 2 B thick may be needed to resist penetration to inside faces. A wall 2 B thick is more than adequate to support the loads of all but heavily loaded structures and for resistance to rain penetration a less thick wall protected with rendering or slate or tile hanging is a more sensible option.

External weathering to walls of brick and block

In exposed positions such as high ground, on the coast and where there is little shelter from trees, high ground or surrounding buildings it may well be advisable to employ a system of weathering on the outer face of both solid and cavity walling to provide protection against wind driven rain. The two systems used are external rendering and slate or tile hanging.

Rendering

The word rendering is used in the sense of rendering the coarse texture of a brick or block wall smooth by the application of a wet mix of lime, cement and sand over the face of the wall, to alter the appearance of the wall or improve its resistance to rain penetration, or both. The wet mix is spread over the external wall face in one, two or three coats and finished with either a smooth, coarse or textured finish while wet. The rendering dries and hardens to a decorative or protective coating that varies from dense and smooth to a coarse and open texture.

Stucco is a term, less used than it was, for external plaster or rendering that was applied as a wet mix of lime and sand, in one or two coats, and finished with a fine mix of lime or lime and sand, generally in the form imitating stone joints and mouldings formed around projecting brick courses as a background for imitation cornices and other architectural decorations. To protect the comparatively porous lime and sand coating, the surface was usually painted.

The materials and application of the various smooth, textured, rough cast and pebble dash rendering are described in Volume 2.

The materials of an external rendering should have roughly the same density and therefore permeability to water as the material of the wall to which it is applied. There are many instances of the application of a dense rendering to the outside face of a wall that is permeable to water, in the anticipation of protecting the wall from rain penetration. The result is usually a disaster.

A dense sand and cement rendering, for example, applied to the face of a wall of porous bricks, will, on drying, shrink fiercely, pull away from the brick face or tear off the face of the soft bricks, and the rendering will craze with many fine hair cracks over its surface. Wind driven rain will then penetrate the many hair cracks through which water will be unable to evaporate to outside air during dry spells and the consequence is that the wall behind will become more water logged than before and the rendering will have a far from agreeable appearance.

Slate and tile hanging

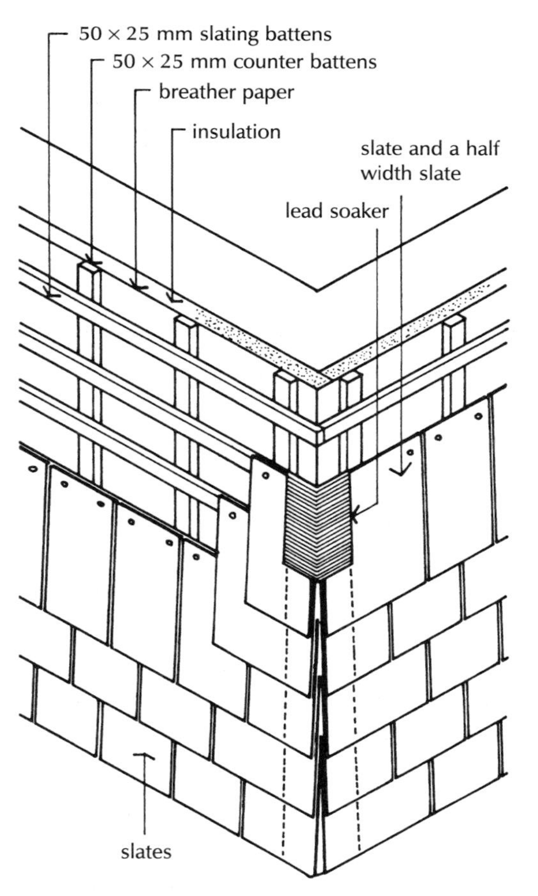

Fig. 85 Slate hanging.

In positions of very severe exposure to wind driven rain, as on high open ground facing the prevailing wind and on the coast facing open sea, it is necessary to protect both solid and cavity walls with an external cladding. The traditional wall cladding is slate or tile hanging in the form of slates or tiles hung double lap on timber battens nailed to counter battens. Slate hanging has generally been used in the north and tile in the south of Great Britain. Either natural or manufactured slates and tiles can be used.

As a fixing for slating or tiling battens, 50 × 25 mm timber counter battens are nailed at 300 mm centres up the face of the wall to which timber slating or tiling battens are nailed at centres suited to the gauge (centres) necessary for double lap slates or tiles, as illustrated in Fig. 85.

As protection against decay, pressure impregnated softwood timber battens should be used and secured with non-ferrous fixings to avoid the deterioration and failure of steel fixings by rusting.

Where slate or tile hanging is used as cladding to a solid wall of buildings normally heated, then the necessary insulation can be fixed to the wall behind the counter battens. Rigid insulation boards of organic or inorganic insulation are fixed with a mechanically operated hammer gun that drives nails through both the counter battens, a breather paper and the insulation boards into the wall.

The continuous layer of breather paper, that is fixed between the counter battens and the insulation, is resistant to the penetration of water in liquid form but will allow water vapour to pass through it. Its purpose is to protect the outer surface of the insulation from cold air and any rain that might penetrate the hanging and to allow movement of vapour through it.

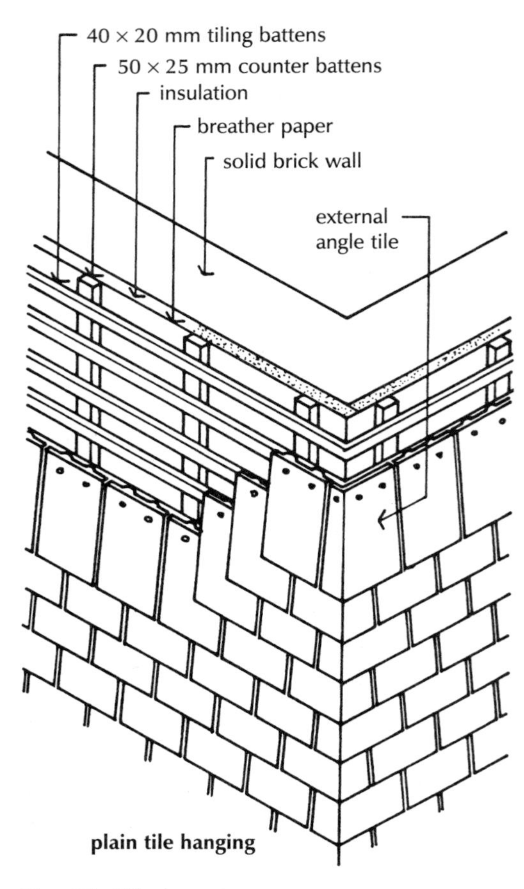

Fig. 86 Tile hanging.

For vertically hung slating it is usual to use one of the smaller slates such as 405 × 205 mm slate which is headnailed to 50 × 25 mm battens and is less likely to be lifted and dislodged in high wind than longer slates would be. Each slate is nailed with non-ferrous nails to overlap two slates below, as illustrated in Fig. 85, and double lapped by overlapping the head of slates two courses below.

At angles and the sides of openings a slate one and a half the width of slates is used to complete the overlap. This width of slate is specifically used to avoid the use of a half width slate that might easily be displaced in wind.

Internal and external angles are weathered by lead soakers – hung over the head of slates – to overlap and make the joint weathertight. Slate hanging is fixed either to overlap or butt to the side of window and door frames with exposed edges of slates pointed with cement mortar or weathered with lead flashings.

At lower edges of slate hanging a projection is formed on or in the wall face by means of blocks, battens or brick corbel courses on to which the lower courses of slates and tiles bell outwards slightly to throw water clear of the wall below.

Tile hanging is hung and nailed to 40 × 20 mm tiling battens fixed at centres to counter battens to suit the gauge of plain tiles. Each tile is hung to battens and also nailed, as security against wind, as illustrated in Fig. 86.

At internal and external angles special angle tiles may be used to continue the bond around the corner, as illustrated in Fig. 86. As an alternative and also at the sides of openings tile and a half width tiles may be used with lead soakers to angles and pointing to exposed edges or weathering to the sides of the openings.

As weather protection to the solid walls of buildings with low or little heat requirements the hanging is fixed directly to walling and to those buildings that are heated the hanging may be fixed to external or internal insulation for solid walling and directly to cavity walling with cavity insulation.

OPENINGS IN SOLID WALLS

For the strength and stability of walling the size of openings in walls is limited by regulations for both solid and cavity walls.

Jambs of openings

The jambs of openings for windows and doors in solid walls may be plain (square) or rebated.

Plain or square jambs are used for small section window or door frames of steel and also for larger section frames where the whole of the external face of frames is to be exposed externally. The bonding of brickwork at square jambs is the same as for stop ends and angles with a closer next to a header in alternate courses to complete the bond.

Rebated jambs

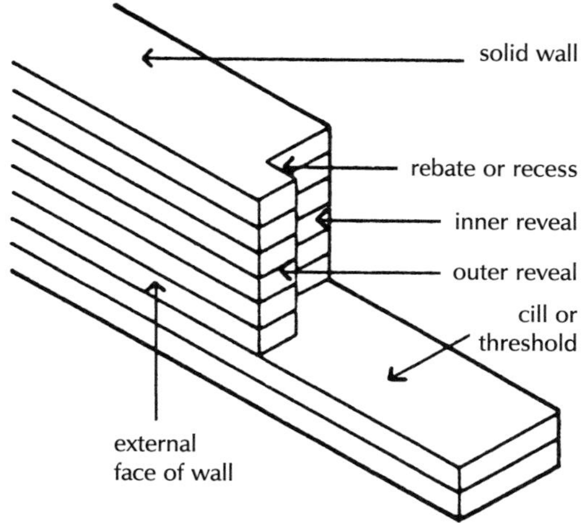

Fig. 87 Rebated jamb.

Window and door frames made of soft wood have to be painted for protection from rain, for if wood becomes saturated it swells and in time may decay. With some styles of architecture it is thought best to hide as much of the window frame as possible. So either as a partial protection against rain or for appearance sake, or for both reasons, the jambs of openings are rebated.

Figure 87 is a diagram of one rebated jamb on which the terms used are noted.

As one of the purposes of a rebated jamb is to protect the frame from rain the rebate faces into the building and the frame of the window or door is fixed behind the rebate.

The thickness of brickwork that shows at the jamb of openings is described as the reveal. With rebated jambs there is an inner reveal and an outer reveal separated by the rebate.

The outer reveal is usually $\frac{1}{2}$B wide for ease of bonding bricks and may be 1 B wide in thick solid walling. The width of the inner reveal is determined by the relative width of the outer reveal and wall thickness.

The depth of the rebate is either $\frac{1}{4}$B (about 51 mm) or $\frac{1}{2}$B (102.5 mm). A $\frac{1}{4}$B rebate is used to protect and mask solid wood frames and the $\frac{1}{2}$B deep rebate to protect and mask the box frames to vertically sliding wood sash windows. The $\frac{1}{2}$B deep rebate virtually covers the external face of cased wood frames (see Volume 2) to the extent that a window opening appears to be glass with a narrow surround of wood.

Bonding of bricks at rebated jambs

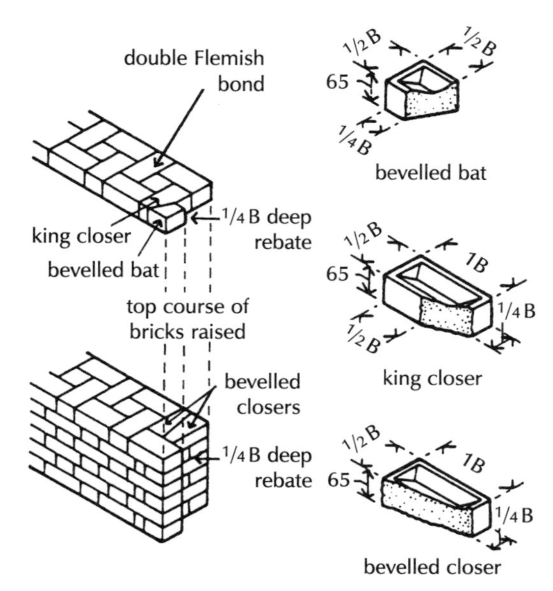

Fig. 88 Bonding at rebated jambs.

Just as at an angle or quoin in brickwork, bricks specially cut have to be used to complete, or close, the $\frac{1}{4}$B overlap caused by bonding, so at jambs special closer bricks $\frac{1}{4}$B wide on face have to be used.

Provided that the outer reveal is $\frac{1}{2}$B wide, the following basic rules will apply irrespective of the sort of bond used or the thickness of the wall. If the rebate is $\frac{1}{4}$B deep the bonding at one jamb will be arranged as illustrated in Fig. 88. In every other course of bricks a header face and then a closer of $\frac{1}{4}$B wide face must appear at the jamb or angle of the opening. To do this and at the same time to form the $\frac{1}{4}$B deep rebate and to avoid vertical joints continuously up the wall, two cut bricks have to be used.

These are a bevelled bat (a 'bat' is any cut part of a brick), which is shaped as shown in Fig. 88, and a king closer, which is illustrated in Fig. 88. Neither of these bricks is made specially to the shape and size shown, but is cut from whole bricks on the site.

In the course above and below, two other cut bricks, called bevelled closers, should be used behind the stretcher brick. These two bricks are used so as to avoid a vertical joint. Figure 88 shows a view of a bevelled closer.

Where the rebate is $\frac{1}{2}$B deep the bonding is less complicated. An arrangement of half bats as quoin header and two bevelled closers in alternate courses for English bond and half bats and king closers in alternate courses for Flemish bond is used.

Head of openings in solid walls

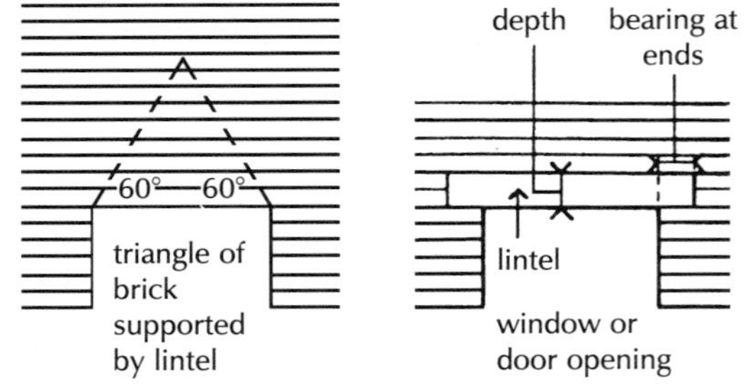

Fig. 89 Head of openings.

Solid brickwork over the head of openings has to be supported by either a lintel or an arch. The brickwork which the lintel or arch has to support is an isosceles triangle with 60° angles, formed by the bonding of bricks, as illustrated in Fig. 89. The triangle is formed by the vertical joints between bricks which overlap $\frac{1}{4}$B. In a bonded wall if the solid brickwork inside the triangle were taken out the load of the wall above the triangle would be transferred to the bricks of each side of the opening in what is termed 'the arching effect'.

Lintel is the name given to any single solid length of timber, stone, steel or concrete built in over an opening to support the wall over it, as shown in Fig. 89. The ends of the lintel must be built into the brick or blockwork over the jambs to convey the weight carried by the lintel to the jambs. The area of wall on which the end of a lintel bears is termed its bearing at ends. The wider the opening the more weight the lintel has to support and the greater its bearing at ends must be to transmit the load it carries to an area capable of supporting it. For convenience its depth is usually made a multiple of brick course height, that is about 75 mm, and the lintels are not usually less than 150 mm deep.

Timber lintels

Up to the beginning of the twentieth century it was common practice to support the brickwork over openings on a timber lintel. Wood lintels are less used today because wood may be damaged during a fire and because timber is liable to rot in conditions of persistent damp.

Concrete lintels

Since Portland cement was first mass produced towards the end of the nineteenth century it has been practical and economic to cast and use concrete lintels to support brickwork over openings.

Concrete is made from reasonably cheap materials, it can easily be moulded or cast when wet and when it hardens it has very good strength in resisting crushing and does not lose strength or otherwise deteriorate when exposed to the weather. The one desirable quality that concrete lacks, if it is to be used as a lintel, is tensile strength, that is strength to resist being pulled apart. To provide the necessary tensile strength to concrete steel reinforcement is cast into concrete.

For a simple explanation for the need and placing of reinforcement in concrete lintels suppose that a piece of india rubber were used as a lintel. Under load any material supported at its ends will deflect, bend, under its own weight and loads that it supports. India rubber has very poor compressive and tensile strength so that under load it

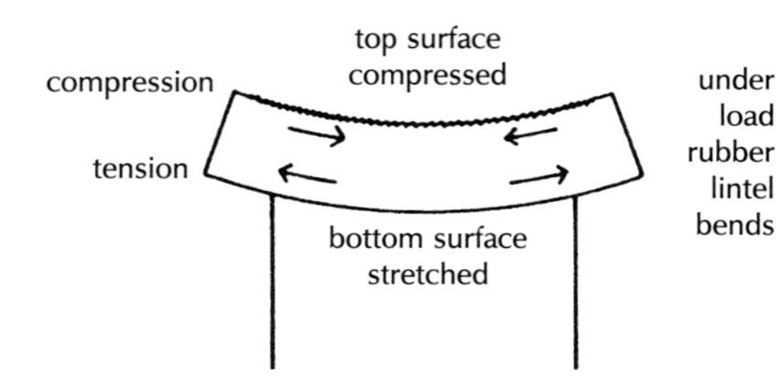

Fig. 90 Bending under load.

will bend very noticeably, as illustrated in Fig. 90. The top surface of the rubber becomes squeezed, indicating compression, and the lower surface stretched, indicating tension. A close examination of the india rubber shows that it is most squeezed at its top surface and progressively less to the centre, and conversely most stretched and progressively less up from its bottom surface to the centre of depth.

A concrete lintel will not bend so obviously as india rubber, but it will bend and its top surface will be compressed and its bottom surface stretched or in tension under load. Concrete is strong in resisting compression but weak in resisting tension, and to give the concrete lintel the strength required to resist the tension which is maximum at its lower surface, steel is added, because steel is strong in resisting tension. This is the reason why rods of steel are cast into the bottom of a concrete lintel when it is being moulded in its wet state.

Lengths of steel rod are cast into the bottom of concrete lintels to give them strength in resisting tensile or stretching forces. As the tension is greatest at the underside of the lintel it would seem sensible to cast the steel rods in the lowest surface. In fact the steel rods are cast in some 15 mm or more above the bottom surface. The reason for this is that steel very soon rusts when exposed to air and if the steel rods were in the lower surface of the lintel they would rust, expand and rupture the concrete around them, and in time give way and the lintel might collapse. Also if a fire occurs in the building the steel rods would, if cast in the surface, expand and come away from the concrete and the lintel collapse. The rods are cast at least 15 mm up from the bottom of the lintel and 15 mm or more of concrete below them is called the concrete cover.

Reinforcing rods

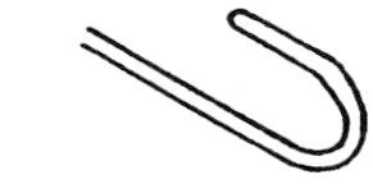

Fig. 91 Ends of reinforcing rods.

Reinforcing rods are usually of round section mild steel 10 or 12 mm diameter for lintels up to 1.8 m span. The ends of the rods should be bent up at 90° or hooked as illustrated in Fig. 91.

The purpose of bending up the ends is to ensure that when the lintel does bend the rods do not lose their adhesion to the concrete around them. After being bent or hooked at the ends the rods should be some 50 or 75 mm shorter than the lintel at either end. An empirical rule for determining the number of 12 mm rods required for lintels of up to, say, 1.8 m span is to allow one 12 mm rod for each half brick thickness of wall which the lintel supports.

Casting lintels

The word 'precast' indicates that a concrete lintel has been cast inside a mould, and has been allowed time to set and harden before it is built into the wall.

The words 'situ-cast' indicate that a lintel is cast in position inside a timber mould fixed over the opening in walls. Whether the lintel is precast or situ-cast will not affect the finished result and which method is used will depend on which is most convenient.

It is common practice to precast lintels for most normal door and window openings, the advantage being that immediately the lintel is placed in position over the opening, brickwork can be raised on it, whereas the concrete in a situ-cast lintel requires a timber mould or formwork and must be allowed to harden before brickwork can be raised on it.

Lintels are cast in situ, that is in position over openings, if a precast lintel would have been too heavy or cumbersome to have been easily hoisted and bedded in position.

Precast lintels must be clearly marked to make certain that they are bedded with the steel reinforcement in its correct place, at the bottom of the lintel. Usually the letter 'T' or the word 'Top' is cut into the top of the concrete lintel whilst it is still wet.

Prestressed concrete lintels

Prestressed, precast concrete lintels are used particularly over internal openings. A prestressed lintel is made by casting concrete around high tensile, stretched wires which are anchored to the concrete so that the concrete is compressed by the stress in the wires. (See also Volume 4.) Under load the compression of concrete, due to the stressed wires, has to be overcome before the lintel will bend.

Two types of prestressed concrete lintel are made, composite lintels and non-composite lintels.

Composite and non-composite lintels

Composite lintels are stressed by a wire or wires at the centre of their depth and are designed to be used with the brickwork they support which acts as a composite part of the lintel in supporting loads. These comparatively thin precast lintels are built in over openings and brickwork is built over them. Prestressed lintels over openings more than 1200 mm wide should be supported to avoid deflection, until the mortar in the brickwork has set. When used to support blockwork the composite strength of these lintels is considerably less than when used with brickwork.

Non-composite prestressed lintels are made for use where there is insufficient brickwork over to act compositely with the lintel and also where there are heavy loads.

These lintels are made to suit brick and block wall thicknesses, as illustrated in Fig. 92. They are mostly used for internal openings, the inner skin of cavity walls and the outer skin where it is covered externally.

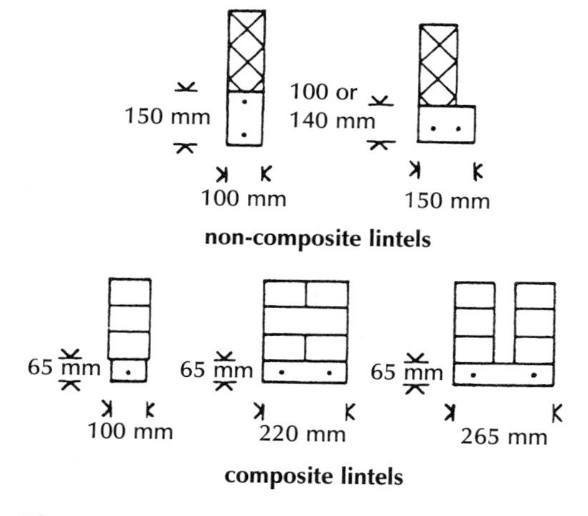

Fig. 92 Prestressed lintels.

Precast, or prestressed lintels may be used over openings in both internal and external solid walls. In external walls prestressed lintels are used where the wall is to be covered with rendering externally and for the inner leaf of cavity walls where the lintel will be covered with plaster.

Precast reinforced concrete lintels may be exposed on the external face of both solid and cavity walling where the appearance of a concrete surface is acceptable.

Boot lintels

When concrete has dried it is a dull, light grey colour. Some think that a concrete lintel exposed for its full depth on the external face of brick walls is not attractive. In the past it was for some years common practice to hide the concrete lintel behind a brick arch or brick lintel built over the opening externally.

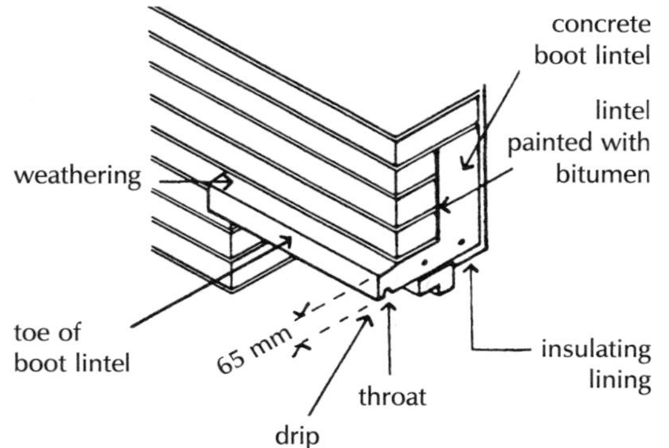

Fig. 93 Boot lintel.

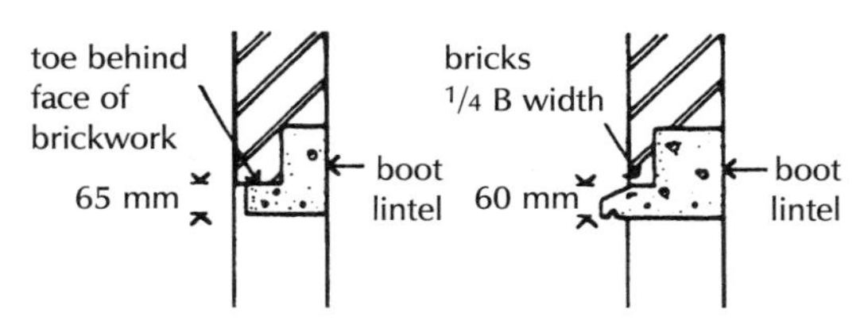

Fig. 94 Boot lintels.

A modification of the ordinary rectangular section lintel, known as a boot lintel, was often used to reduce the depth of the lintel exposed externally. Figure 93 is an illustration of a section through the head of an opening showing a boot lintel in position. The lintel is boot-shaped in section with the toe part showing externally. The toe is usually made 65 mm deep. The main body of the lintel is inside the wall where it does not show and it is this part of the lintel which does most of the work of supporting brickwork. Some think that the face of the brickwork looks best if the toe of the lintel finishes just 25 or 40 mm back from the external face of the wall, as in Fig. 94. The brickwork built on the toe of the lintel is usually $\frac{1}{2}$B thick for openings up to 1.8 m wide. The 65 mm deep toe, if reinforced as shown, is capable of safely carrying the two or three courses of $\frac{1}{2}$B thick brickwork over it. The brickwork above the top of the main part of the lintel bears mainly on it because the bricks are bonded. If the opening is wider than 1.8 m the main part of the lintel is sometimes made sufficiently thick to support most of the thickness of the wall over, as in Fig. 94.

The brickwork resting on the toe of the lintel is built with bricks cut in half. When the toe of the lintel projects beyond the face of the brickwork it should be weathered to throw rainwater out from the wall face and throated to prevent water running in along soffit or underside, as shown in Fig. 93.

When the external face of brickwork is in direct contact with concrete, as is the brickwork on the toe of these lintels, an efflorescence of salts is liable to appear on the face of the brickwork. This is caused by soluble salts in the concrete being withdrawn when the wall dries out after rain and being left on the face of the brickwork in the form of unsightly white dust. To prevent the salts forming, the faces of the lintel in direct contact with the external brickwork should be painted with bituminous paint as indicated in Fig. 93. The bearing at ends where the boot lintel is bedded on the brick jambs should be of the same area as for ordinary lintels.

Pressed steel lintels

Galvanised pressed steel lintels may be used as an alternative to concrete as a means of support to both loadbearing and non-loadbearing internal walls.

Mild steel strip is pressed to shape, welded as necessary and galvanised. The steel lintels for support over door openings in loadbearing internal walls are usually in hollow box form, as illustrated in Fig. 95. A range of lengths and sections is made to suit standard openings, wall thicknesses, course height for brickwork and adequate bearing at ends. For use over openings in loadbearing concrete block internal walls it is usually necessary to cut blocks around the bearing ends of these shallow depth lintels.

Over wide openings it may be necessary to fill the bearing ends of these hollow lintels with concrete to improve their crushing resistance.

The exposed faces of these lintels are perforated to provide a key for plaster.

To support thin, non-loadbearing concrete blocks over narrow door openings in partition walls, a small range of corrugated, pressed steel lintels is made to suit block thickness. These shallow depth, galvanised lintels are made to match the depth of horizontal mortar joints to avoid cutting of blocks.

The corrugations provide adequate key for plaster run over the face of partitions and across the soffit of openings, as illustrated in Fig. 96.

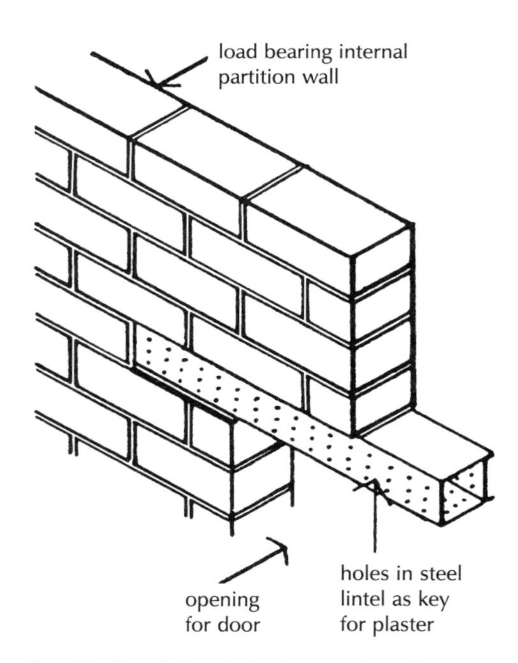

Fig. 95 Steel lintels in internal walls.

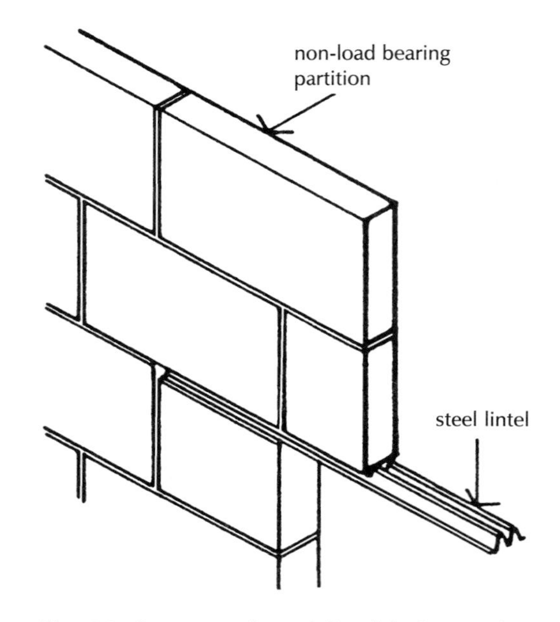

Fig. 96 Corrugated steel lintel in internal wall.

These shallow depth lintels act compositely with the blocks they support. To prevent sagging they should be given temporary support at mid-span until the blocks above have been laid and the mortar hardened.

Brick lintels

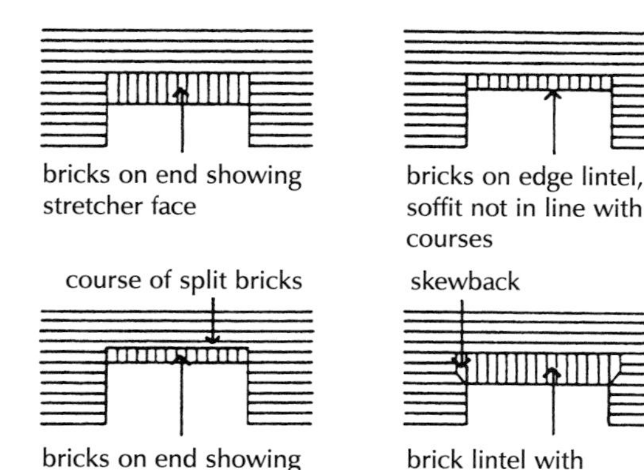

Fig. 97 Brick lintels.

A brick lintel may be formed as bricks on end, bricks on edge or coursed bricks laid horizontally over openings. The small units of brick, laid in mortar, give poor support to the wall above and usually need some form of additional support.

A brick on-end lintel is generally known as a 'soldier arch' or 'brick on end' arch. The word arch here is wrongly used as the bricks are not arranged in the form of an arch or curve but laid flat. The brick lintel is built with bricks laid on end with stretcher faces showing, as illustrated in Fig. 97. In building a brick lintel, mortar should be packed tightly between bricks.

A brick on end or soldier arch was a conventional method of giving the appearance of some form of support over openings in fairface brickwork.

For openings up to about 900 mm wide it was common to provide some support for soldier arches by building the lintel on the head of timber window and door frames. The wood frame served as temporary support as the bricks were laid, and support against sagging once the wall was built.

A variation was to form skew back bricks at each end of the lintel with cut bricks so that the slanting surface bears on a skew brick in the jambs, as illustrated in Fig. 97. The skew back does give some little extra stability against sagging.

For openings more than 900 mm wide a brick on end lintel may be supported by a 50×6 mm iron bearing bar, the ends of which are built into jambs as illustrated in Fig. 98A. The bearing bar provides little effective support and may in time rust. As a more effective alternative a steel 50 or 75 mm angle is built into jambs to give support to the lintel. The 50 mm flange of the angle supports the back edge of the bricks and may be masked by the window or door frame.

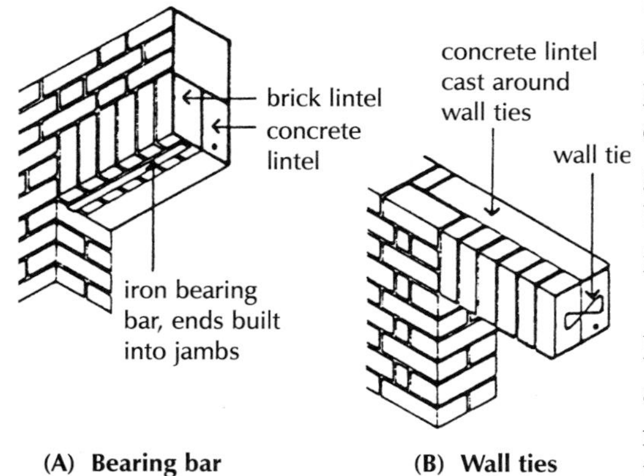

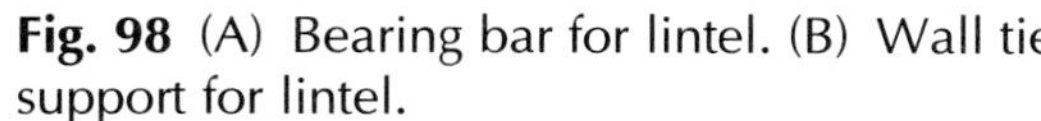

Fig. 98 (A) Bearing bar for lintel. (B) Wall tie support for lintel.

Another method of support was to drill a hole in each brick of the lintel. This can only successfully be done with fine grained bricks such as marls or gaults. Through the holes in the bricks a round-section mild steel rod is threaded and the ends of the rod are built into the brickwork either side of the lintel. This method of supporting the lintel is quite satisfactory but is somewhat expensive because of the labour involved.

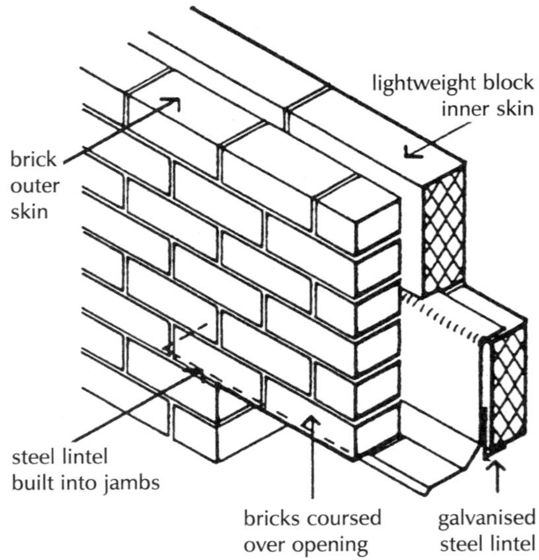

Fig. 99 Steel lintel support.

A more satisfactory method of providing support for brick on edge lintels is by wall ties cast into a concrete lintel. The lintel bricks are laid on a temporary supporting soffit board. As the bricks are laid wall ties are bedded between joints. An in situ reinforced concrete lintel is then cast behind the brick lintel so that when the concrete has set and hardened the ties give support, as illustrated in Fig. 98B.

Bricks laid on edge, showing a header face, were sometimes used as a lintel. Where the soffit of the lintel is in line with a brick course there has to be an untidy split course of bricks, some 37 mm deep above. Alternatively, the top of the lintel may be in line with a course, as illustrated in Fig. 97.

As support for coursed brickwork over openings a galvanised, pressed steel lintel is used. The lintel illustrated in Fig. 99 is for use with cavity walling to provide support for both the brick outer leaf and the block inner.

Brick arches

Semi-circular arch

An arch, which is the most elegant and structurally efficient method of supporting brickwork, has for centuries been the preferred means of support for brickwork over the small openings for doors and windows and for arcades, viaducts and bridges. The adaptability and flexibility of the small units of brick, laid in mortar, is demonstrated in the use of accurately shaped brick in ornamental brickwork and large span, rough archwork for railway bridges.

The traditional skills of bricklaying have for many years been in decline. Of recent years the use of brick arching has, to an extent, come back into fashion in the form of arched heads to openings in loadbearing walls, brick facework to framed buildings and arcading.

The most efficient method of supporting brickwork over an opening is by the use of a semi-circular arch which transfers the load of the wall it supports most directly to the sides of the opening through the arch. Figure 100 is an illustration of a semi-circular brick arch with the various terms used noted.

A segmented arch, which takes the form of a segment (part) of a circle is less efficient in that it transmits loads to the jambs by both vertical and outward thrust.

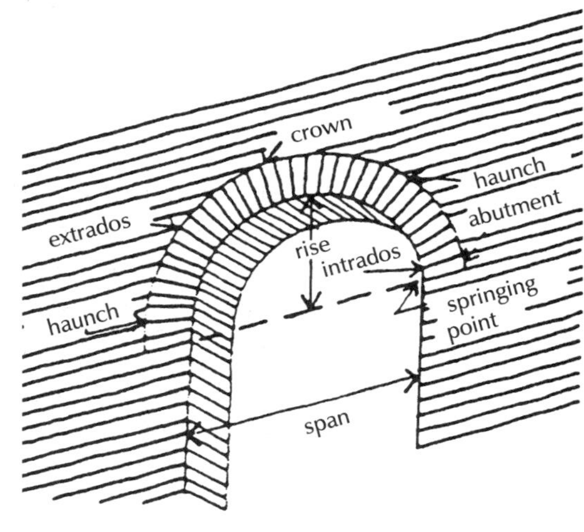

Fig. 100 Semi-circular brick arch.

Rough and axed arches

The two ways of constructing a curved brick arch are with bricks laid with wedge shaped mortar joints or with wedge-shaped bricks with mortar joints of uniform thickness, as illustrated in Fig. 101.

An arch formed with uncut bricks and wedge shaped mortar joints is termed a rough brick arch because the mortar joints are irregular

Fig. 101 Rough and axed arches.

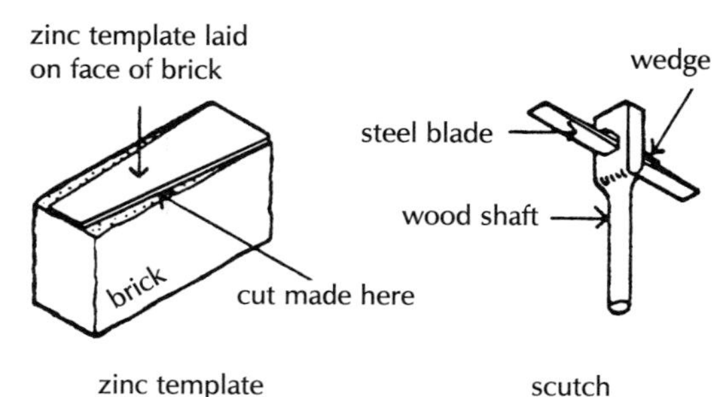

Fig. 102 Axed brick.

and the finished effect is rough. In time the joints, which may be quite thick at the crown of the arch, may tend to crack and emphasise the rough appearance. Rough archwork, which may be used for its rugged appearance with irregularly shaped bricks, is not generally used for fairface work.

Arches in fairface brickwork are usually built with bricks cut to wedge shape with mortar joints of uniform width. The bricks are cut to the required wedge shape by gradually chopping them to shape, hence the name 'axed bricks'.

Any good facing brick, no matter how hard, can be cut to a wedge shape either on or off the building site. A template, or pattern, is cut from a sheet of zinc to the exact wedge shape to which the bricks are to be cut. The template is laid on the stretcher or header face of the brick as illustrated in Fig. 102. Shallow cuts are made in the face of the brick either side of the template. These cuts are made with a hacksaw blade or file and are to guide the bricklayer in cutting the brick. Then, holding the brick in one hand the bricklayer gradually chops the brick to the required wedge shape. For this he uses a tool called a scutch, illustrated in Fig. 102. When the brick has been cut to a wedge shape the rough, cut surfaces are roughly levelled with a coarse rasp, which is a steel file with coarse teeth.

From the description this appears to be a laborious operation but in fact the skilled bricklayer can axe a brick to a wedge shape in a few minutes. The axed wedged-shaped bricks are built to form the arch with uniform 10 mm mortar joints between the bricks.

Gauged bricks

The word gauge is used in the sense of measurement, as gauged bricks are those that have been so accurately prepared to a wedge shape that they can be put together to form an arch with very thin joints between them. This does not improve the strength of the brick arch and is done entirely for reasons of appearance. Hard burned clay facing bricks cannot be cut to the accurate wedge shape required for this work because the bricks are too coarse grained, and bricks which are to be gauged are specially chosen. One type of brick used for gauged brickwork is called a rubber brick because its composition is such that it can be rubbed down to an accurate shape on a flat stone.

Rubber bricks are made from fine grained sandy clays. The bricks are moulded and then baked to harden them, and the temperature at which these bricks are baked is lower than that at which clay bricks are burned, the aim being to avoid fusion of the material of the bricks so that they can easily be cut and accurately rubbed to shape. Rubber bricks have a fine sandy texture and are usually 'brick red' in colour, although grey, buff and white rubber bricks are made. These bricks are usually somewhat larger than most clay bricks.

Sheet zinc templates, or patterns, are cut to the exact size of the wedge-shaped brick voussoirs. These templates are placed on the stretcher or header face of the brick to be cut and the brick is sawn to a wedge shape with a brick saw. A brick saw consists of an 'H' shaped wooden frame across which is strung a length of twisted steel wires. Because rubber bricks are soft this twisted wire quickly saws through them.

After the bricks have been cut to a wedge shape they are carefully rubbed down by hand on a large flat stone until they are the exact wedge shape required as indicated by the sheet zinc template.

The gauged rubber bricks are built to form the arch with joints between the bricks as thin as 1.5 mm thick. A mortar of coarse sand and lime or cement, is too coarse for narrow joints and the mortar used between the gauged bricks is composed of either fine sand and cement and lime or lime and water, depending on the thickness of joint selected. The finished effect of accurately gauged red bricks with thin white joints between them was considered very attractive. Gauged bricks are used for flat camber arches.

A disadvantage of thin, lime mortar joints with fine grained rubber bricks is that bricks may become saturated with rainwater and crumble due to the effect of frost and the lime mortar joints may break up.

Two ring arch

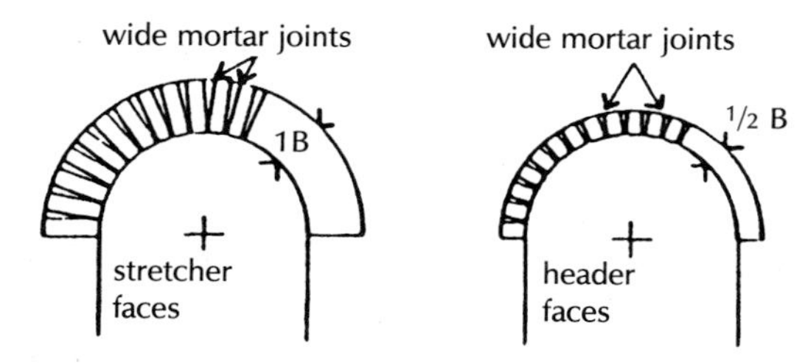

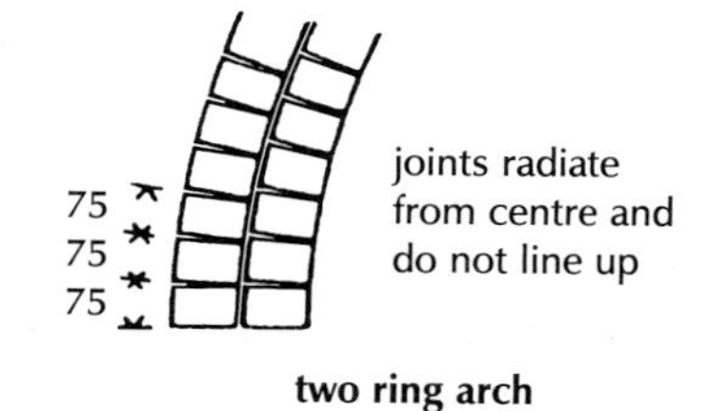

Fig. 103 Two ring arch.

Rough and axed bricks are used for both semi-circular and segmental arches and gauged brick for segmental and flat camber arches to avoid the more considerable cutting necessary with semi-circular arches.

Rough, axed or gauged bricks can be laid so that either their stretcher or their header face is exposed. Semi-circular arches are often formed with bricks showing header faces to avoid the excessively wedge-shaped bricks or joints that occur with stretcher faces showing. This is illustrated by the comparison of two arches of similar span first with stretcher face showing and then with header face showing, as illustrated in Fig. 103. If the span of the arch is of any considerable width, say 1.8 m or more, it is often practice to build it with what is termed two or more rings of bricks, as illustrated in Fig. 103.

An advantage of two or more rings of bricks showing header faces is that the bricks bond into the thickness of the wall. Where the wall over the arch is more than 1 B thick it is practical to effect more bonding of arch bricks in walls or viaducts by employing alternate snap headers (half bricks) in the face of the arch.

Segmental arch

The curve of this arch is a segment, that is part of a circle, and the designer of the building can choose any segment of a circle that he thinks suits his design. By trial and error over many years bricklayers have worked out methods of calculating a segment of a circle related to the span of this arch, which gives a pleasant looking shape, and which at the same time is capable of supporting the weight of brickwork over the arch. The recommended segment is such that the rise of the arch is 130 mm for every metre of span of the arch.

Centering

As temporary support for brick arches it is necessary to construct a rough timber framework to the profile of the underside of the arch on which the arch bricks are laid and jointed with mortar.

The timber frame is described as centering. It is fixed and supported in the opening while the bricks of the arch are being built and the coursed brickwork over the arch laid. Once the arch and brickwork above are finished the centering is removed.

A degree of both skill and labour is involved in arch building, in setting out the arch, cutting bricks for the arch and the abutment of coursed brickwork to the curved profile of the arch so that an arched opening is appreciably more expensive than a plain lintel head.

Flat camber arch

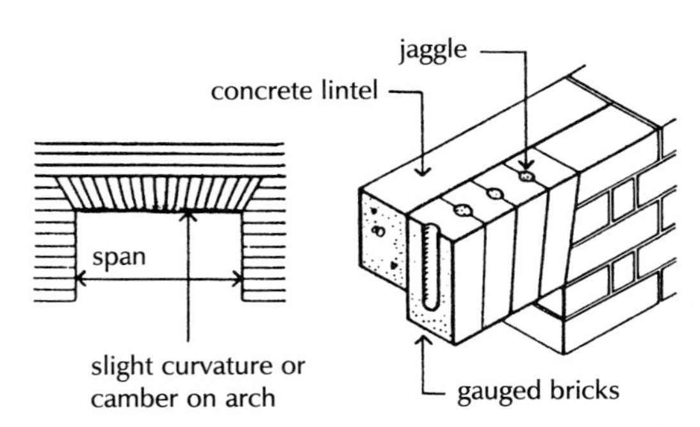

Fig. 104 Flat gauged camber arch.

This is not a true arch as it is not curved and might well be more correctly named flat brick lintel with voussoirs radiating from the centre, as illustrated in Fig. 104.

The bricks from which the arch is built may be either axed or gauged to the shape required so that the joints between the bricks radiated from a common centre and the widths of voussoirs measured horizontally along the top of the arch are the same. This width will be 65 mm or slightly less so that there are an odd number of voussoirs, the centre one being a key brick.

The centre from which the joints between the bricks radiate is usually determined either by making the skew or slating surface at the end of the arch 60° to the horizontal or by calculating the top of this skew line as lying 130 mm from the jamb for every metre of span.

If the underside or soffit of this arch were made absolutely level it would appear to be sagging slightly at its centre. This is an optical illusion and it is corrected by forming a slight rise or camber on the soffit of the arch. This rise is usually calculated at 6 or 10 mm for every metre of span and the camber takes the form of a shallow curve.

The camber is allowed for when cutting the bricks to shape. In walls built of hard coarse grained facing bricks this arch is usually built of axed bricks. In walls built of softer, fine grained facing bricks the arch is usually of gauged rubber bricks and is termed a flat gauged camber arch. This flat arch must be of such height on face that it bonds in with the brick course of the main walling. The voussoirs of this arch, particularly those at the extreme ends, are often longer overall than a normal brick and the voussoirs have to be formed with two bricks cut to shape.

Flat gauged camber arch

The bricks in this arch are jointed with lime and water, and the joints are usually 1.5 mm thick. Lime is soluble in water and does not adhere strongly to bricks as does cement. In time the jointing material, that is lime, between the bricks in this arch may perish and the bricks may slip out of position. To prevent this, joggles are formed between the bricks. These joggles take the form of semi-circular grooves cut in both bed faces of each brick, as shown in Fig. 104, into which mortar is run.

Solid walls

Thermal insulation

A requirement of the Building Regulations is that measures be taken, in new buildings, for the conservation of fuel and power. There is no requirement for particular forms of construction to meet the requirement. The practical guidance to the regulation, contained in Approved Document L for dwellings, is based on assumed levels of heating to meet the expectation of indoor comfort of the majority of the largely urban population of this country who are engaged in sedentary occupations.

The advice in the Approved Document is based on an assumption that walls will be of cavity construction with the insulation in the cavity, which is the optimum position for insulation. In consequence it is likely that insulated cavity wall construction will be the first choice for the walls of dwellings for some time to come.

The regulations do make allowance for the use of any form of construction providing the calculated energy use of such buildings is no greater than that of a similar building with recommended insulated construction.

To provide the insulation required to meet the standard for conservation with a solid wall it is necessary to fix a layer of some lightweight insulating material to either the external or the internal face of the wall.

For external insulation it is necessary to cover the insulation material with either rendering, tile, slate or some sheet metal covering as protection against weather. Internal insulation has to be protected with plasterboard or some other solid material to provide an acceptable finish. The cost of the additional materials and the very considerable labour involved is so great that it is an unacceptable alternative to the more straightforward, less expensive and more satisfactory use of cavity wall insulation for new buildings.

Internal insulation

Internal insulation may be fixed to the solid brick walls of existing buildings where, for example, there is to be a change of use from warehouse to dwelling to enhance the thermal insulation of the external walls.

Insulating materials are lightweight and do not generally have a smooth hard finish and are not, therefore, suitable as the inside face of the walls of most buildings. It is usual to cover the insulating layer with a lining of plasterboard or plaster so that the combined thickness of the inner lining and the wall have a U value of 0.45 W/m^2K, or less.

Internal linings for thermal insulation are either of preformed, laminated panels that combine a wall lining of plasterboard glued to an insulation board or of separate insulation material that is fixed to the wall and then covered with plasterboard or wet plaster. The method of fixing the lining to the inside wall surface depends on the surface to which it is applied.

Adhesive fixing

Adhesive fixing directly to the inside wall face is used for preformed, laminate panels and for rigid insulation boards. Where the inside face of the wall is clean, dry, level and reasonably smooth, as, for example, a sound plaster finish or a smooth and level concrete, brick or block face, the laminate panels or rigid insulation boards are secured with organic based, gap filling adhesive that is applied in dabs and strips to the back of the boards or panels or to both the boards and wall. The panels or boards are then applied and pressed into position against the wall face and their position adjusted with a foot lifter.

Where the surface of the wall to be lined is uneven or rough the laminated panels or insulation boards are fixed with dabs of plaster bonding, applied to both the wall surface and the back of the lining. Dabs are small areas of wet plaster bonding applied at intervals on the surface with a trowel, as a bedding and adhesive. The lining is applied and pressed into position against the wall. The wet dabs of bonding allow for irregularities in the wall surface and also serve as an adhesive. Some of the lining systems use secondary fixing in addition to adhesive. These secondary fixings are non-ferrous or plastic nails or screws driven or screwed through the insulation boards into the wall.

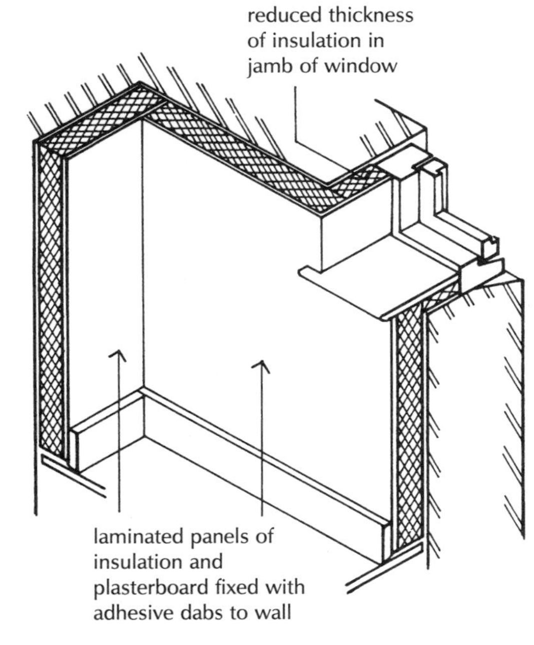

Fig. 105 Internal insulation.

Figure 105 is an illustration of laminated insulation panels fixed to the inside face of a solid wall.

Internal insulation is used where solid walls have sufficient resistance to the penetration of rain, an alteration to the external appearance is not permitted or is unacceptable and the building is not occupied. A disadvantage of internal insulation is that as the insulation is at, or close to, the internal surface, it will prevent the wall behind from acting as a heat store where constant, low temperature heating is used.

The principal difficulty with both external and internal insulation to existing buildings is that it is not usually practical to continue the insulation into the reveals of openings to avoid thermal bridges, because the exposed faces of most window and door frames are not wide enough to take the combined thickness of the insulation and rendering or plaster finish.

Mechanical fixing

As alternative to adhesive fixing, the insulating lining and the wall finish can be fixed to wood battens that are nailed to the wall with packing pieces as necessary, to form a level surface. The battens should be impregnated against rot and fixed with non-ferrous fixings. The insulating lining is fixed either between the battens or across the battens and an internal lining of plasterboard is then nailed to the battens, through the insulation.

The thermal resistance of wood is less than that of most insulating materials. When the insulating material is fixed between the battens there will be cold bridges through the battens that may cause staining on wall faces.

Details of some insulating materials used for internal lining are given in Table 6.

Table 6 Internal insulating materials.

Solid wall internal insulation	Thickness	U value W/m^2K
Glass fibre		
laminated panel glass fibre slab and plasterboard	30, 40, 60	0.031
Rockwool		
laminated panel rockwool slab and plasterboard	25, 32, 40, 50	0.033
EPS		
laminated panel EPS board and plasterboard	25, 32, 40, 50	0.037
XPS		
boards keyed for plaster	25, 50, 75, 110	0.033
PIR		
boards reinforced with glass fibre tissue both sides	25, 30, 35, 40	0.022
PUR		
laminated panel PUR board and plasterboard	12.5, 15, 20, 25, 30	0.022

EPS expanded polystyrene
XPS extruded polystyrene
PIR rigid polyisocyanurate
PUR rigid polyurethane

Internal finish

An inner lining of plasterboard can be finished by taping and filling the joints or with a thin skim coat of neat plaster. A plaster finish of lightweight plaster and finishing coat is applied to the ready keyed surface of some insulating boards or to expanded metal lathing fixed to battens. (For details of plaster, see Volume 2.)

Laminated panels of insulation, lined on one side with a plasterboard finish are made specifically for the insulation of internal walls. The panels are fixed with adhesive or mechanical fixings to the inside face of the wall. For internal lining the organic insulants such as XPS, PIR and PUR have the advantage of least thickness of material necessary due to their low U value.

Vapour barrier

Vapour check

The moisture vapour pressure from warm moist air inside insulated buildings may find its way through internal linings and condense to water on cold outer faces. Where the condensation moisture is absorbed by the insulation it will reduce the efficiency of the insulation and where condensation saturates battens, they may rot. With insulation that is permeable to moisture vapour, a vapour check should be fixed on the room side of insulation. A vapour barrier is one that completely stops the movement of vapour through it and a vapour check is one that substantially stops vapour. As it is difficult to make a complete seal across the whole surface of a wall including all overlaps of the barrier and at angles, it is in effect impossible to form a barrier and the term vapour check should more properly be used. Sheets of polythene with edges overlapped are commonly used as a vapour check, providing the edges of panels or boards of these materials can be tightly butted together.

External insulation

Insulating materials by themselves do not provide a satisfactory external finish to walls against rain penetration or for appearance sake and have to be covered with a finish of cement rendering, paint or a cladding material such as tile, slate or weatherboarding. For rendered finishes, one of the inorganic insulants, rockwool or cellular glass in the form of rigid boards, is most suited. For cladding, one of the organic insulants such as XPS, PIR or PUR is used because their low U values necessitate least thickness of board.

As a base for applied rendering the insulation boards or slabs are first bedded and fixed in line on dabs of either gap filling organic adhesive or dabs of polymer emulsion mortar and secured with corrosion resistant fixings to the wall. As a key for the render coats, either the insulation boards have a keyed surface or expanded metal lath or glass fibre mesh is applied to the face of the insulation. The weather protective render is applied in two coats by traditional wet

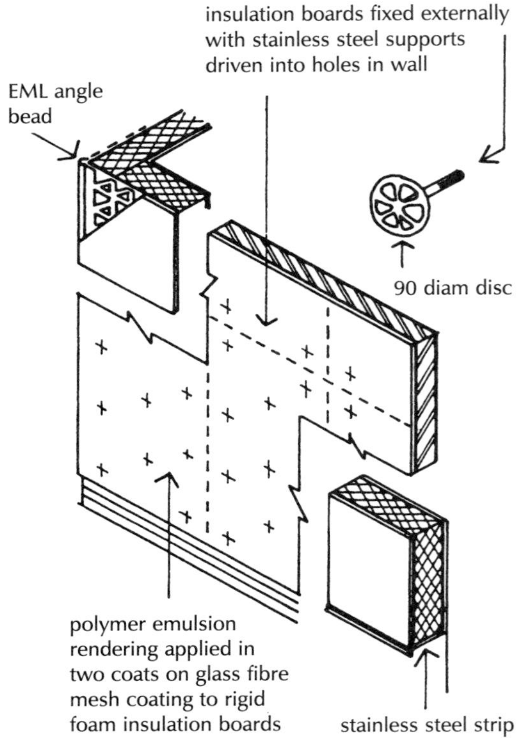

Fig. 106 External insulation.

render application, by rough casting or by spray application and finished smooth, coarse or textured. Coarse, spatter dash or textured finishes are preferred as they disguise hair cracks that are due to drying shrinkage of the rendering.

Because the rendering is applied over a layer of insulation it will be subject to greater temperature fluctuations than it would be if applied directly to a wall, and so is more liable to crack. To minimise cracking due to temperature change and moisture movements, the rendering should be reinforced with a mesh securely fixed to the wall, and movement joints should be formed at not more than 6 m intervals. The use of a light coloured finish and rendering incorporating a polymer emulsion will reduce cracking.

As the overall thickness of the external insulation and rendering is too great to be returned into the reveals of existing openings it is usual to return the rendering by itself, or fix some non-ferrous or plastic trim to mask the edge of the insulation and rendering. The reveals of openings will act as thermal bridges to make the inside face of the wall around openings colder than the rest of the wall. Figure 106 is an illustration of insulated rendering applied externally.

Tile and slate hanging, timber weatherboarding and profiled sheets can be fixed over a layer of insulating material behind the battens or sheeting rails to which these cladding materials are fixed.

Slabs of compressed rockwool are cut and shaped with bevel edges to simulate the appearance of masonry blocks. The blocks are secured to the external face of the wall with stainless steel brackets, fixed to the wall to support and restrain the blocks that are arranged with either horizontal, bonded joints or vertical and horizontal continuous joints. An exterior quality paint is then applied to the impregnated surface of the blocks. At openings, non-ferrous or plastic trim is fixed around outer reveals.

Details of insulating materials are given in Table 7.

Resistance to the passage of sound

The requirement of Part E of Schedule 1 to the Building Regulations is that walls which separate a dwelling from another building or from another dwelling shall have reasonable resistance to airborne sound.

Where solid walls of brick or block are used to separate dwellings the reduction of airborne sound between dwellings depends mainly on the weight of the wall and its thickness. A cavity wall with two leaves of brick or block does not afford the same sound reduction as a solid wall of the same equivalent thickness because the stiffness of the two separate leaves is less than that of the solid wall and in consequence is more readily set into vibration.

The joints between bricks or blocks should be solidly filled with mortar and joints between the top of a wall and ceilings should be filled against airborne sound transmission.

Table 7 External insulating materials.

Solid wall external insulation	Thickness mm	U value W/m²K
Rockwool		
rigid slab with polymer cement finish	30, 40, 50, 60, 75, 100	0.033
Cellular glass		
boards for rendering on EML	30, 35, 40, 50, 60, 70	0.042
XPS		
boards T & G on long edges for cladding	25, 50	0.025
PIR		
boards behind cladding	25, 30, 35, 40	0.022
PUR		
boards behind cladding	20, 25, 30, 35, 40, 50	0.022

XPS extruded polystyrene
PIR rigid polyisocyanurate
PUR rigid polyurethane

In Approved Document E, giving practical guidance to meeting the requirements of the Building Regulations in relation to walls between dwellings, is a table giving the minimum weight of walls to provide adequate airborne sound reduction. For example, a solid brick wall 215 mm thick, plastered both sides, should weigh at least 300 kg/m^2 including plaster, and a similar cavity wall 255 mm thick, plastered both sides, should weigh at least 415 kg/m^2 including plaster, and a cavity block wall 250 mm thick, plastered both sides, should weigh at least 425 kg/m^2, including plaster.

STONE MASONRY WALLS

Before the Industrial Revolution, many permanent buildings in hill and mountain districts and many large buildings in lowland areas in this country were built of natural stone. At that time the supply of stone from local quarries was adequate for the buildings of the small population of this country. The increase in population that followed the Industrial Revolution was so great that the supply of sound stone was quite inadequate for the new buildings being put up. Coal was cheap, the railway spread throughout the country and cheap mass produced bricks largely replaced stone as the principal material for the walls of all but larger buildings.

Because natural stone is expensive it is principally used today as a facing material bonded or fixed to a backing of brickwork or concrete (see Volume 4). Many of the larger civic and commercial buildings are faced with natural stone because of its durability, texture, colour and sense of permanence. Natural stone is also used as the outer leaf of cavity walls for houses in areas where local quarries can supply stone at reasonable cost.

In recent years much of the time consuming and therefore expensive labour of cutting, shaping and finishing building stone has been appreciably reduced by the use of power operated tools, edged or surfaced with diamonds. This facility has improved output in the continuing and extensive work of repair and maintenance to stone buildings and encourages the use of natural stone as a facing material for new buildings.

Because natural stone is an expensive material, cast stone has been used as a cheaper substitute. Cast stone is made from either crushed natural stone or natural aggregate and cement and water which is cast in moulds. The cast stone blocks are made to resemble natural stone.

Natural stone

The natural stones used in building may be classified by reference to their origin as:

(1) igneous
(2) sedimentary
(3) metamorphic.

Igneous stone

Igneous stones were formed by the cooling of molten magma as the earth's crust cooled, shrank and folded to form beds of igneous rock. Of the igneous stones that can be used for building such as granite, basalt, diorite and serpentine, granite is most used for walls of buildings.

Granite consists principally of crystals of felspar, which is made up of lime and soda with other minerals in varying proportions and small grains of quartz and mica which give a sparkle to the surface of the stone. The granite that is native to these islands that is most used for walling is sometimes loosely described as Aberdeen granite as it is mined from deep beds of igneous rock near that town in Scotland. The best known Aberdeen granites are Rubislaw which is blue grey, Kemnay which is grey and Peterhead which is pink in colouring. All of these granites are fine grained, hard and durable and can be finished to a smooth polished surface. Aberdeen granites have been much used for their strength and durability as a walling material for large buildings and are now used as a facing material.

Cornish and Devon granites are coarse grained, light grey in colour with pronounced grains of white and black crystals visible. The stone is very hard and practically indestructible. Because these granites are coarse grained and hard they are laborious to cut and shape and cannot easily be finished with a fine smooth face. These granites have been principally used in engineering works for bridges, lighthouses and docks and also as a walling material for buildings in the counties of their origin.

Sedimentary stone

Sedimentary stone was formed gradually over thousands of years from the disintegration of older rocks which were broken down by weathering and erosion or from accumulations of organic origin, the resulting fine particles being deposited in water in which they settled in layers, or being spread by wind in layers that eventually consolidated and hardened to form layers of sedimentary rocks and clays. Because sedimentary stone is formed in layers it is said to be stratified. The strata or layers make this type of stone easier to split and cut than hard, igneous stones that are not stratified. The strata also affect the way in which the stone is used, if it is to be durable, as the divisions between the layers or strata are, in effect, planes of weakness. A general subdivision of sedimentary stones is

limestone
sandstone

The limestones used for walling consist mainly of grains of shell or sand surrounded by calcium carbonate, which are cemented together with calcium carbonate. The limestones most used for walling are quarried from beds of stone in the south-west of England, those most used being Portland and Bath stone. Because limestone is a stratified rock, due to the deposit of layers, it must be laid on its natural bed in walls.

Portland stone

Portland stone is quarried in Portland Islands on the coast of Dorset. There were extensive beds of this stone which is creamy white in colour, weathers well and used to be particularly popular for walling for larger buildings in towns. Many large buildings have been built in Portland stone because an adequate supply of large stone was available, the stone is fine grained and delicate mouldings can be cut on it and it weathers well even in industrial atmospheres.

Among the buildings constructed with this stone are the great banqueting hall in Whitehall (1639), St Paul's Cathedral (1676), the British Museum (1753) and Somerset House (1776). More recently, many large buildings have been faced with this stone.

In the Portland stone quarries are three distinct beds of the stone, the base bed, the whit bed and the roach. The base bed is a fine, even grained stone which is used for both external and internal work to be finished with delicate mouldings and enrichment. The whit bed is a hard, fairly fine grained stone which weathers particularly well, even in towns whose atmosphere is heavily polluted with soot and it was extensively used as a facing material for large buildings.

The roach is a tough, coarse grained stone which has principally been used for marine construction such as piers and lighthouses.

The stones from the different beds of Portland limestone look alike to the layman. It is sometimes difficult for even the trained stonemason to distinguish base bed from whit bed. Roach can be distinguished by its coarse grain and by the remains of fossil shells embedded in it. When taken from the quarry the stone is moist and comparatively soft, but gradually hardens as moisture (quarry sap) dries out.

Bath stone

Many of the buildings in the town of Bath were built with a limestone quarried around the town. This limestone is one of the great oolites and a similar stone was also quarried in Oxfordshire. Bath stone from the Tayton (Oxfordshire) quarry was extensively used in the construction of the early colleges in Oxford (St Johns, for example) during the twelfth, thirteenth and fourteenth centuries. Many of the permanent buildings in Wiltshire and Oxfordshire were built of this stone, which varies from fine grained to coarse grained in texture and light cream to buff in colour. Most of the original quarries are no longer being worked.

The durability of Bath stone varies considerably. Some early buildings constructed with this stone are well preserved to this day, but others have so decayed over the years and been so extensively repaired that little of the original stone remains. Extensive repair of the Bath stone fabric of several of the colleges in Oxford has been carried out and continuing repair is necessary.

Sandstone

Sandstone was formed from particles of rock broken down over thousands of years by the action of wind and rain. The particles were washed into and settled to the beds of lakes and seas in combination with clay, lime and magnesia and gradually compressed into strata of sandstone rock. The particles of sandstone are practically indestructible and the hardness and resistance to the weather of this stone depends on the composition of the minerals binding the particles of sand. If the sand particles are bound with lime the stone often does not weather well as the soluble lime dissolves and the stone disintegrates. The material binding the sand particles should be insoluble and crystalline. Sandstones are generally coarse grained and cannot be worked to fine mouldings.

The stratification of most sandstones is visible as fairly close spaced divisions in the sandy mass of the stone. It is essential that this type of stone be laid on its natural bed in walls.

Most sandstones have been quarried in the northern counties of England where for centuries this stone has been the material commonly used for the walls of buildings. Some of the sandstones that have been used are:

Crosland Hill (Yorkshire). A light brown sandstone of great strength which weathers well and is used for masonry walls as a facing material and for engineering works. It is one of the stones known as hard York stone, a general term used to embrace any hard sandstone not necessarily quarried in Yorkshire.

Blaxter stone (Northumberland). A hard, creamy coloured stone used for wall and as a facing.

Doddington (Northumberland). A hard, pink stone used for walling.

Darley Dale (Derbyshire). A hard, durable stone of great strength much used for engineering works and as walling. It is hard to work and generally used in plain, unornamented wall. Buff and white varieties of this stone were quarried.

Forest of Dean (Gloucestershire). A hard, durable, grey or blue grey stone which is hard to work but weathers well as masonry walling.

Metamorphic stone

Metamorphic stones were formed from older stones that were changed by pressure or heat or both. The metamorphic stones used in building are slate and marble.

Slate

Slate was formed by immense pressure on beds of clay that were compressed to hard, stratified slate which is used for roofing and as cills and copings in building. Riven, split, Welsh slate has for centuries been one of the traditional roofing materials used in this country. The stone can be split to comparatively thin slates that are hard and very durable.

Marble

The description marble is used to include many stones that are not true metamorphic rocks, such as limestones, that can take a fine polish. In the British Isles true marble is only found in Ireland and Scotland. Marble is principally used as an internal facing material in this country.

Durability of natural stone

Natural stone has been used in the construction of buildings because it was thought that any hard, natural stone would resist the action of wind and rain for centuries. Many natural stones have been used in walling and have been durable for a hundred or more years and are likely to have a comparable life if reasonably maintained.

There have been some notable failures of natural stone in walling, due in the main to a poor selection of the material and poor workmanship. The best known example of decay in stonework occurred in the fabric of the Houses of Parliament, the walls of which were built with a magnesian limestone from Ancaster in Yorkshire. A Royal Commission reported in 1839 that the magnesian limestone quarried at Bolsover Moor in Yorkshire was considered the most durable stone for the Houses of Parliament. After building work had begun it was discovered that the quarry was unable to supply sufficient large stones for the building and a similar stone from the neighbouring quarry at Anston was chosen as a substitute. The quarrying, cutting and use of the stone was not supervised closely and in consequence many inferior stones found their way into the building and many otherwise sound stones were incorrectly laid.

Decay of the fabric has been continuous since the Houses of Parliament were first completed and extensive, costly renewal of stone has been going on for many years. At about the same time that the Houses of Parliament were being built, the Museum of Practical Geology was built in London of Anston stone from the same quarry that supplied the stone for the Houses of Parliament, but the quarrying, cutting and use of the stones was closely supervised for the museum, whose fabric remained sound.

The variability of natural stone that may appear sound and durable, but some of which may not weather well, is one of the disadvantages of this material which can, when carefully selected and used, be immensely durable and attractive as a walling material.

Seasoning natural stone

Some natural stones are comparatively soft and moist when first quarried but gradually harden. Building stones should be seasoned (allowed to harden) for periods of up to a few years, depending on the size of the stones. Once stone has been seasoned it does not revert to its original soft moist state on exposure to rain, but on the contrary hardens with age.

Bedding stones

Natural stones that are stratified, limestone and sandstone, must be used in walling so that they lie on their natural bed to support compressive stress. The bed of a stone is its face parallel to the strata (layer) of the stones in the quarry and the stress that the stone suffers in use should be at right angles to the strata or bed which otherwise might act as a plane of weakness and give way under compressive stress. The stones in an arch are laid with the bed or strata radiating roughly from the centre of the arch so that the bed is at right angles to the compressive stress acting around the curve of the arch.

Cast stone

Cast stone is one of the terms used to describe concrete cast in moulds to resemble blocks of natural stone. When the material first came into use some 50 years ago it was called artificial stone. To avoid the use of the pejorative term artificial, the manufacturers now prefer the description reconstructed stone.

Reconstructed stone

Reconstructed stone is made from an aggregate of crushed stone, cement and water. The stone is crushed so that the maximum size of the particles is 6 mm and it is mixed with cement in the proportions of 1 part cement to 3 or 4 parts of stone. Either portland cement, white cement or coloured cement may be used to simulate the colour of a natural stone as closely as possible. A comparatively dry mix of cement, crushed stone and water is prepared and cast in moulds. The mix is thoroughly consolidated inside the moulds by vibrating and left to harden in the moulds for at least 24 hours. The stones are then taken out of the moulds and allowed to harden gradually for 28 days.

Well made reconstructed stone has much the same texture and colour as the natural stone from which it is made and can be cut, carved and dressed just like natural stone. It is not stratified, is free from flaws and is sometimes a better material than the natural stone from which it is made. The cost of a plain stone, cast with an aggregate of crushed natural stone, is about the same as that of a similar natural stone. Moulded cast stones can often be produced more cheaply by repetitive casting than similar natural stones that have to be cut and shaped.

A cheaper, inferior, form of cast stone is made with a core of ordinary concrete, faced with an aggregate of crushed natural stone and cement. This material should more properly be called cast concrete.

The core is made from clean gravel, sand and Portland cement and the facing from crushed stone and cement to resemble the texture and colour of a natural stone. The crushed stone, cement and water is first spread in the base of the mould to a thickness of about 25 mm, the core concrete is added and the mix consolidated. If the stone is to be exposed on two or more faces the natural stone mix is spread up the sides and the bottom of the mould. This type of cast stone obviously cannot be carved as it has only a thin surface of natural looking stone.

As an alternative to a facing of reconstructed stone, the facing or facings can be made of cement and sand pigmented to look somewhat like the colour of a natural stone.

Cast stones made with a surface skin of material to resemble stone do not usually weather in the same way that natural stone does, by a gradual change of colour. The material tends to have a lifeless, mechanical appearance and may in time tend to show irregular, unsightly dirt stains at joints, cracks and around projections.

Reconstructed stone is used as an ashlar facing to brick or block backgrounds for both solid and the outer leaf of cavity walls and as facings (see Volume 4).

Functional requirements

Strength and stability

The strength of sound building stone lies in its very considerable compressive strength. The ultimate or failing stress of stone used for walling is about 300 to 100 N/mm^3 for granite, 195 to 27 N/mm^3 for sandstone and 42 to 16 N/mm^3 for limestone. The considerable compressive strength of building stone was employed in the past in the construction of massive stone walls for fortifications and in other large structures. The current use of stone as a facing material makes little use of the inherent compressive strength of the material.

The stability of a stone wall is affected by the same limitations that apply to walls of brick or block. The construction of foundations and the limits of slenderness ratio, the need for buttressing walls, piers and chimneys along the length of walls and the requirements for lateral support from floors and roofs up the height of walls apply to stone walls as they do for brick and block walls.

Resistance to weather and ground moisture

To prevent moisture rising from the ground through foundation walls it is necessary to form a continuous horizontal dpc some 150 mm above ground level. One way of achieving this is to construct foundation walls of dense stone, such as granite, that does not readily absorb moisture. More usually one of the damp-proof materials described for use with brickwork is used. A sheet lead dpc is commonly used as it is less likely to be squeezed out and forms a comparatively thin and therefore less unsightly joint than a bitumen felt dpc.

The resistance to the penetration of wind driven rain was not generally a consideration in the construction of solid masonry walls. The very considerable thickness of masonry walls of traditional large buildings was such that little, if any, rain penetrated to the inside face.

With the use of stone, largely as a facing material for appearance sake, it is necessary to construct walls faced with stone as cavity walling with a brick or block inner leaf separated by a cavity from the stone faced outer leaf, as illustrated in Fig. 107.

The outer leaf illustrated in Fig. 107 is built with natural stone blocks bonded to a brick backing, with full width stones in every other course and the stones finished on face in ashlar masonry. This is an expensive form of construction because of the considerable labour costs in preparing the ashlared stones. As alternatives the outer leaf of small buildings may be constructed with stone blocks by themselves for the full thickness of the outer leaf, or with larger buildings the outer leaf may be constructed of brick to which a facing of stone slabs is fixed (see Volume 4).

The leaves of the cavity are tied with galvanised steel or stainless steel wall ties in the same way that brick and block walls are constructed and the cavity is continued around openings, or dpcs are formed to resist rain penetration at head, jambs and cills of openings.

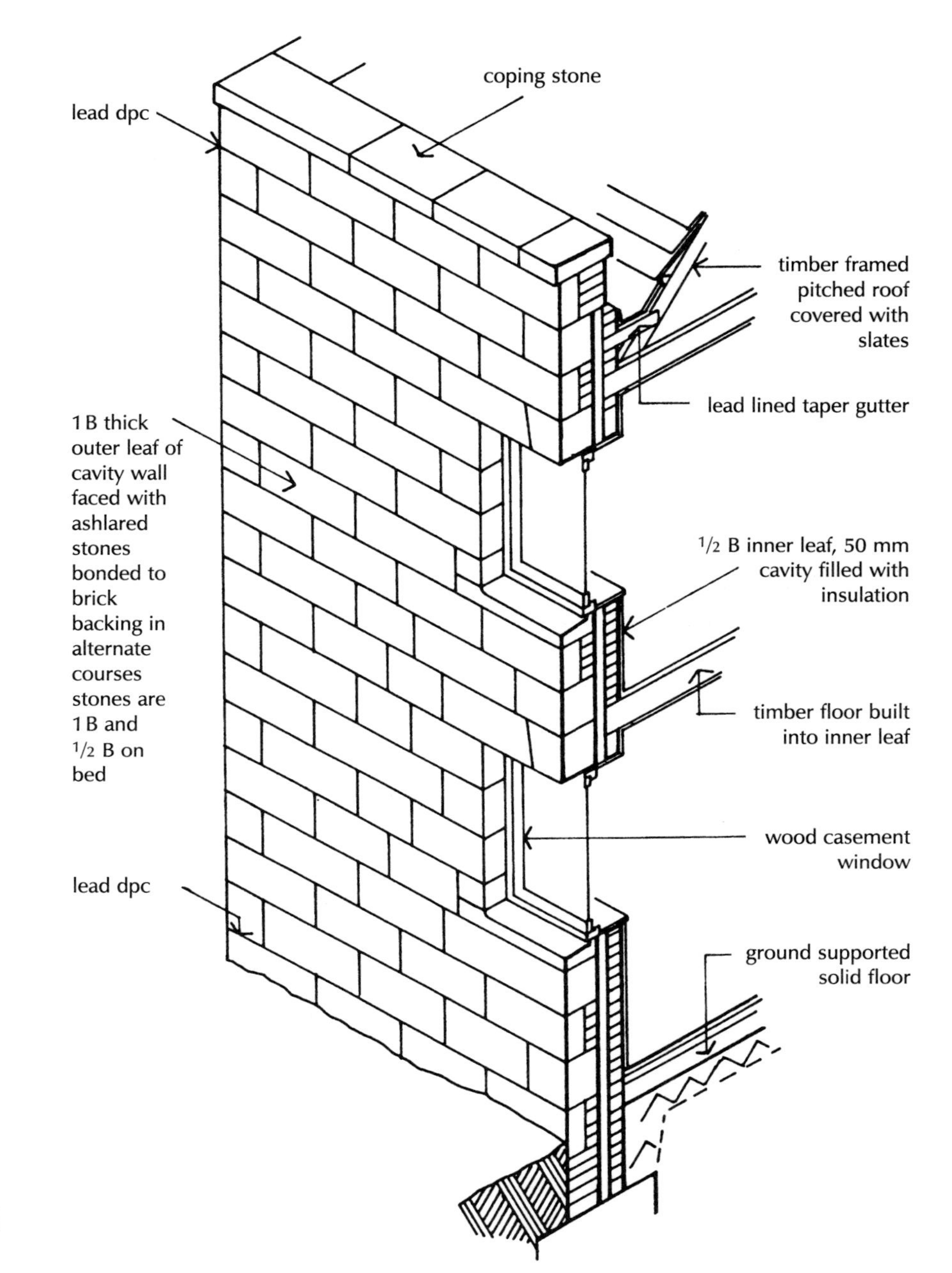

Fig. 107 Cavity wall faced with ashlared stone and brick backing.

Durability and freedom from maintenance

Sound natural stone is highly durable as a walling material and will have a useful life of very many years in buildings which are adequately maintained.

Granite is resistant to all usual weathering agents, including highly polluted atmospheres, and will maintain a high natural polished surface for a hundred years or more. The lustrous polish will be enhanced by periodic washing.

Hard sandstones are very durable and inert to weathering agents but tend to dirt staining in time, due to the coarse grained texture of the material which retains dirt particles. The surface of sandstone may be cleaned from time to time to remove dirt stains by abrasive blasting with grit or chemical processes and thorough washing.

Sound limestone, sensibly selected and carefully laid, is durable for the anticipated life of the majority of buildings. In time the surface weathers by a gradual change of colour over many years, which is commonly held to be an advantage from the point of view of appearance. Limestones are soluble in rainwater that contains carbon dioxide so that the surface of a limestone wall is to an extent self-cleansing when freely washed by rain, while protected parts of the wall will collect and retain dirt. This effect gives the familiar black and white appearance of limestone masonry. The surface of limestone walls may be cleaned by washing with a water spray or by steam and brushing to remove dirt encrustations and the surface brought back to something near its original appearance.

In common with the other natural walling material, brick, a natural stone wall of sound stone sensibly laid will have a useful life of very many years and should require little maintenance other than occasional cleaning.

Fire resistance

Natural stone is incombustible and will not support or encourage the spread of flame. The requirements of Part B of Schedule 1 to the Building Regulations for structural stability and integrity and for concealed spaces apply to walls of stone as they do for walls of block or brick masonry.

Resistance to the passage of heat

The natural stones used for walling are poor insulators against the transfer of heat and will contribute little to thermal resistance in a wall. It is necessary to use some material with a low U value as cavity insulation in walls faced with stone in the same way that insulation is used in cavity walls of brick or blockwork.

Resistance to the passage of sound

Because natural building stone is dense it has good resistance to the transmission of airborne sound and will provide a ready path for impact sound.

Ashlar walling

Ashlar walling is constructed of blocks of stone that have been very accurately cut and finished true square to specified dimensions so that the blocks can be laid, bedded and bonded with comparatively thin mortar joints, as illustrated in Fig. 107. The very considerable labour involved in cutting and finishing individual stones is such that this type of walling is very expensive. Ashlar walling has been used for the larger, more permanent buildings in towns, and on estates where the formal character of the building is pronounced by the finish to the walling. Ashlar walling is now used principally as a facing material (see Volume 4).

Openings to stone walls

Stone walls over door and window openings are supported by flat stone lintels or by segmental or semi-circular arches.

Lintels

A stone lintel for small openings of up to about a metre wide can be formed of one whole stone with its ends built into jambs and its depth corresponding to one or more stone courses. The poor tensile strength of stone limits the span of single stone lintels unless they are to be disproportionately deep.

Over openings wider than about a metre it is usual to form lintels with three or five stones cut in the form of a flat arch. The stones are cut so that the joints between the ends of stones radiate from a common centre so that the centre, or key stone, is wedge-shaped, as illustrated in Fig. 108. The stones are cut so that the lower face of each stone occupies a third or a fifth of the width of the opening.

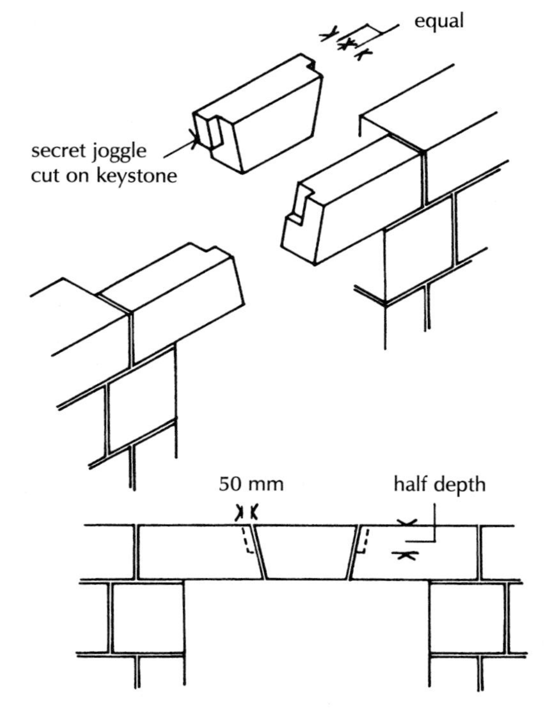

Fig. 108 Stone lintel with secret joggle joints.

To prevent the key stone sinking due to settlement and so breaking the line of the soffit, it is usual to cut half depth joggles in the ends of the key stone to fit to rebates cut in the other stones. The joggles and rebates may be cut the full thickness of each stone and show on the face of the lintel or more usually the joggles and rebates are cut on the inner half of the thickness of stones as secret joggles, which do not show on the face, as illustrated in Fig. 108. The depth of the lintel corresponds to a course height, with the ends of the lintel built in at jambs as end bearing. Stone lintels are used over both ashlar and rubble walling.

The use of lintels is limited to comparatively small openings due to the tendency of the stones to sink out of horizontal alignment. For wider openings some form of arch is used.

Arches

A stone arch consists of stones specially cut to a wedge shape so that the joints between stones radiate from a common centre, the soffit is arched and the stones bond in with the surrounding walling. The individual stones of the arch are termed 'voussoirs', the arched soffit the 'intrados' and the upper profile of the arch stones the 'extrados'.

Figure 109 is an illustration of a stone arch whose soffit is a segment of a circle. The choice of the segment of a circle that is selected is to an extent a matter of taste, which is influenced by the appearance of strength. A shallow rise is often acceptable for small openings and a greater rise for larger, as the structural efficiency of the arch increases the more nearly the segment approaches a full half circle. The voussoirs of the segmental arch illustrated in Fig. 109 are cut with steps that correspond in height with stone courses, to which the stepped extrados is bonded.

The stones of an arch are cut so that there is an uneven number of voussoirs with a centre or key stone. The key stone is the last stone to be put in place as a key to the completion and the stability of the arch, hence the term key stone.

The majority of semi-circular arches are formed with stones cut to bond in with the surrounding stonework in the form of a stepped extrados similar to that shown for a segmental arch in Fig. 109.

Crossetted arch

The semi-circular arch, illustrated in Fig. 110, is formed with stones that are cut to bond into the surrounding walling to form a stepped extrados and also to bond horizontally into the surrounding stones. The stones, voussoirs, are said to be crossetted, or crossed. This extravagant cutting of stone is carried out purely for appearance sake. This is not a structurally sound idea as a very slight settlement might cause the crossetted end of a stone to crack away from the main body of the stone, whereas with plain voussoirs the slight settlement would be taken up by the joints.

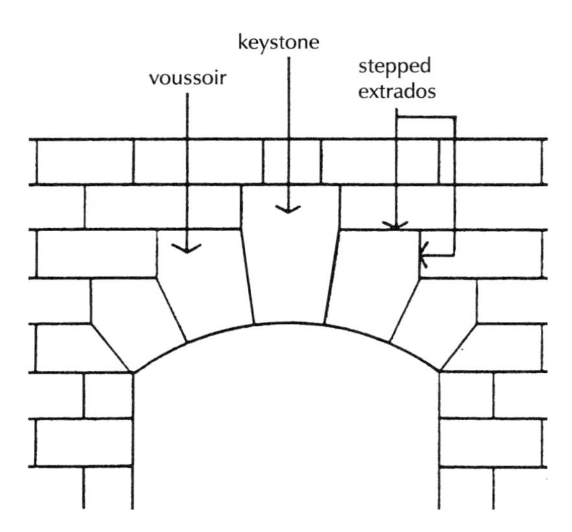

Fig. 109 Segmental stone arch.

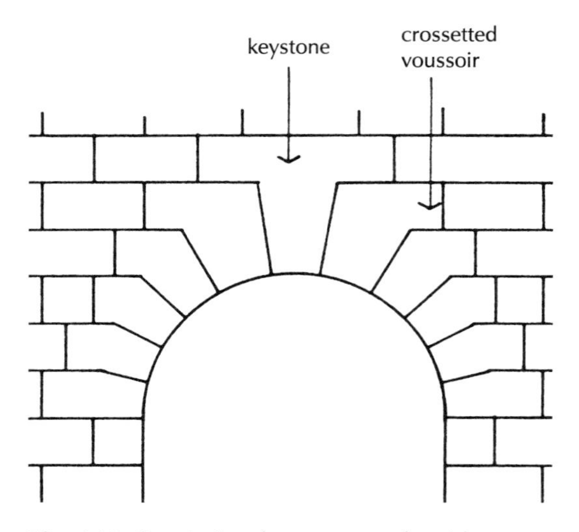

Fig. 110 Semi-circular stone arch with crossetted voussoirs.

The stones radiate from a common centre with an odd number of stones arranged around the half circle so that there is a central key stone. The extent that the crossetted top of each stone extends into the surrounding masonry is not necessarily dictated by the stretcher bond of the main walling. Any degree of bond into the main wall may be chosen and the bond of masonry adjusted accordingly. No matter which dimension of crossetted end is chosen the key stone has to be wastefully cut from one large stone.

The effect of the crossetted voussoir is to emphasise the arch as structurally separate from the coursed masonry and is chosen for that effect.

Ashlar masonry joints

Ashlar stones may be finished with smooth faces and bedded with thin joints, or the stones may have their exposed edges cut to form a channelled or 'V' joint to emphasise the shape of each stone and give the wall a heavier, more permanent appearance. The ashlar stones of the lower floor of large buildings are often finished with channelled or V joints and the wall above with plain ashlar masonry to give the base of the wall an appearance of strength. Ashlar masonry finished with channelled or V joints is said to be rusticated. A channelled joint (rebated joint) is formed by cutting a rebate on the top and one side edge of each stone, so that when the stones are laid, a channel rebate appears around each stone, as illustrated in Fig. 111A. The rebate is cut on the top edge of each stone so that when the stones are laid, rainwater which may run into the horizontal joint will not penetrate the mortar joint.

A V joint (chamfered joint) is formed by cutting all four edges of stones with a chamfer so that when they are laid a V groove appears on face, as illustrated in Fig. 111B. Often the edges of stones are cut with both V and channelled joints to give greater emphasis to each stone.

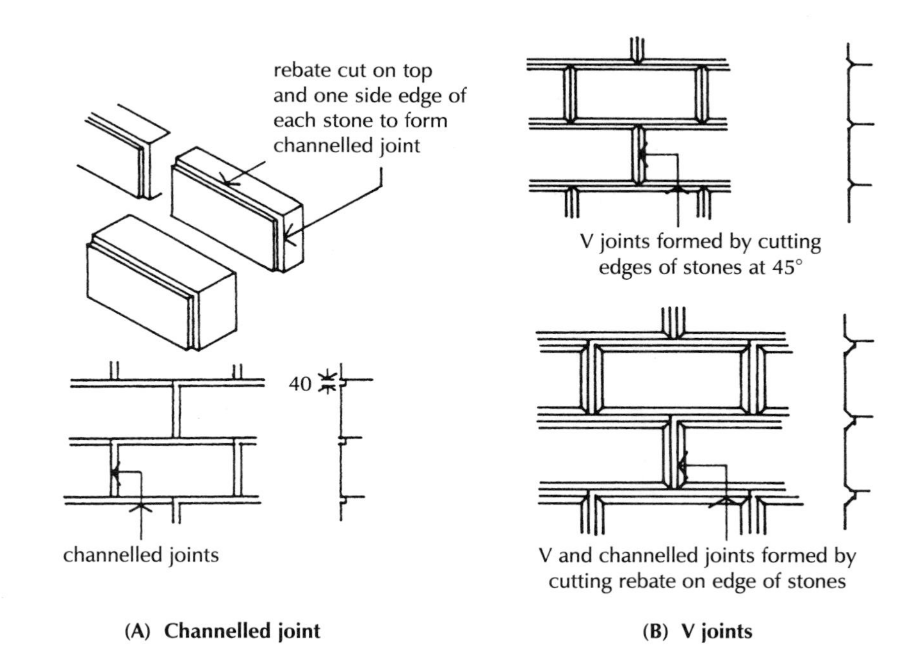

Fig. 111 (A) Channelled joint. (B) V joint.

Tooled finish

Plain ashlar stones are usually finished with flat faces to form plain ashlar facing. The stones may also be finished with their exposed faces tooled to show the texture of the stone. Some of the tooled finishes used with masonry are illustrated in Fig. 112. It is the harder stones such as granite and hard sandstone that are more commonly finished

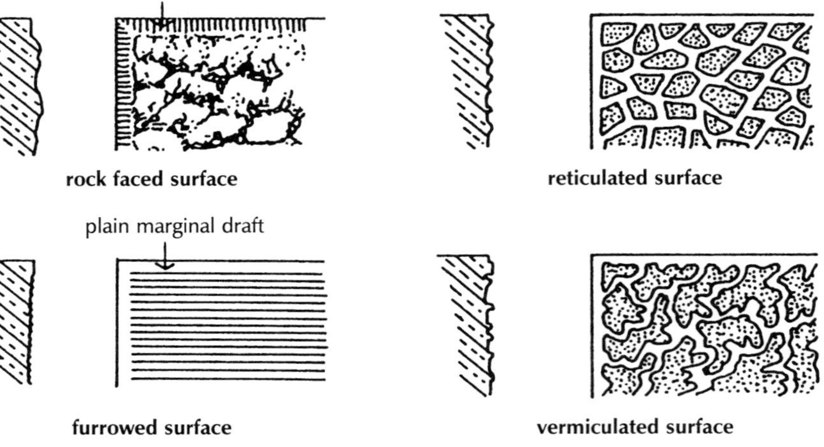

Fig. 112 Tooled finishes.

with rock face, pitched face, reticulated or vermiculated faces. The softer, fine grained stones are usually finished as plain ashlar.

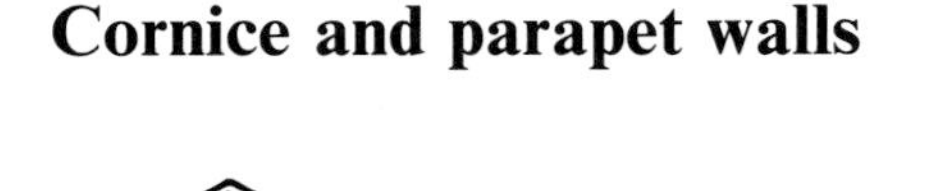

Cornice and parapet walls

It is common practice to raise masonry walls above the levels of the eaves of a roof, as a parapet. The purpose of the parapet is partly to obscure the roof and also to provide a depth of wall over the top of the upper windows for the sake of appearance in the proportion of the building as a whole.

In order to provide a decorative termination to the wall, a course of projecting moulded stones is formed. This projecting stone course is termed a cornice and it is generally formed some one or more courses of stone below the top of the parapet. Figure 113 is an illustration of a cornice and a parapet wall to an ashlar faced building. An advantage of the projecting cornice is that it affords some protection against rain to the wall below.

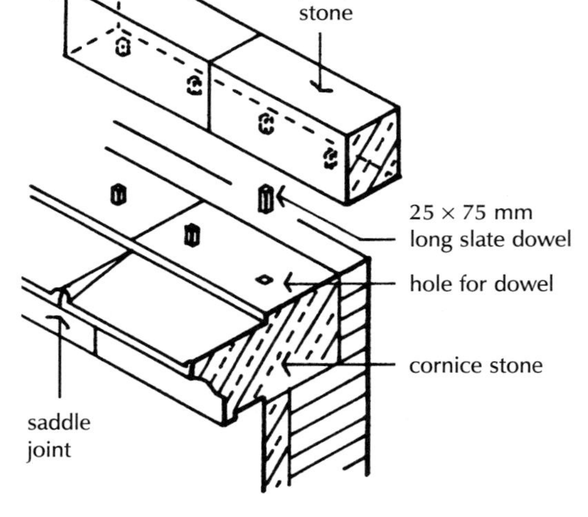

Fig. 113 Cornice and parapet.

The parapet wall usually consists of two or three courses of stones capped with coping stones bedded on a dpc of sheet metal. The parapet is usually at least 1 B thick or of such thickness that its height above roof is limited by the requirements of the Building Regulations as described in Chapter 4 for parapet walls. The parapet may be built of solid stone or stones bonded to a brick backing.

The cornice is constructed of stones of about the same depth as the stones in the wall below, cut so that they project and are moulded for appearance sake. Because the stones project, their top surface is weathered (slopes out) to throw water off.

Saddle joint

The projecting, weathered top surface of coping stones is exposed and rain running off it will in time saturate the mortar in the vertical joints between the stones. To prevent rain soaking into these joints it is usual to cut the stones to form a saddle joint as illustrated in Fig. 113.

The exposed top surface of the stones has to be cut to slope out (weathering) and when this cutting is executed a projecting quarter circle of stone is left on the ends of each stone. When the stones are laid, the projections on the ends of adjacent stones form a protruding semi-circular saddle joint which causes rain to run off away from the joints.

Weathering to cornices

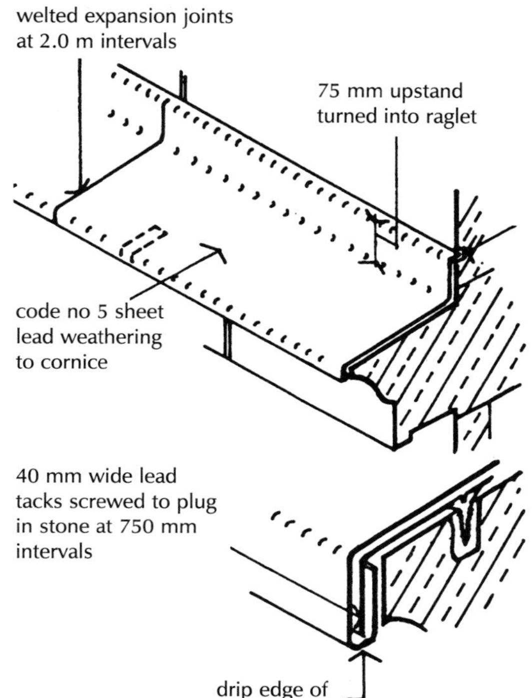

Fig. 114 Lead weathering to cornice.

Because cornices are exposed and liable to saturation by rain and possible damage by frost, it is good practice to cover the exposed top surface of cornice stones cut from limestone or sandstone with sheet metal. The sheet metal covering is particularly useful in urban areas where airborne pollutants may gradually erode stone.

Sheet lead is preferred as a non-ferrous covering because of its ductility, that facilitates shaping, and its impermeability.

Sheets of lead, code No 5, are cut and shaped for the profile of the top of the cornice, and laid with welted (folded) joints at 2 m intervals along the length of the cornice. The purpose of these comparatively closely spaced joints is to accommodate the inevitable thermal expansion and contraction of the lead sheet. The top edge of the lead is dressed up some 75 mm against the parapet as an upstand, and turned into a raglet (groove) cut in the parapet stones and wedged in place with lead wedges. The joint is then pointed with mortar.

The bottom edge of the lead sheets is dressed (shaped) around the outer face of the stones and welted (folded). To prevent the lower edge of the lead sheet weathering being blow up in high winds, 40 mm wide strips of lead are screwed to lead plugs set in holes in the stone at 750 mm intervals, and folded into the welted edge of the lead, as illustrated in Fig. 114.

Where cornice stones are to be protected with sheet lead weathering there is no purpose in cutting saddle joints.

Cement joggle

Cornice stones project and one or more stones might in time settle slightly so that the decorative line of the mouldings cut on them would be broken and so ruin the appearance of the cornice. To prevent this possibility shallow V-shaped grooves are cut in the ends of each stone so that when the stones are put together these matching V grooves form a square hole into which cement grout is run. When the cement hardens it forms a joggle which locks the stones in their correct position.

Dowels

To maintain parapet stones in their correct position in a wall, slate dowels are used. The stones in a parapet are not kept in position by the weight of walling above and these stones are, therefore, usually fixed with slate dowels. These dowels consist of square pins of slate that are fitted to holes cut in adjacent stones, as illustrated in Fig. 113.

Cramps

Coping stones are bedded on top of a parapet wall as a protection against water soaking down into the wall below. There is a possibility that the coping (capping) stones may suffer some slight movement and cracks in the joints between them open up. Rain may then saturate the parapet wall below and frost action may contribute to some movement and eventual damage.

To keep coping stones in place a system of cramps is used. Either slate or non-ferrous metal is used to cramp the stones together.

A short length of slate, shaped with dovetail ends, is set in cement grout (cement and water) in dovetail grooves in the ends of adjacent stones, as illustrated in Fig. 115A.

As an alternative a gunmetal cramp is set in a groove and mortice in the end of each stone and bedded in cement mortar, as illustrated in Fig. 115B.

For coping stones cut from limestone or sandstone a sheet metal weathering is sometimes dressed over coping stones. The weathering of lead is welted and tacked in position over the stones.

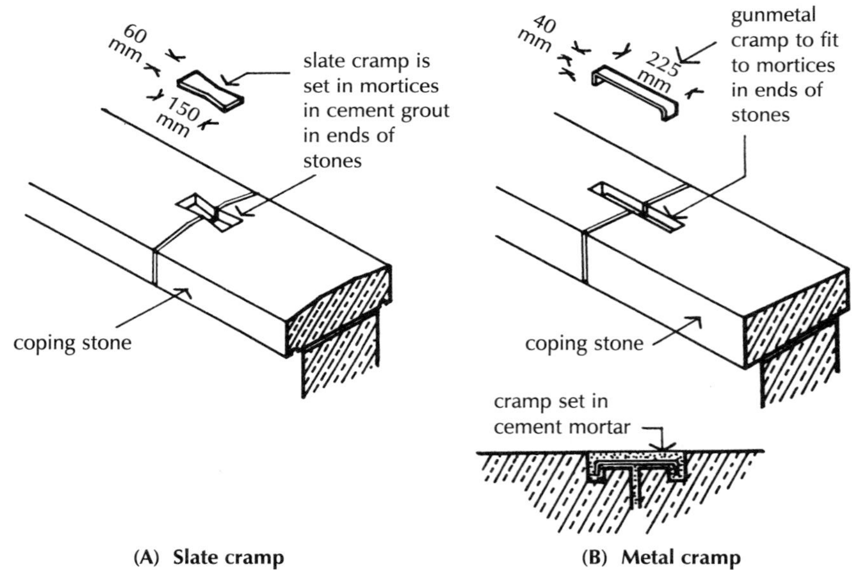

Fig. 115 (A) Slate cramp. (B) Metal cramp.

Rubble walling

Rubble walling has been extensively used for agricultural buildings in towns and villages in those parts of the country where a local source of stone was readily available. The term rubble describes blocks of stones as they come from the quarry. The rough rubble stones are used in walling with little cutting other than the removal of inconvenient corners. The various types of rubble walling depend on the nature of the stone used. Those stones that are hard and laborious to cut or shape are used as random rubble and those sedimentary stones that come from the quarry roughly square are used as squared rubble.

The various forms of rubble walling may be classified as random rubble and squared rubble.

Random rubble

Uncoursed random rubble

Uncoursed random rubble stones of all shapes and sizes are selected more or less at random and laid in mortar, as illustrated in Fig. 116A. No attempt is made to select and lay stones in horizontal courses. There is some degree of selection to avoid excessively wide mortar joints and also to bond stones by laying some longer stones both along the face and into the thickness of the wall, so that there is a bond stone in each square metre of walling. At quoins, angles and around openings selected stones or shaped stones are laid to form roughly square angles.

Random rubble brought to course

Random rubble brought to course is similar to random rubble uncoursed except that the stones are selected and laid so that the walling is roughly levelled in horizontal courses at vertical intervals of from 600 to 900 mm, as illustrated in Fig. 116B. As with uncoursed rubble, transverse and longitudinal bond stones are used.

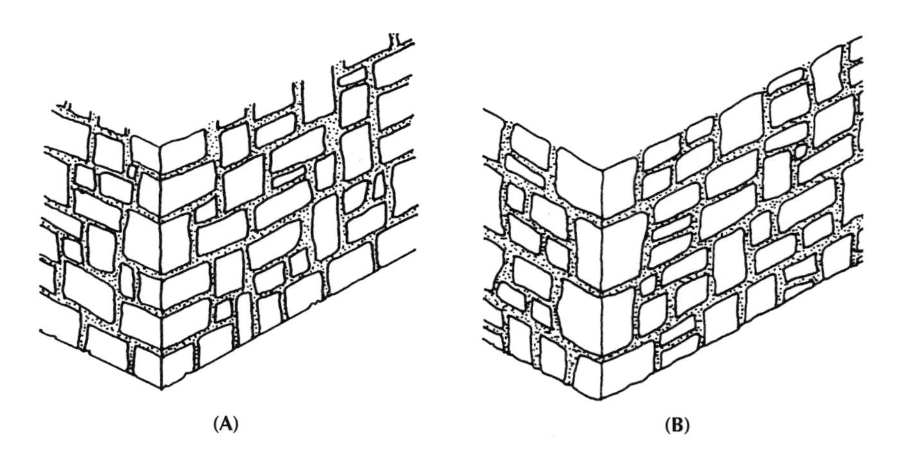

Fig. 116 (A) Random rubble uncoursed. (B) Random rubble coursed.

Squared rubble

Squared rubble uncoursed

Squared rubble uncoursed is laid with stones that come roughly square from the quarry in a variety of sizes. The stones are selected at random, are roughly squared with a walling hammer and laid without courses, as illustrated in Fig. 117A. As with random rubble, both transverse and longitudinal bond stones are laid at invervals.

Snecked rubble

Snecked rubble is a term for squared rubble in which a number of small squared stones, snecks, are laid to break up long continuous vertical joints. Snecked rubble is often difficult to distinguish from squared random rubble.

Squared rubble brought to course

Squared rubble brought to course is constructed from roughly square stone rubble, selected and squared so that the work is brought to courses every 300 to 900 mm intervals, as illustrated in Fig. 117B.

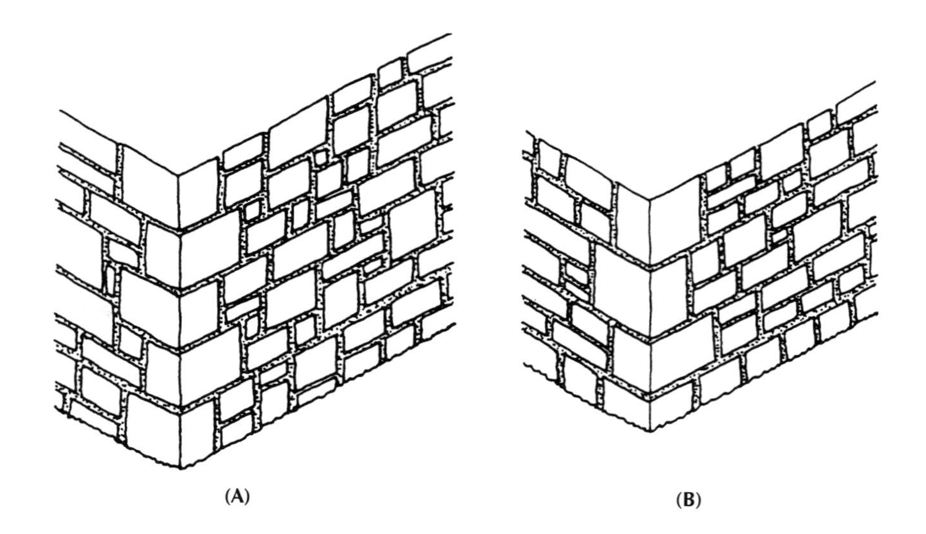

Fig. 117 (A) Squared rubble uncoursed. (B) Squared rubble coursed.

Squared rubble coursed is built with stones that are roughly squared so that the stones in each course are roughly the same height and the courses vary in height, as illustrated in Fig. 118A. The face of the stones may be roughly dressed to give a rock faced appearance or dressed smooth to give a more formal appearance.

Polygonal walling

Stones that are taken from a quarry where the stone is hard, have no pronounced laminations and come in irregular shapes can be laid as polygonal walling. The stones are selected and roughly dressed to fit when laid, to an irregular pattern, with no attempt at regular courses or vertical joints. At corners and as a base, roughly square edged stones are used, as illustrated in Fig. 118B.

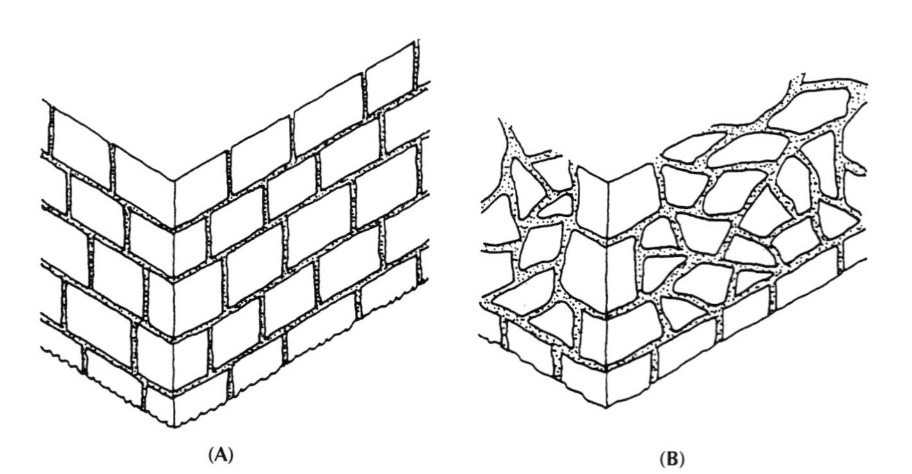

Fig. 118 (A) Squared rubble coursed. (B) Polygonal walling.

Flint walling

Flint walling is traditional to East Anglia and the south and south-east of England. Both field and shore flints are used. The flints used for walling are up to 300 mm in length and from 75 to 250 mm in width and thickness. The flints or cobbles may be used whole or split to show the heart of the flint, and also knapped or snapped so that they show a roughly square face.

Flint walling is built with a dressing of stone or brick at angles and in horizontal lacing courses that level the wall at intervals. Figure 119A is an illustration of whole flints laid without courses in brick dressing to angles and as lacing and Fig. 119B an illustration of knapped flints laid to courses in stone dressing.

Rubble walling is generally considerably thicker than a wall of square bricks or blocks because of the inherent instability of a wall built of irregular shaped blocks. Rubble walling for buildings is usually at least 400 mm thick.

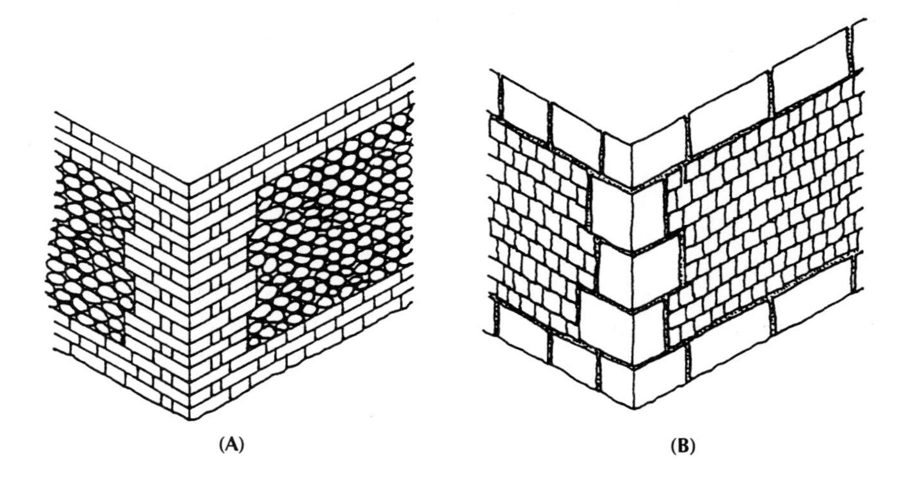

Fig. 119 (A) Whole flint wall. (B) Knapped flint wall.

TIMBER FRAMED WALLS

TIMBER

The word timber describes wood which has been cut for use in building.

Timber has many advantages as a building material. It is a light-weight material that is easy to cut, shape and join by relatively cheap and simple hand or power operated tools in the production of wall, floor and roof panels, timber joists, and for rafters, walls, floors and roofs and joinery generally.

As a structural material it has favourable weight to cost, weight to strength and weight to modulus of elasticity ratios and coefficients of thermal expansion, K values, density and specific heat. With sensible selection, fabrication and fixing and adequate protection it is a reasonably durable material in relation to the life of most buildings.

Wood burns at temperatures of about 350°C and chars, the charred outer faces of the wood protecting the unburnt inner wood for periods adequate for escape during fires, in most buildings.

In this age of what the layman calls 'plastics' and 'synthetic' materials some people remark that timber is old-fashioned and suggest that more modern materials such as reinforced concrete should be used for floors and roofs. At present the cost of a timber upper floor for a house is about half that of a similar reinforced concrete floor and as the timber floor is quite adequate for its purpose it seems senseless to double the cost of the floor just to be what is called modern.

Much of the timber used in buildings today is cut from the wood of what are called 'conifers'. A coniferous tree is one that has thin needle-like leaves that remain green all year round and whose fruit is carried in woody cones. Examples of this type of tree are fir and pine.

Softwood

Hardwood

In North America and northern Europe are very extensive forests of coniferous trees, the wood from which can be economically cut and transported. The wood of coniferous trees is generally less hard than that of other sorts of trees, for example oak, and the wood from all conifers is classified as softwood, whilst the wood from all other trees which have broad leaves is termed hardwood.

The cost of a typical softwood used today is about half that of a typical hardwood.

Softwood is made up of many very thin, long cells with their long axis along the length of the trunk or branch. These cells are known as tracheids and they give softwood its strength and texture, and serve to convey sap. The structure of hardwood is more complicated and in most hardwoods the bulk of the wood consists of long fibres. In addition to the fibres there are long cells, called vessels, which conduct water from the roots of the tree to its crown.

Trees whose wood is used for building are all exogens, meaning growing outwards, as each year most of these trees form new layers of wood under the bark. Beneath the protective bark around tree trunk and branches is a slimy light green skin. This is termed the cambium and it consists of a layer of wood cells which begin each spring to divide several times to form a layer of new wood cells. All these new wood cells are formed inside the cambium and the first wood cells formed each year are termed spring wood, and the cells formed later are termed summer wood.

The wood cells which are formed first are thinner walled than those formed in the summer and in any cross section cut of wood there are distinct circular rings of light coloured spring wood then darker coloured and thinner rings of summer wood. As a new spring and summer ring is formed each year they are called annual rings, which

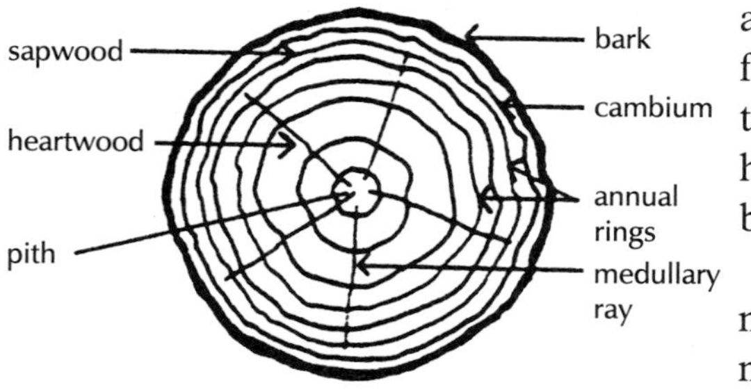

Fig. 120 Section through growing wood.

are shown in Fig. 120. The wood of the annual rings most recently formed close to the bark is less dense than that close to the centre of the tree. The outer wood is sapwood and that towards the centre is heartwood. In the conversion of wood to timber there should ideally be as little sapwood as possible.

In the diagram are shown a number of irregular radial lines marked medullary rays. These rays consist of specialised wood cells whose main purpose is to store food in readiness for it to be conveyed to any part of the tree that may require it.

Seasoning of timber

Up to two-thirds of the weight of growing wood is due to water in the cells of the wood. When the tree is felled and the wood is cut into timber this water begins to evaporate to the air around the timber, and the wood gradually shrinks as water is removed from the cell walls. As the shrinkage in timber is not uniform the timber may lose shape and it is said to warp.

It is essential that before timber is used in buildings, either it should be stacked for a sufficient time in the open air for most of the water in it to dry out, or it should be artificially dried out. If wet timber is used in building it will dry out and shrink and cause cracking of plaster and twisting of doors and windows. The process of allowing, or causing, newly cut wood to dry out is called seasoning, and timber which is ready for use in building is said to have been properly seasoned. The amount of water in wood varies, and it is not sufficient to allow all timber to dry out for some specific length of time, as one piece of timber may be well seasoned and dried out, whilst another similar piece stacked for the same length of time may still be too wet to use immediately.

Moisture content of timber

It is necessary to specify that there shall be a certain amount of water, and no more or less, in timber suitable for building. It is said that timber shall have a certain moisture content, and the moisture content is stated as a percentage of the dry weight of the timber. The dry weight of any piece of timber is its weight after it has been so dried that further drying causes it to lose no more weight. This dry weight is reasonably constant for a given cubic measure of each type of wood and is used as the constant against which the moisture content can be assessed. Table 8 sets out moisture contents for timber.

The moisture content of timber should be such that the timber will not appreciably gain or lose moisture in the position in which it is fixed in a building.

Table 8 Moisture content of timber.

Position of timber in building	1 %	2 %
External uses fully exposed	20 or more	—
Covered and generally unheated	18	24
Covered and generally heated	15	20
Internal and continuously heated building	12	20

Column 1 Average moisture content likely to be attained in service conditions
Column 2 Moisture content which should not be exceeded in individual pieces at time of erection
Taken from BS 5268 : Part 2:1996 (issue 2, May 1997)

Natural dry seasoning

When logs have been cut into timber it is stacked either in the open or in a rough open sided shed. The timbers are stacked with battens between them to allow air to circulate around them. The timbers are left stacked for a year or more, until most of the moisture in the wood has evaporated. Softwoods have to be stacked for a year or two before they are sufficiently dried out or seasoned, and hardwoods for up to ten years. The least moisture content of timber that can be achieved by this method of seasoning is about 18%.

Artificial or kiln seasoning

Because of the great length of time required for natural dry seasoning and because sufficiently low moisture contents of wood cannot be achieved, artificial seasoning is largely used today. After the wood has been converted to timber it is stacked with battens between the timber and they are then placed in an enclosed kiln. Air is blown through the kiln, the temperature and humidity of the air being regulated to effect seasoning more rapidly than with natural seasoning, but not so rapidly as to cause damage to the timber. If the timber is seasoned too quickly by this process it shrinks and is liable to crack and lose shape badly. To avoid this it is common practice to allow timber to season naturally for a time and then complete the process artificially as described.

Conversion of wood into timber

The method of cutting a log into timber will depend on the ultimate use of the timber.

Most large softwood logs are converted into timbers of different sizes so that there is the least wastage of wood. Smaller softwood logs are usually converted into a few long rectangular section timbers. Most hardwood today is converted into boards.

The method of converting wood to timber affects the timber in two ways: (a) by the change of shape of the timber during seasoning and (b) in the texture and differences in colour on the surface of the wood. Because the spring wood is less dense than the summer wood the

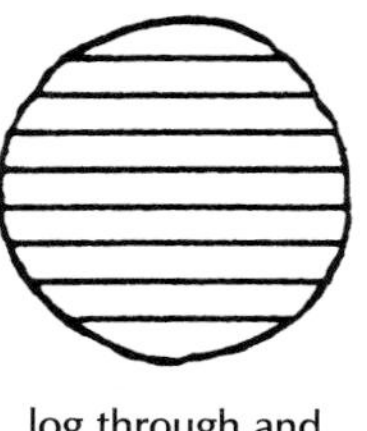

Fig. 121 Conversion of log to timber.

shrinkage caused when the wood is seasoned (dried) occurs mainly along the line of the annual rings. The circumferential shrinkage is greater than the radial shrinkage. Because of this the shrinkage of one piece of timber cut from a log may be quite different from that cut from another part of the log.

This can be illustrated by showing what happens to the planks of a log converted by the 'through and through' cut method shown in Fig. 121. When the planks have been thoroughly seasoned their deformation due to shrinkage can be compared by putting them together in the order in which they were cut from the log as in Fig. 121. From this it will be seen that the plank which was cut with its long axis on the radius of the circle of the log, lost shape least noticeably and was the best timber after seasoning.

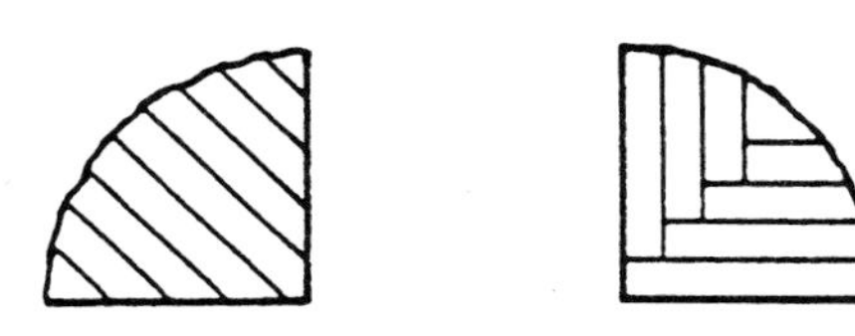

Fig. 122 Log quartered and converted into boards.

It is apparent that timber which is required to retain its shape during seasoning, such as good quality boarding, must somehow be cut as nearly as possible along the radius of the centre of the log. As it is not practicable to cut a log in the way we cut a slice out of an apple, logs which are to be cut along their radius are first cut into quarters. Each quarter of the log is then cut into boards or planks. This can be done in a variety of ways. Two of the most economical ways of doing this are shown in Fig. 122. It will be seen from these diagrams that one or two boards or planks are cut very near a radius of the circle, whilst the rest are cut somewhat off the radius and the former will lose shape least.

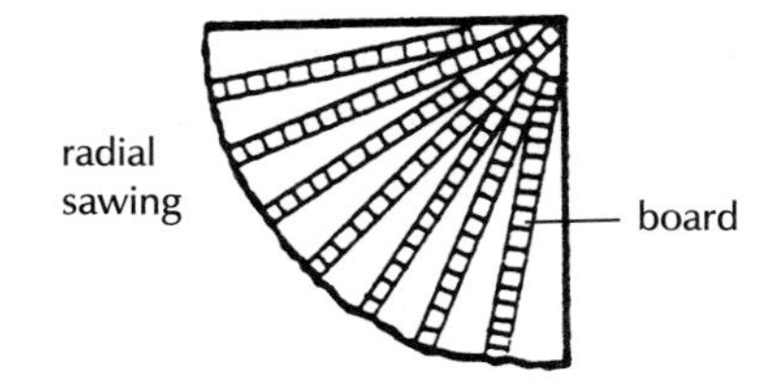

Fig. 123 Radial sawn boards.

In describing the structure and growth of a tree the medullary rays were described as being narrow radial lines of wood cells of different shape and structure than the main wood cells. If the face of a timber is cut on a radius of the circle of the log, the cells of the medullary rays may be exposed where the cut is made. With many woods this produces very pleasing texture and colour on the surface of the wood and it is said that the 'figure' of the wood has been exposed. To expose the figure of the wood by cutting along the medullary rays a quarter of a log has to be very wastefully cut, as shown in Fig. 123. The radial cutting of boards as shown is very expensive and is employed only for high-class cabinet making and panneling timbers where the exposed figure of the wood will be used decoratively.

Decay in timber

Any one of a number of wood-destroying fungi may attack timber that is persistently wet and has a moisture content of over 20%.

Fungal decay

Dry rot

This is the most serious form of fungal decay and is caused by Seupula lacrymans which can spread and cause extensive destruction of timber. The description dry rot derives from the fact that timber which has been attacked appears dry and powdery.

The airborne spores of this fungus settle on timber and if its moisture content is greater than 20% they germinate. (If the moisture content is less than 20%, germination does not occur.) The spore forms long thread-like cells which pierce the wood cells and use the wood as a food. The thread-like cells multiply, spreading out long white thread-like arms called mycelium which feed on other wood cells. This fungus can spread many tens of feet from the point where the spore first began to thrive, and is capable of forming thick greyish strands which can find their way through lime mortar and softer bricks.

Timber which is affected by this fungus turns dark brown and shrinks and dries into a cracked powdery dry mass which looks as though it has been charred by fire.

It is generally accepted that there is little risk of fungal decay in softwood if the timber is maintained at a moisture content of 20% or less. Fungal decay will only occur when the moisture content is above 20%.

Prevention of dry rot

Do not use unseasoned timber in buildings. Prevent seasoned timber becoming so wet that it can support the fungus by:

(1) building in a good horizontal dpc;
(2) either ventilating the space below or around timber floors or by designing the building so that these spaces do not become damp;
(3) immediately repairing all leaking water, rainwater and drain pipes which otherwise might saturate timber to such an extent as to make it liable to dry rot.

Replacement of timber affected by dry rot

The cause of the persistent dampness that has raised the moisture content of timber above 20%, such as by leaking gutters or water pipes, must be corrected. Timber which has been affected by the fungus or is in close proximity to it should be taken out or cut out of the building, and this timber should be burnt immediately. The purpose of burning the affected timber is to ensure that none of it is used in the repairs and to kill any spore that might cause further rot.

All walls on which, or against which, the fungus grew must be thoroughly cleaned and sterilised by application of a fungicidal solution. Any old lime plaster on which or through which the rot has spread should be hacked off and renewed. New timber used to replace affected timer should be treated with a wood preservative before it is fixed or built in.

Wet rot

Wet rot is caused principally by Coniophora puteana, the cellar fungus, which occurs more frequently, but is less serious, than dry rot. Decay of timber due to wet rot is confined to timber that is in damp situations such as cellars, ground floors without dpcs and roofs. The rot causes darkening and longitudinal cracking of timber and there is often little or no visible growth of fungus on the surface of timber.

Prevention of wet rot

Timber should not be built into or in contact with any part of the structure that is likely to remain damp. Damp-proof courses and damp-proof membranes above and at ground level and sensibly detailed flashings and gutters to roofs and chimneys will prevent the conditions suited to the growth of wet rot fungus.

Replacement of timber affected by wet rot

Affected timber should be cut out and replaced by sound new timber treated with a preservative. It is not necessary to sterilise brickwork around the area of affected timber.

Insect attack on wood

In this country the three sorts of insect which most commonly cause damage to timber are the furniture beetle, the death-watch beetle and the house longhorn beetle. Insect attack on wood in occupied buildings is much less common than is generally supposed.

Furniture beetle

The common furniture beetle (Anobium punctatum) is the beetle whose larvae most commonly attack timber and furniture and its attack is generally known as 'wood worm'. Between June and August the beetles fly around and after mating the female lays small eggs in the cracks and crevices of any dry softwood or hardwood. The eggs hatch and a small white grub emerges. These grubs (larvae) have powerful biting jaws with which they tunnel into the wood, using the wood as food. The grubs tunnel along the grain of the wood for a year or two until they are fully grown, then they bore to just below the surface of the wood where they pupate or change into beetles. These emerge in June to August and the life cycle of the beetle is repeated. The hole which each grub makes is very small and timber is structurally weakened only if a great number of these holes tunnelled by a number of grubs are formed over many years. The holes are very unsightly, particularly if the beetle infests furniture or panelling.

Death-watch beetle

The grubs of the death-watch beetle (Xestobium rufovillosum) thrive particularly well on very old timbers that have suffered decay and rarely attack softwoods. The beetle lays eggs which hatch into grubs (larvae) which, in turn, do as the larvae of the furniture beetle, tunnel along the wood for from one to several years. The name death-watch beetle derives from the ticking sound made by the beetles as they tap their heads on timber during the mating season in May and June. Again only after years of infestation by this beetle is the strength of the timber seriously affected.

House longhorn beetle

House longhorn beetle (Hylotrupes bajulus) is a large insect that, in recent years, has been active in London and the home counties and that attacks mainly softwood. It has a life cycle of up to 11 years and can cause such extensive damage that only a thin shell of sound timber is left. There is little external evidence of the infestation except some unevenness of wood surfaces due to boring just below the surface. The Building Regulations specify areas of the south of England in which the softwood timbers of roofs are required to be treated with a suitable preservative against attack by the longhorn beetle.

Control of attack by beetles

Timber which has been affected by the larvae of these beetles should be sprayed or painted with a preservative that contains an insecticide during early summer and autumn. These preservatives prevent the larvae changing to beetles at the surface of the wood and so arrest further infestation.

Wood preservatives

Wood may be preserved as a precaution against fungal decay or insect attack. There is very little likelihood of softwood timber in buildings, which are maintained at a moisture content below 20 to 22%, being affected by fungal attack and little possibility of attack by beetles.

Over recent years it has become fairly common practice to preserve softwood used in buildings, where the risks barely warrant the expenditure, because of lurid accounts of the very few instances of attack that have occurred. Fungal decay has occurred where simple, commonsense care was not taken to avoid persistent damp from leaking gutters and pipes. The few instances of beetle attack have been blown up out of all proportion.

The consequence is that regulations have become over cautious and builders and makers of preservative promote preservation as a sales advantage.

Where it is decided to adopt preservation the two types of preservative in general use are water borne or organic solvent formulations where water or a volatile solvent serves as a vehicle for the active fungicide or insecticide components.

Water borne preservatives

The most commonly used water borne preservative is based on solutions of copper sulphate, sodium dichromate and arsenic pentoxide, abbreviated to CCA. The liquid preservative is applied by pressure in a pressure tank.

The degree to which a water borne preservative will penetrate timber depends on the species. A lateral penetration of 6 to 19 mm in about 2 to 3 hours occurs under pressure with European Redwood and 3 to 6 mm with Canadian Douglas Fir for example.

To gain the maximum advantage of preservation, all timber should be cut and notched as necessary before preservative treatment. Any cutting after treatment, particularly where the cutting penetrates below the surface penetration, should be treated with preservative.

Preservation with a water borne preservative causes a considerable increase in the moisture content of the timber, an increase in the cross section and a rise in the surface grain. After preservation timber has to be dried to the required moisture content.

Water borne CCA preservatives are most suited for use with sawn structural timbers such as floor joists, roof rafters and stud framing where the raised grain of the wood is of no consequence.

Organic solvent preservation

These preservative solutions comprise an organic solvent such as white spirits with fungicides such as pentachlorophenol and zinc naphthenate. An insecticide is added as required. The liquid preservative is applied by double vacuum process and by immersion. The organic solvent evaporates to air, leaving the fungicide in the wood to the depth of penetration depending on the species of wood.

An organic solvent is commonly used for cut joinery sections both plain and moulded because the solvent dries quickly, does not cause an increase in section and does not noticeably affect the surface of the prepared joinery sections.

Tar oil preservatives

Tar oil preservatives, such as creosote, are used for the preservation of wood used for fences where the strong smell of this material is not offensive. Because of the appreciable additional cost in the preservative treatment of timber it should be used sparingly where there is a real risk of fungal or insect attack.

Surface finishes for timber

There are three types of surface finish for wood, paint, varnish and stains. The traditional finishes paint and varnish are protective and decorative finishes which afford some protection against water externally and provide a decorative finish which can easily be cleaned internally. Paints are opaque and hide the surfaces of the wood whereas varnishes are sufficiently transparent for the texture and grain of the wood to show.

Of recent years stains have been much used on timber externally. There is a wide range of stains available, from those that leave a definite film on the surface to those that penetrate the surface and range from gloss through semi-gloss to matt finish. The purpose of this finish is to give a selected uniform colour to wood without masking the grain and texture of the wood. Most stains contain a preservative to inhibit fungal surface growth. These stains are most effective on rough sawn timbers.

TIMBER WALLS

The construction of a timber framed wall is a rapid, clean, dry operation. The timbers can be cut and assembled with simple hand or power operated tools and once the wall is raised into position and fixed it is ready to receive wall finishes. A timber framed wall has adequate stability and strength to support the floors and roof of small buildings, such as houses. Covered with wall finishes it has sufficient resistance to damage by fire, good thermal insulating properties and reasonable durability providing it is sensibly constructed and protected from decay. In North America, timber is as commonly used for walls as brick is in the United Kingdom. A timber framed house can be constructed on site by two men in a matter of a few days.

There has been a prejudice against timber buildings in this country for many years. For some years after the end of the Second World War (1945) timber framed houses were built as a means of satisfying the need for new housing, by the rapid building possible with this form of construction. Comparatively few such houses were built, partly because of an inherent feeling that timber was not as solid or durable a material as brick and through the unwillingness of building societies to lend money for the construction or purchase of these houses.

More recently, timber framed houses have been constructed and once again the timber wall frame has been given a bad name because of a few problems of condensation in the fabric of walls due to ill considered design.

It is irrational to construct a timber frame that has by itself adequate strength to support the floors and roof of a house and then enclose it with a brick or block wall purely for the sake of appearance. The combination of these two systems of construction impedes the rapid construction possible with timber and confuses the 'wet trades' system of the construction of solid walling with that of the 'dry system' of timber framed construction.

Strength of timber

The strength of timber varies with species and is generally greater with dense hardwoods than less dense softwoods. Strength is also affected by defects in timber such as knots, shakes, wane and slope of the grain of the wood.

Stress grading of timber

There is an appreciable variation in the actual strength of similar pieces of timber which had led in the past to very conservative design in the use of timber as a structural material. Uncertain of the strength of individual timbers, it was practice to over design, that is, make allowance for the possibility of weakness in a timber and so select timbers larger than necessary.

Of recent years systems of stress grading have been adopted, with the result that a more certain design approach is possible with a reduction of up to 25% in the section of structural timbers. Stress grading of structural timbers, which was first adopted in the Building Regulations 1972, is now generally accepted in selecting building timber.

There are two methods of stress grading: visual grading and machine grading.

Visual grading

Trained graders determine the grade of a timber by a visual examination from which they assess the effect on strength of observed defects such as knots, shakes, wane and slope of grain. There are two visual grades, general structural (GS) and special structural (SS), the allowable stress in SS being higher than in GS.

Machine grading

Timbers are subjected to a test for stiffness by measuring deflection under load in a machine which applies a specified load across overlapping metre lengths to determine the stress grade. This mechanical test, which is based on the fact that strength is proportional to stiffness, is a more certain assessment of the true strength of a timber than a visual test. The machine grades, which are comparable to the visual grades, are machine general structural (MGS) and machine special structural (MSS). There are in addition two further machine grades, M50 and M75. The stress of M50 lies between MGS and MSS and M75 is the highest stress grade in the series.

Stress graded timbers are marked GS and SS at least once within the length of each piece for visually graded timber together with a mark to indicate the grader or company. Machine graded timber is likewise marked MGS, M50, MSS and M75 together with the BS kitemark and the number of the British Standard, 4978.

Approved Document A, which gives practical guidance to meeting the requirements of the Building Regulations for small buildings, includes tables of the sizes of timber required for floors and roofs, related to load and span.

Stability

The stability of a timber framed wall depends on a reasonably firm, stable foundation on which a stable structure can be constructed. In common with other systems of walling the foundation to a timber framed wall should serve equally for the timber framed walls of small buildings, depending on ground conditions.

Because timber framing is a lightweight form of construction it will depend less on support from the foundation than will a similar brick construction and more on being firmly anchored to the foundation against uplift due to wind forces. Figure 124 is an illustration of the base of a timber framed wall set on a brick upstand raised from a strip foundation. The 150 × 50 mm timber sole plate is bedded on a horizontal dpc, with 13 mm bolts at 2 m centres built into the wall to

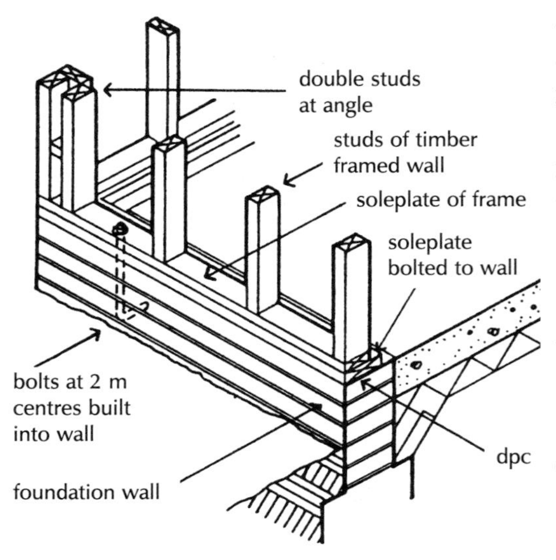

Fig. 124 Base of timber framed wall.

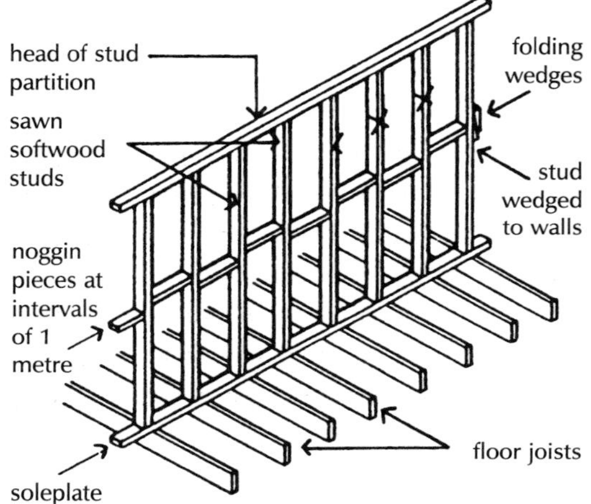

Fig. 125 Timber stud frame.

anchor the plate against wind uplift. As an alternative the bolts may be shaped so that the bottom flange is built into the wall, run up on the inside face of the wall with a top flange turned over the top of the plate.

Where a concrete raft serves as foundation the upstand kerb of the raft serves as a base for the timber wall with the anchor bolts set into the concrete curbs and turned over the top of the sole plate.

The vertical 100 × 50 mm studs are nailed to the sole plate at 400 to 600 mm centres with double studs at angles to facilitate fixing finishes.

A timber stud wall consists of small section timbers fixed vertically between horizontal timber head and sole plates, as illustrated in Fig. 125. The vertical stud members are usually spaced at centres of 400 to 600 mm to support the anticipated loads and to provide fixing for external and internal linings. The horizontal noggins fixed between studs are used to stiffen the studs against movement that might otherwise cause finishes to crack.

By itself a timber stud wall has poor structural stability along its length because of the non-rigid, nailed connection of the studs to the head and sole plate which will not strongly resist racking deformation. A timber stud wall must, therefore, be braced (stiffened) against racking.

As an internal wall or partition a timber stud frame may be braced by diagonal timbers framed in the stud wall or by being wedged between solid brick or block walls, as illustrated in Fig. 125, or a combination of the two systems of bracing.

As an external wall a timber stud frame may be braced between division walls and braced at angles where one wall butts to another, as illustrated in Fig. 124. In addition an external stud frame wall is braced by diagonally fixed boarding or plywood sheathing fixed externally as a background for finishes.

Because of its small mass, a timber frame wall has poor lateral stability against forces such as wind that tend to overturn the wall. For stability along the length of the wall, connected external and internal walls or partitions will serve as buttresses.

For stability up the height of the wall, timber upper floors and roof connected to the wall will serve as buttresses. Buttressing to timber walls that run parallel to the span of floor joists and roof frames is provided by steel straps that are fixed across floor joists and roof rafters and fixed to timber walls in the same way that straps are used to buttress solid walls as previously described.

The usual method of supporting and fixing the upper floor joists to the timber wall frame is by using separate room height wall frames. The heads of the ground floor frames provide support for the floor joists on top of which the upper floor wall frame is fixed, as illustrated in Fig. 126. The roof rafters are notched and fixed to the head of the upper floor wall frame.

As an alternative a system of storey height wall frames may be used with the top of the head of the lower frame in line with the top of the floor joists which are supported by a timber plate nailed to the studs, as illustrated in Fig. 127. The upper frame is formed on the lower frame. The advantage of this system is that there is continuity of the wall frame and the disadvantage a less secure connection and therefore lateral bracing.

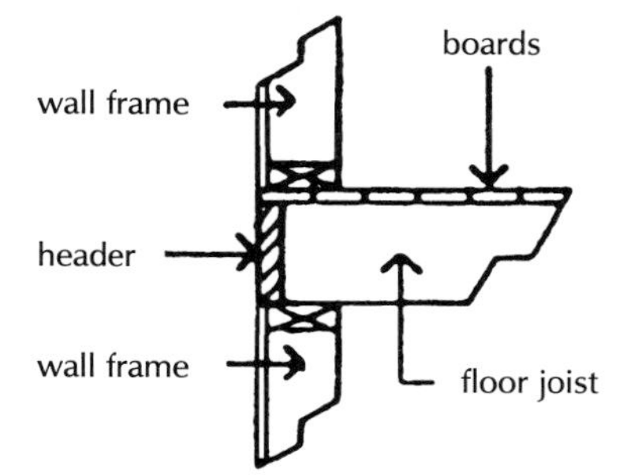

Fig. 126 Support for floor joists.

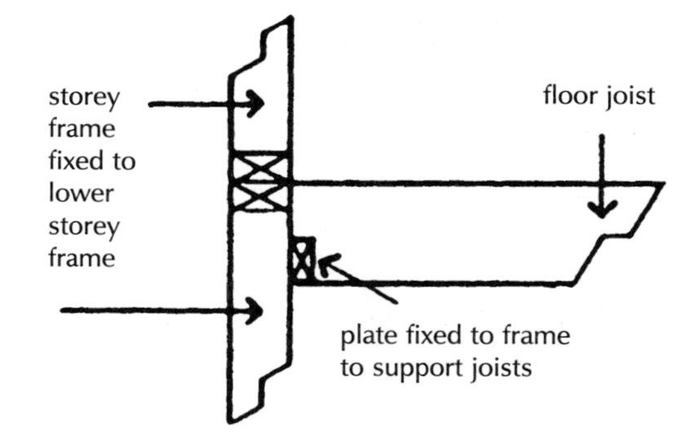

Fig. 127 Storey frame.

Resistance to weather

The traditional weather envelope for timber walls is timber weatherboarding nailed horizontally across the stud frame. The weather boards are shaped to overlap to shed water. Some typical sections of boarding are illustrated in Fig. 128.

The wedge section, feather edge boarding, is either fixed to a simple overlap or rebated to lie flat against the studs as illustrated. The shaped chamfered and rebated and tongued and grooved shiplap boarding is used for appearance sake, particularly when the boarding is to be painted for protection and decoration.

To minimise the possibility of boards twisting it is practice to use boards of narrow widths of as little as 100 and usually 150 mm.

As protection against rain and wind penetrating the weatherboarding it is usual to fix sheets of roofing underlay or breather paper behind the weatherboarding. Breather paper serves to act as a barrier to water and at the same time allow the release of moisture vapour, under pressure to move through the sheet.

Instead of nailing weatherboarding directly to the studs of the wall frame it is usual to fix either diagonally fixed boarding or sheets of plywood across the external faces of the stud frame. The boarding and ply sheets serve as a brace to the frame and as a sheath to seal the frame against weather. Figure 129 is an illustration of weatherboarding fixed to plywood sheathing with insulation fixed between studs.

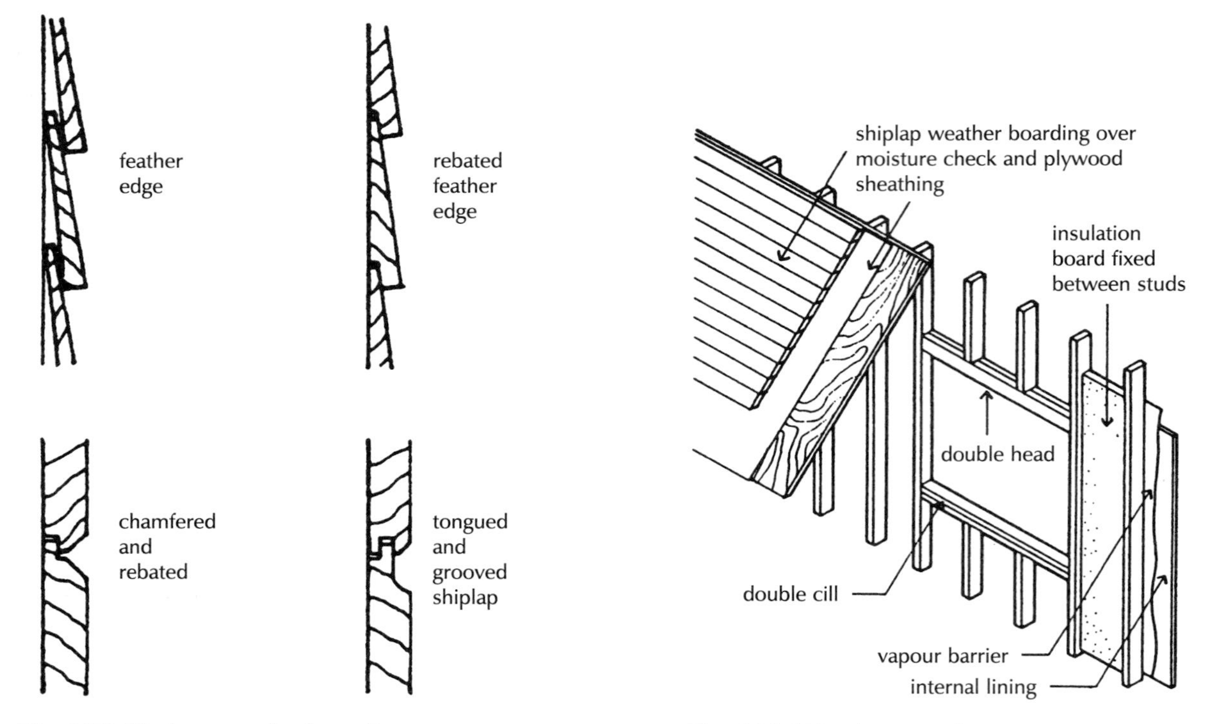

Fig. 128 Timber weatherboarding.

Fig. 129 Weather envelope.

Around openings to windows and doors the weatherboarding and ply sheath may be butted to the back of window and door frames fixed to project beyond the stud frames for the purpose. At the head of the opening the head of the frame may be reduced in depth so that the boarding runs down over the face of the frame. The weatherboarding butts up to the underside of a projecting cill. For extra protection sheet lead may be fixed behind the weatherboarding and nailed and welted to window and door frames.

At external angles the weatherboarding may be mitred or finished square edged. As a seal and finish to the joint between the weatherboarding at external angles timber cover mouldings have been used without much success as the mouldings soon become saturated, swell and defeat the object of their use. A more straightforward and effective weathering is to fix a strip of lead behind the weatherboarding to form a sort of secret gutter.

In exposed positions weatherboarding may not provide adequate protection. Tile or slate hanging may be used to provide more satisfactory protection.

Brick and timber framed wall

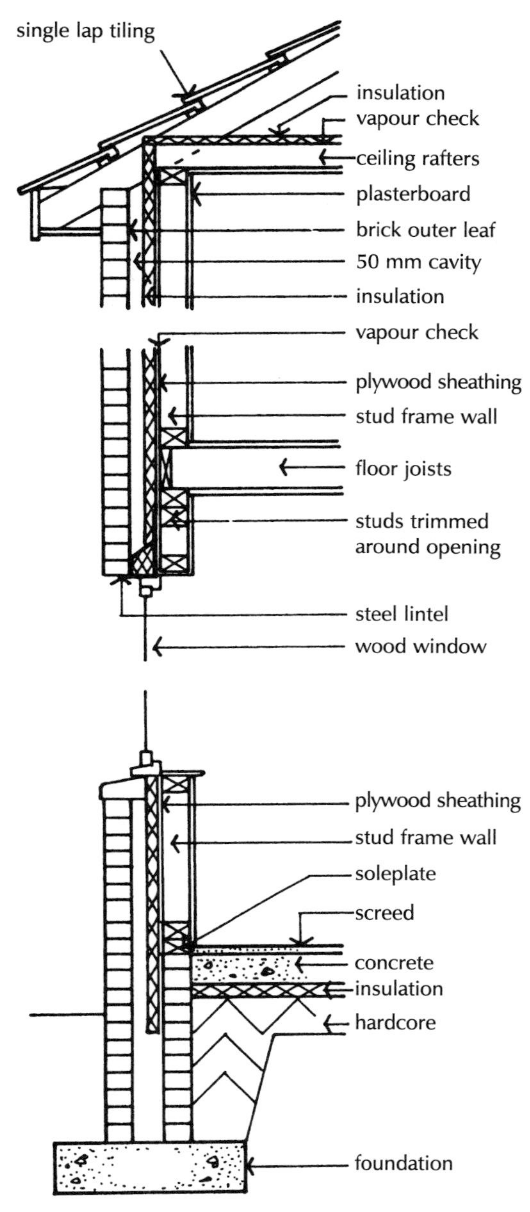

Fig. 130 Brick outer and timber framed cavity wall.

The external walls of small buildings such as houses have been constructed as a cavity wall with a brick outer leaf and a timber framed inner leaf or frame. This seemingly perverse form of construction which combines the 'wet trade' form of construction with a 'dry trade' form of construction may be justified by the permanence and appearance of an outer brick wall.

The external brick leaf may well overcome the prejudice of buyer and building society surveyor that timber is a temporary building form, by providing the sense of weather resistance and durability that brickwork gives.

A sensible argument for this odd form of construction could be speed of erection and completion of building work by combining the rapid framing of a timber wall, floor and roof structure that could be completed and covered in a matter of a few days, with a brick outer leaf and speedy installation of electrical, water and heating services and dry linings.

Figure 130 is an illustration of a two storey house with timber walls, floor and roof with a brick outer leaf.

For strength the timber inner leaf, floor and roof are adequate to the small loads. For stability the upper floor and roof are adequate to stiffen the walls. It could be demonstrated that the external brick outer leaf, buttressed at angles, has sufficient stability by itself, or the use of ties across the cavity at first floor level and roof to the timber structure could be used to augment stability if need be.

For resistance to weather a brick outer leaf is generally accepted as being thick enough to prevent penetration of rain to the inside.

For thermal resistance one of the thermal insulation boards is fixed to a vapour check and plywood sheathing nailed to the stud frame. The thermal insulation is carried up in the cavity to unite with the roof insulation laid on a vapour check. The plywood sheathing is used to diagonally brace the stud frame.

Internal plasterboard linings to the timber framed walls, the soffit of the first floor and the ceiling will provide a sufficient period of fire resistance to meet the requirements for a two floor house.

The requirement for barriers in external cavity walls to small houses applies only to the junction of a cavity and a wall separating buildings.

Prefabricated timber frames

With the use of a wide range of wood working tools that are available it is practice to prefabricate timber wall frames, particularly in North America and Scandinavia where there is a plentiful supply of timber and a traditional use of timber for small buildings.

The advantage of using frames that are fabricated either on or off site complete with outer and inner finishes is speed of erection. Where

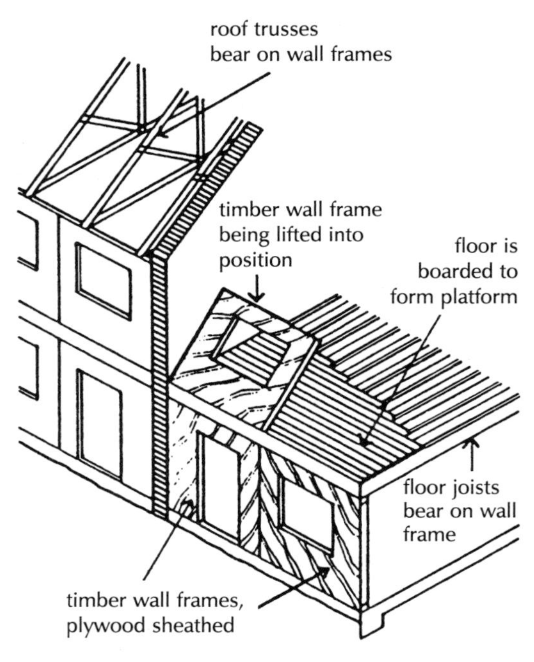

Fig. 131 Platform frame.

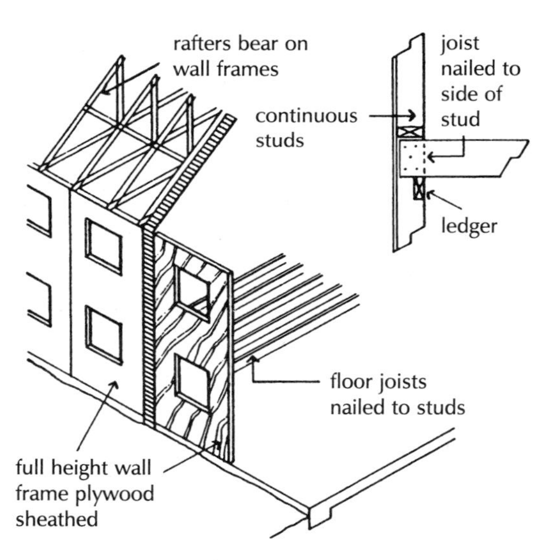

Fig. 132 Balloon frame.

a number of houses is to be built it is possible to complete a building in a matter of days.

Prefabricated timber wall frames have been little used in England mainly because of a prejudice against timber. Prefabrication has been used to some extent in the external walls of terrace housing where the walls can readily be erected as timber framed front and back walls, complete with windows and doors, between solid brick separating walls.

The systems most used are either platform or storey frames.

The platform frame system of construction employs prefabrication frames that are floor to ceiling level high, with the sole of the lower stud frame bearing on the foundation and the head of the frame supporting first floor joists, as illustrated in Fig. 131. The first floor can then be used as a working platform from which the upper frames are set on top of the lower.

The wall frames or panels may be the full width of the front and rear walls of narrow terrace houses or made in two or more panels. The first floor joists and roof provide sufficient bracing up the height and the separating wall will brace across the width of the wall.

The wall frames may be prefabricated as stud frames sheathed with plywood or made complete with finishes both sides.

Storey frames are made the height of a storey, floor to floor so that the top of the head of a frame is level with the top of floor joists. The joists are supported by a bearer fixed to the stud frame. This arrangement provides continuity of the stud framing up the height of the wall at the expense of some loss of secure anchor of floor joists to wall.

A balloon wall frame is fabricated as one continuous panel the height of the two floors of small houses, as illustrated in Fig. 132. This system is most used in North America and Scandinavia where timbers of the required length are more available than they are in England.

The advantage of the balloon frame is speed of fabrication and erection, and the least number of joints between frames that have to be covered and weathered externally.

Fire safety

The requirements for means of escape from one or two storey houses are that each habitable room either opens directly on to a hallway or stair leading to the entrance, or that it has a window or door through which escape could be made and that means are provided for giving early warning in the event of fire.

With increased height and size, where floors are more than 4.5 m above ground, it is necessary to protect internal stairways or provide alternative means of escape. Where windows and doors may be used as a means of escape their minimum and maximum size and the minimum and maximum height of the window cill are defined.

All new houses should be fitted with self-contained smoke alarms permanently wired to a separate fused circuit at the distribution board.

Internal fire spread (structure)

The premature failure of the structural stability of a building is restricted by specifying a minimum period of fire resistance for the elements of structure. A timber framed wall covered with plaster-board internally satisfies the requirement for houses of up to two storeys.

External fire spread

To prevent the spread of fire between buildings, limits to the size of 'unprotected areas' of walls and finishes to roofs close to boundaries are set out in the Building Regulations. By reference to the boundaries of the site the control will limit spread of fire. Unprotected areas are those parts of external walls that may contribute to spread of fire and include glazed windows, doors and those parts of a wall that may have less than a notional fire resistance.

Limits are set on the use of roof coverings that will not provide adequate protection against the spread of fire across their surface to adjacent buildings.

Resistance to the passage of heat

Timber is a comparatively good insulator having a U value of 0.13 for softwood and 0.15 for hardwood. The sections of a timber frame do not by themselves generally afford sufficient insulation to meet the requirements of the Building Regulations and a layer of some insulating material has to be incorporated in the construction.

The layer of insulation is fixed either between the vertical studs of the frame or on the outside or inside of the framing.

The disadvantage of fixing the insulation between the studs is that there may be a deal of wasteful cutting of insulation boards to fit them between studs and to the extent that the U value of the timber stud is less than that of the insulation material, there will be a small degree of thermal bridge across the studs.

The advantage of fixing the insulation across the outer face of the timber frame is simplicity in fixing and the least amount of wasteful cutting and that the void space between the studs will augment insulation and provide space in which to conceal service pipes and cables.

The disadvantage of external insulation is that the weathering finish such as weatherboarding has to be fixed to vertical battens screwed or nailed through the insulation to the studs. Unless the insulation is one of the rigid boards it may be difficult to make a fixing for battens sufficiently firm to nail the battens to.

Internal insulation is usually in the form of one of the insulation boards that combine insulation with a plasterboard finish.

Vapour check

A high level of insulation required for walls may well encourage moisture vapour held by warm inside air, particularly in bathrooms and kitchens, to find its way due to moisture vapour pressure into a timber framed wall and condense to water on the cold side of the insulation. The condensation moisture may then damage the timber frame.

As a barrier to warm moist air there should be some form of vapour check fixed on the warm side of the insulation. Closed cell insulating materials such as extruded polystyrene, in the form of rigid boards, are impermeable to moisture vapour and will by themselves act as a vapour check. The boards should either be closely butted together or supplied with rebated or tongued and grooved edges so that they fit tightly and serve as an efficient vapour check.

Where insulation materials that are pervious to moisture vapour, such as mineral fibre, are used for insulation between studs, a vapour check of polythene sheet must be fixed right across the warm side of the insulation. The polythene sheet should be lapped at joints and continued up to unite with any vapour check in the roof and should, as far as practical, not be punctured by service pipes.

Electrical cables that are run through the members of a timber wall and the insulation between the studs may overheat due to the surrounding insulation, with a risk of short circuit or fire. To prevent overheating of cables run through insulation, the cables should be derated by a factor of 0.75 by using larger cables than specified, which will generate less heat. So that cables are not run through insulation it is wise to fix the inside dry lining to timber frames that are filled with insulation, on timber battens nailed across the frame so that there is a void space in which cables can be safely run.

Insulation for timber walls

The inorganic materials glass fibre and rockwool are most used for insulation between studs as there is no advantage in using the more expensive organic materials, as the thickness of insulation required is not usually greater than the width of the studs. Either rolls of loosely felted fibres or compressed semi-rigid batts or slabs of glass fibre or rockwool are used. The material in the form of rolls is hung between the studs where it is suspended by top fixing and a loose friction fit between studs, which generally maintains the insulating material in position for the comparatively small floor heights of domestic buildings. The friction fit of semi-rigid slabs or batts between studs is generally sufficient to maintain them, close butted, in position.

For insulating lining to the outside face of studs one of the organic insulants such as XPS or PIR provides the advantage of least thickness of insulating material for given resistance to the transfer of heat. The more expensive organic insulants, in the form of boards, are fixed across the face of studs for ease of fixing and to save wasteful cutting.

A vapour check should be fixed on to or next to the warm inside face of insulants against penetration of moisture vapour. Organic insulants, such as XPS, which are substantially impervious to moisture vapour can serve as a vapour check, particularly when rebated edge boards are used and the boards are close butted together.

Table 9 lists some of the insulants suited for use in timber framed walls.

Table 9 Insulating materials.

	Thickness mm	**U value W/m²K**
Timber framed wall (insulation between studs)		
Glass fibre		
rolls	50, 80, 90, 100	0.04
semi-rigid batts	80, 90, 100, 120, 140, 160	0.04
Rockwool		
rolls	60, 80, 90, 100, 150	0.037
semi-rigid slabs	60, 80, 90, 100	0.037
Timber framed wall (insulation fixed to face of studs)		
XPS		
boards T & G on long edges	25, 50	0.025
PIR		
boards with heavy duty aluminium facings	20, 25, 30, 35, 50	0.02

XPS extruded polystyrene
PIR rigid polyisocyanurate

Resistance to the passage of sound

The small mass of a timber framed wall affords little resistance to airborne sound and does not readily conduct impact sound. The insulation necessary for the conservation of heat will give some reduction in airborne sound and the use of a brick or block outer leaf will appreciably reduce the intrusion of airborne sound.

3: Floors

FUNCTIONAL REQUIREMENTS

The functional requirements of a floor are:

Strength and stability
Resistance to weather and ground moisture
Durability and freedom from maintenance
Fire safety
Resistance to passage of heat
Resistance to the passage of sound

Strength

The strength of a floor depends on the characteristics of the materials used for the structure of the floor, such as timber, steel or concrete. The floor structure must be strong enough to safely support the dead load of the floor and its finishes, fixtures, partitions and services and the anticipated imposed loads of the occupants and their movable furniture and equipment. BS 6399: Part 1 is the Code of Practice for dead and imposed loads for buildings.

Where imposed loads are small, as in single family domestic buildings of not more than three storeys, a timber floor construction is usual. The lightweight timber floor structure is adequate for the small loads over small spans and appreciably cheaper than a reinforced concrete floor.

For larger imposed loads and wider spans a reinforced concrete floor is used both for strength in support and also for resistance to fire.

Approved Document A to the Building Regulations includes tables of recommended sizes and spacing for softwood timber floor joists of two strength classes, for various dead loads and spans.

Stability

A floor is designed and constructed to serve as a horizontal surface to support people and their furniture, equipment or machinery. The floor should have adequate stiffness to remain reasonably stable and horizontal under the dead load of the floor structure and such partitions and other fixtures it supports and the anticipated static and live loads it is designed to support. The floor structure should also support and accommodate, either in its depth, or below or above, electrical, water, heating and ventilating services without affecting its stability. For stability there should be adequate support for the floor structure and the floor should have adequate stiffness against gross deflection under load.

Upper or suspended floors are supported by walls or beams and should have adequate stiffness to minimise deflection under load. Under load a floor will deflect and bend and this deflection or bending should be limited to avoid cracking of rigid finishes such as plasterboard and to avoid the sense of apprehension in those below the floor that they might suffer, if the deflection or bending were obvious. A deflection of about 1/300 of the span is generally accepted as a maximum in the design of floors.

Solid ground and basement floors are usually built off the ground from which they derive support. The stability of such floors depends, therefore, on the characteristics of the concrete under them. For small domestic loads the site concrete, without reinforcement, provides adequate stability. For heavier loads, such as heavy equipment or machinery, a reinforced concrete slab is generally necessary with, in addition, a separate foundation under heavy machinery.

On shrinkable clay soils it may be necessary to use a suspended reinforced concrete slab against differential expansion or contraction of the soil, especially where there are deep rooted trees near the building.

Resistance to weather and ground moisture

The ground floor of a building, especially a heated building, will tend to encourage moisture from the ground below to rise and make the floor damp and feel cold and uncomfortable. This in turn may require additional heating to provide reasonable conditions of comfort. An appreciable transfer of moisture from the ground to the floor may promote conditions favourable to wood rot and so cause damage to timber ground floors and finishes.

Obviously, the degree of penetration of moisture from the ground to a floor will depend on the nature of the subsoil, the water table and whether the site is level or sloping. On a gravel or coarse grained sand base, where the water table throughout the year is well below the surface, there will be little penetration whereas on a clay base, with the water table close to the surface, there will be appreciable penetration of moisture from the ground to floors. In the former instance a concrete slab alone may be sufficient barrier and in the latter a waterproof membrane on, in, or under the concrete slab will be necessary to prevent moisture rising to the surface of the floor.

The requirements of the Building Regulations for the resistance of the passage of moisture to the inside of buildings are described in Chapter 1.

Durability and freedom from maintenance

Ground floors on a solid base protected against rising moisture from the ground, and suspended upper floors solidly supported and adequately constructed and protected inside a sound envelope of walls and roof, should be durable for the expected life of the building and require little maintenance or repair.

The durability and freedom from maintenance of floor boards and finishes to solid floors will depend on the nature of the materials used and the wear that they are subject to.

Fire safety

Suspended upper floors should be so constructed as to provide resistance to fire for a period adequate for the escape of the occupants from the building. The notional periods of resistance to fire, from $\frac{1}{2}$ to 4 hours, depending on the size and use of the building, are set out in the Building Regulations. In general a timber floor provides a lesser period of resistance to fire than a reinforced concrete floor. In consequence timber floors will provide adequate resistance to fire in small domestic buildings, and concrete floors the longer periods of resistance to fire required in large buildings.

Resistance to the passage of heat

A floor should provide resistance to transfer of heat where there is normally a significant air temperature difference on the opposite sides of the floor, as, for example, where an open car port is formed under a building and the floor over the port is exposed to outside air, the floor over should be insulated and have a U value the same as an exposed wall.

Obviously a ground floor should be constructed to minimise transfer of heat from the building to the ground or the ground to the building. Both hardcore and a damp-proof membrane on, under or sandwiched in the oversite concrete will assist in preventing the floor being damp and feeling cold and so reduce heating required for comfort and reduce transfer of heat. The use of insulation under solid ground floors is described in Chapter 1.

Where under floor heating is used it is essential to introduce a layer of insulation below and around the edges of the floor slab to reduce transfer of heat to the ground.

Resistance to the passage of sound

Upper floors that separate dwellings, or separate noisy from quiet activities, should act as a barrier to the transmission of sound. The comparatively low mass of a timber floor will transmit airborne sound more readily than a high mass concrete floor, so that floors between dwellings, for example, are generally constructed of concrete.

The resistance to sound transmission of a timber floor can be improved by filling the spaces between the timber joists with either lightweight insulating material or a dense material. The additional cost of filling to a new floor for the comparatively small reduction in sound transmission may not be worthwhile where, for a modest increase in cost, a concrete floor will be more effective.

Where existing buildings, with timber floors, are to be converted into flats the only reasonable way of improving sound insulation between floors is the use of filling between joists or some form of floating floor.

The reduction of impact sound is best effected by a floor covering such as carpet or a resilient layer under the floor surface, that deadens the sound of footsteps on either a timber or a concrete floor.

The hard surfaces of the floor and ceiling of both timber and concrete floors will not appreciably absorb airborne sounds which will be reflected and may build up to an unacceptable level. The sound absorption of a floor can be improved by carpet or felt, and a ceiling by the use of one of the absorbent 'acoustic' tile or panel finishes.

CONCRETE GROUND FLOORS

Ground supported slab

The majority of ground floors are constructed as ground supported in-situ cast concrete slabs on a hardcore bed with a damp-proof membrane and insulation, as described in Chapter 1.

Suspended concrete slabs

Where ground under a floor is sloping, has poor or uncertain bearing capacity, or is liable to volume change due to seasonal loss or gain of moisture and a ground supported slab might sink or crack due to settlement, it may be wise to form the ground floor as a suspended reinforced concrete slab, supported by external and internal load-bearing walls, independent of the ground.

Suspended concrete slabs are constructed with one of the pre-cast reinforced concrete plank, slab or beam and block floor systems described later for upper floors, because there is no ready means of constructing centering on which to cast an in-situ concrete floor. The one way spanning, pre-cast concrete floor bears on internal and external loadbearing foundation walls with endbearing of at least 90 mm, and is built into the walls. The depth of the plank, slab or beams depends on the loads to be carried and the span between supporting walls.

Damp-proof membrane

Where the suspended ground floor slab is formed above the ground level inside the building, with an air space of at least 75 mm below the underside of the slab and the air space is ventilated to outside air, then it may not be necessary to use a dpm. The purpose of ventilating the space below a suspended floor is to prevent the build-up of stagnant moist air in the space, which would otherwise tend to make the slab damp.

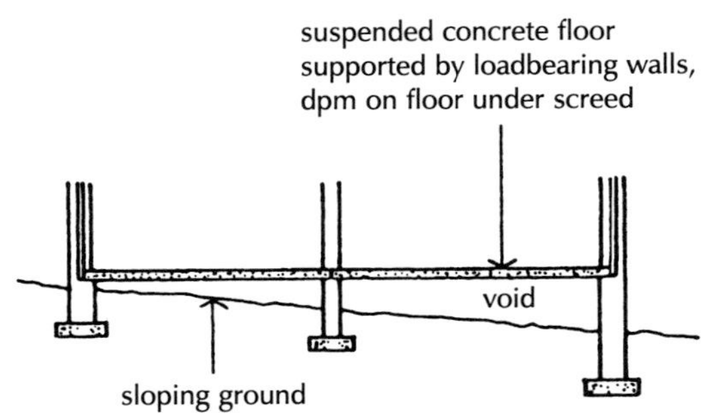

Fig. 133 Suspended concrete floor and damp-proof membrane.

Where there is a likelihood of an accumulation of gas building up in the space below the floor, which might lead to an explosion, then the space below the floor should be at least 150 mm clear and there should be ventilation on opposite sides of the space. The ventilation openings should be at least equivalent to 1500 mm for each metre run of wall.

Where the ground level inside a building under a suspended slab is below the surrounding level or the site slopes, it may be necessary to provide drainage to prevent a build-up of water. In these situations it is wise to form a dpm in the suspended slab under the screed, as illustrated in Fig. 133.

Insulation

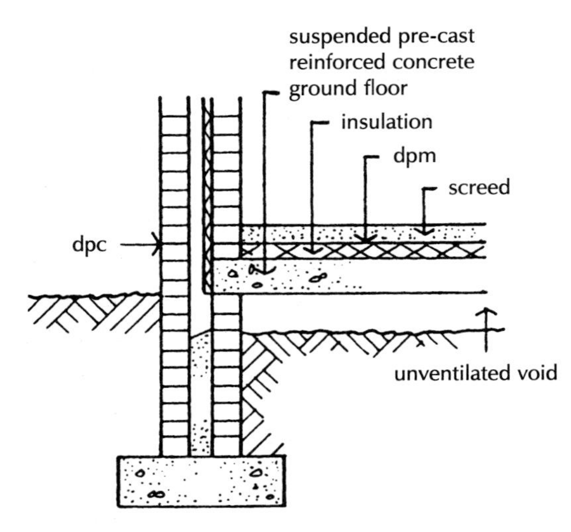

Fig. 134 Suspended concrete ground floor, insulation.

Suspended concrete ground floor slabs that are constructed above the ground on well drained subsoils, with an unventilated void below the slab, may not require insulation.

Where there is a likelihood of the void space below a suspended concrete floor becoming damp or causing a build-up of gases it is necessary to ventilate the space. For the ventilation to be effective there should be a reasonably vigorous flow of air across the space which in winter will cause an appreciable transfer of heat through the floors.

In this situation the floor slab should be insulated to reduce heat transfer and with a built in dpm to maintain a reasonably comfortable dry floor temperature. The best place for the layer of floor insulation is on top of the concrete floor either over the dpm or below it on to which the floor screed is laid, as illustrated in Fig. 134. The dpm should be laid at the same level as the dpc in the internal leaf of the cavity wall.

Where the dpm is under the insulation, the top surface of the concrete slab should be covered with a thin layer of cement and sand levelled off to protect the dpm from damage by irregularities in the top of the slab.

Floor surface

The term floor surface is generally used to describe the top surface of a floor structure. The level top surface of a structural concrete floor and the top surface of the floor boards that are an integral part of a structural timber floor are the floor surface. The concrete and timber board surface may be used as a finished floor surface.

The term floor finish is generally used to describe the material or materials that are applied to a floor surface as a finished surface, such as tiles and thin sheets of plastic to concrete.

For sheds, workshops, stores and garages, the finished top surface of the oversite concrete is sometimes used as the finished floor surface to save the cost of an applied floor finish. Concrete is not generally satisfactory as a finished floor surface, because even though it can be given a smooth finish with a power float, many of the fine particles of sand and cement are brought to the surface. These particles have poor resistance to wear and in a short time the surface of the concrete 'dusts' and requires frequent vigorous brushing. Being a coarse grained material, concrete cannot be washed clean and if it becomes stained the stains are permanent.

Extensive areas of concrete floor may be levelled and finished by power floating as a satisfactory base for the thicker floor finishes such as mastic asphalt, tiles and wood blocks. For the thin finishes such as plastic, linoleum, rubber sheet and tile, the more precisely level, smooth surface of a screeded base is necessary.

Floor screeds

The purpose of a floor screed is to provide a level surface to which a floor finish can be applied.

The word screed is used to describe the narrow strips of the wet mix that are first laid in bays across the length and width of the floor. The screed strips are carefully levelled in both directions to set out a precise level finish. The main bulk of the wet mix is then spread and levelled between the screeds.

The usual materials for a floor screed are cement, sand and water which are thoroughly mixed, spread over the surface of the concrete base, compacted, levelled and trowelled to a smooth finish. The thickness of the screed and the mix of cement and sand depends on the surface on which the screed is laid. The cement-rich mix used in a screed will shrink as it dries out and the thinner the screed the more rapidly it will dry and the more it will shrink and crack.

On the majority of building sites the concrete ground and upper floors are cast and roughly levelled as a working platform for subsequent building operations. To avoid damage to screeded surfaces that will serve as a finished floor surface or as a level base or substrate to applied floor finishes it is usual to lay a screed after the concrete floor has dried and hardened.

Where it is practical to lay a screed on a concrete base, within 3 hours of placing the concrete it will bond strongly to the concrete and dry slowly with the concrete so that drying shrinkage and cracking of the screed relative to that of the concrete will be minimised. For this monolithic construction of screed a thickness of 12 mm of screed will suffice.

A screed laid on a concrete base that has set and hardened should be at least 40 mm thick. To provide a good bond between the screed and the concrete, the surface of the concrete should be hacked by mechanical means, cleaned and dampened and then covered by a thin grout, or wet mix, of water and cement before the screed is laid. With a good bond to the concrete base a separate screed at least 40 mm thick will dry sufficiently slowly to avoid serious shrinkage cracking.

Where a screed is laid on an impermeable dpm there will be no bond between the screed and the concrete base so that drying shrinkage of the screed is unrestrained. So that the screed does not dry too rapidly and suffer shrinkage cracking, the screed in this unbonded construction should be at least 50 mm thick.

A screed laid on a layer of compressible thermal or sound insulating material should be at least 65 mm thick for domestic and 75 mm for other buildings, if this floating construction is not to crack due to drying shrinkage and the deflection under loads on the floor.

For screeds up to 40 mm thick, a mix of Portland cement and clean sand in the proportions by weight of 1:3 to 1:4½ is used. The lower the proportion of cement to sand the less the drying shrinkage. For screeds over 40 mm thick a mix of fine concrete is often used in the proportions of 1:1½:3 of cement, fine aggregate and coarse aggregate with a maximum of 10 mm for the coarse aggregate.

Screed should be mixed with just sufficient water for workability. The material is spread over the surface of the base and thoroughly compacted by tamping to the required thickness and level and then it is finished with a wood or steel float. A wood float finish is used for wood block and thick tile floors and a steel finish for the thin sheet and tile finishes. The screed should be cured, that is allowed to dry out slowly over the course of several days, by covering it with sheeting, such as polythene, to minimise rapid drying shrinkage and cracking.

Premixed cement screed materials, dry bagged ready for use on site, are used to avoid messy, wasteful site mixing. Just sufficient water is added to the dry mix for workability. The mix may include polymer fibre for reinforcement.

FLOOR SURFACE FINISHES

The traditional floor surface and finish for ground floors was natural stone slabs, thick clay tiles or brick laid on a bed of lime and sand on a consolidated earth base. These natural materials which provided a reasonably level, solid surface required considerable effort to keep clean and were cold underfoot because they were ready conductors of cold from the soil below.

To provide some insulation against cold and damp rising from the ground, a timber boarded floor was often used. The timber boards were nailed to timber battens set in the consolidated earth base or raised on small walls which supported timber joists to which the floor boards were nailed. This raised timber floor construction was in use, particularly for houses, for many years before the introduction of the concrete ground slab in the late nineteenth century.

The disadvantage of the timber boarded ground floor was that the battens laid on the ground would in time rot and the raised timber floor required ventilation of the space below the joists to clear moist air, which made the floor cold under foot.

The use of a solid concrete floor slab as a barrier to rising damp provided a solid surface on which a variety of materials could be laid as a floor finish. The later use of a continuous membrane and a layer of insulation provided a barrier against rising damp and insulation against transfer of heat.

The materials that were used as a finished floor surface in the early days were clay tiles, sheets of linoleum or timber boards nailed to battens. More recently a wide range of plastic sheet and tiles has been used to provide a floor surface to meet the demands for an easily cleaned surface that is reasonably durable.

Most recently plastic floor finishes have to an extent lost favour because of the dull, insipid colours that were used initially with the material, in favour of traditional materials such as linoleum, clay tiles, natural stone slabs and timber.

Floor finishes for concrete floors

A floor finish should generally be level, reasonably resistant to the wear it will be subject to and capable of being easily cleaned. For specific areas the surface should be non-slip, smooth for cleaning and polishing, resistant to liquids likely to be spilled, seamless for hygiene or substantially dust free. There is no one finish that will satisfy the possible range of general and specific requirements. There is a wide range of finishes available from which one may be selected as best suited to a particular requirement.

For the small floor areas of rooms in houses and flats the choice of floor finish is dictated largely by appearance and ease of cleaning. For the larger floor areas of offices, public and institutional buildings ease of cleaning is a prime consideration where power operated cleaning and polishing equipment is used.

It is convenient to make a broad general classification of floor finishes as:

Jointness
Flexible thin sheet and tile
Rigid tiles and stone slabs
Wood and wood based

Jointless floor finishes

This group includes the cement and resin based screeds and mastic asphalt that are laid while plastic and do not show joints and seams other than movement joints where necessary.

Cement screeds

A cement and sand screed finish to a concrete floor may be an acceptable, low cost finish to small area floors of garages, stores and outhouses where the small area does not justify the use of a power float and considerations of ease of cleaning are not of prime importance.

Fibre reinforced cement screed

Premixed, dry bagged cement and sand screed material reinforced with polymer fibre is available. The fibre reinforces against drying, shrinkage and cracking.

Surface hardeners

To produce improved surface resistance to wear and resistance to the penetration of oils and grease a dry powder of titanium alloy with cements may be sprinkled on to the wet surface of concrete or screed and trowelled in.

Granolithic paving

The traditional cement screed finish to the floors of factories, stores, garages and other large floor areas, which have to withstand heavy wear, is granolithic paving or screed.

Granolithic paving consists of a mixture of crushed granite which has been carefully sieved so that the particles are graded from coarse to very fine in such proportions that the material, when mixed, will be particularly free of voids or small spaces, and when mixed with cement will be a dense mass. The usual proportions of the mix are $2\frac{1}{2}$ of granite chippings to 1 of Portland cement by volume. These materials are mixed with water and the wet mix is spread uniformly and trowelled to a smooth flat surface. When this paving has dried and hardened it is hard wearing.

Every material which has a matrix (binding agent) of cement shrinks with quite considerable force as it dries out and hardens. Granolithic paving is rich in cement and when it is spread over a concrete base it shrinks as it dries out and hardens. This shrinkage is resisted by the concrete. If the concrete is dense and hard, and its surface has been thoroughly brushed to remove all dust or loose particles, it will successfully restrain the shrinkage in the granolithic paving. If, however, the concrete on which the granolithic paving is spread is of poor quality, or if the surface of the concrete is covered with dust and loose particles, the shrinkage of the granolithic paving will be unrestrained and it will crack and, in time, break up.

If the granolithic paving is laid as soon as the oversite concrete is hard enough to stand on, then the paving can be spread 15 mm thick. This at once economises in the use of the granolithic material, which is expensive, and at the same time, because the wet granolithic binds firmly to the still damp concrete, there is little likelihood of the paving cracking as it dries out.

On the surface of newly laid concrete that has dried out it is necessary to clean the surface thoroughly by mechanical hacking to remove the surface layer to expose aggregate, thoroughly wetting overnight and covering the surface with a thin grout of cement and water before the granolithic is laid. The granolithic is laid to a thickness of 40 mm, consolidated, levelled and trowelled smooth.

Granolithic paving laid on surfaces such as poor concrete, surfaces fouled with oil or grease and on a waterproof membrane to which the material will not bond, should be laid to a thickness of up to 75 mm. Because of the cost of the material and the skilled labour needed such a thickness is not a sensible or economic floor finish. It would be as economic and more satisfactory to break up the old concrete, lay new and then apply a floor finish.

There are a number of additives variously described as 'sealers' or 'hardeners' that may be added to the granolithic mix to produce improved resistance to surface wear. A thin surface dressing of carborundum granules trowelled into the surface improves resistance to wear.

With the decline of heavy industry in this country and the availability of epoxy resin finishes and epoxy sealers, granolithic paving is less used than it was.

Anhydrite floor finish

Premixed, dry bagged screed material of anhydrite and sand is used as a floor finish. Anhydrite is a mineral product of heating gypsum which will, when mixed with water, act as a cement to bind the grains into a solid mass as the material dries and hardens. The advantage of anhydrite is that it readily combines with water and does not shrink and crack as it dries out and hardens.

The wet mix of anhydrite and sand may be pumped and spread over the concrete base as a self-levelling screed or spread and trowelled by hand. The material may be pigmented.

A disadvantage of the material is that it fairly readily absorbs water and is not suited to use in damp situations.

Resin based floor finish

A range of resin emulsion finishes is available for use where durability, chemical resistance and hygiene are required in laboratories, hospitals and food preparation buildings. This specialist application finish is composed of epoxy resins as binders with cement, quartz, aggregates and pigments. The material is spread on a power floated or cement screed base by pumping and trowelling to a thickness of up to 12 mm. The aggregate may be exposed on the surface as a non-slip finish and as decoration.

This specialist floor finish is little used for small domestic or office floors. On larger floor areas it is used for the advantage of a seamless finish that can be cleaned by a range of power operated devices designed for the purpose.

Polymer resin floor surface sealers

Polyester, epoxy or polyurethane resin floor sealers are specialist thin floor finishes used for their resistance to water, acids, oils, alkalis and some solvents. The materials are spread and levelled on a level power floated or screed surface to provide a seamless finish to provide an easily cleaned surface.

Polyester resin, the most expensive of the finishes, is spread to a finished thickness of 2 to 3 mm to provide the greatest resistance.

Epoxy resin provides a less exacting resistance. It is sprayed or pumped to a self-levelling or trowelled thickness of 2 to 6 mm.

Polyurethane resin, which has moderate resistance, can be spread on a somewhat uneven base by virtue of its possible thickness. It is pumped to be self-levelling for thin finish and trowelled for the thicker. Thicknesses of 2 to 10 mm are used.

Mastic asphalt floor finish

Mastic asphalt for flooring is made from either limestone aggregate, natural rock or black pitch-mastic in the natural colour of the material or coloured. Mastic asphalt serves both as a floor finish and a dpm. It is a smooth, hardwearing, dust free finish, easy to clean but liable to be slippery when wet. The light duty grade is fairly readily indented by furniture. Mastic asphalt has been less used as a floor finish since the advent of the thin plastic tiles and sheets.

The light duty, non-industrial grade, which is laid in one coat to a finished thickness from 15 to 20 mm, is used for offices, schools and housing. The medium grade is laid in one coat to a thickness of 20 to 25 mm and the heavy duty in one coat to a thickness of 30 to 50 mm. Mastic asphalt can be laid on a level, power floated concrete finish or on a level, smooth cement and sand screed. The asphalt finish can be coloured in one of the red or brown shades available.

Flexible thin sheet and tile

The original thin sheet floor finish was linoleum that was extensively used as a smooth, easily cleaned finish to both boarded timber and solid floors. Linoleum, commonly known as 'cork lino', was laid loose on an underlay of paper felt to accommodate structural, thermal and moisture movement of the finish, relative to the floor surface. The sheets of linoleum were laid loose for at least 48 hours to allow maximum expansion at room temperature and were then laid butt jointed between rolls.

Providing the floor surface was reasonably firm and level the linoleum finish provided moderate resistance to wear. On boarded floors this finish would provide poor wear resistance over the edges of boards that twisted due to drying shrinkage. On damp, solid floors the linoleum would deteriorate fairly rapidly.

Linoleum is made from oxidised linseed oil, rosin, cork or wood flour, fillers and pigments compressed on a jute canvas backing. The sheets are made in 2 m widths, 9 to 27 m lengths and thicknesses of 2.0, 2.5, 3.2 and 4.5 mm in a variety of colours. The usual thickness of sheet is 2.5 mm. Tiles 300 and 500 mm square are 3.2 and 4.5 mm thick.

Linoleum should be laid on a firm level base of plywood or particle board on timber floors or on hardboard over timber boarded floors and on a trowelled screed on concrete floors. The material is laid flat for 48 hours at room temperature and then laid on adhesive and rolled flat with butt joints between sheets.

Linoleum has a semi-matt finish, is quiet and warm underfoot and has moderate resistance to wear for the usual 2.5 mm thick sheets and good resistance to wear for the thicker sheets and tiles.

Of late linoleum has been used instead of vinyl for the advantage of the strong colours available in the form of sheets and also in the form of decorative patterns by combining a variety of colours in various designs from cut sheet material.

Flexible vinyl sheet and tiles

Polyvinylchloride (PVC), generally referred to as vinyl, is a thermoplastic used in the manufacture of flexible sheets and tiles as a floor finish. The material combines PVC as a binder with fillers, pigments and plasticisers to control flexibility. The resistance to wear and flexibility vary with the vinyl content, the greater the vinyl content the better the wear and the poorer the flexibility.

Since it was first introduced, vinyl sheet flooring has become the principal sheet flooring used where consideration of cost and ease of cleaning combine with moderate resistance to wear. Sheet thickness from 1.5 to 4.5 mm in widths from 1200 to 2100 mm wide are produced in lengths of up to 27 m.

A wide range of colours and textures is produced from the early thin sheets coloured to produce an insipid imitation of marbled and other grained finishes to the later thicker, less flexible sheets with bright colours and greater resistance to wear.

Foam backed vinyl sheet is produced to provide a resilient surface for the advantage of resilience and quiet underfoot at the expense of the material being fairly easily punctured.

The material is extensively used in domestic kitchens and bathrooms and offices where a low cost, easily cleaned surface is suited to moderate wear.

The thin sheet material should be laid on a smooth, level screeded surface particularly free from protruding hard grains that might otherwise cause undue wear. The thicker, less flexible sheet may be laid on a power floated concrete finish. The sheets are bonded on a thin bed of epoxy resin adhesive and rolled to ensure uniformity of adhesion.

For large areas of flooring the sheets may be heat welded to provide a seamless finish for the sake of hygiene.

A range of flexible vinyl tiles is produced in a variety of colours and textures in 225, 250 or 300 mm squares by 1.5 to 3 mm thicknesses.

Various shapes of cut sheet may be used to provide single or multi-coloured designs.

Vinyl sheets and tiles may be coated with a water–wax emulsion polish for appearance and ease of cleaning with a damp mop and occasional polishing.

Flexible rubber sheet

Before the introduction of vinyl sheet, rubber was extensively used in lieu of linoleum. Natural or synthetic, vulcanised rubber with fillers and pigments was used in the production of sheets 3.8 to 12.7 mm thick, 910 to 1830 mm wide and up to 30 m long. This thick, comparatively expensive floor finish was used because of its resilience and quiet underfoot, good wear resistance and ease of cleaning.

It is made in a wide range of colours and textures from plain black to white.

Since the introduction of vinyl, sheet rubber is less used than it was. It is often preferred as an easily cleaned finish with good resistance to wear in common access corridors and changing rooms. The surface of the sheet may be textured with ribs or studs to provide a non slip, hard wearing surface.

It is bonded to a screeded or power floated concrete surface with an epoxy resin adhesive on to which the sheet is laid under the pressure of a roller.

Rigid tiles and stone slabs

Clay floor tiles

Natural clay floor tiles have been used for centuries as a hard, durable floor surface and finish for both domestic and agricultural ground floors. Before the advent of concrete the thicker tiles were often bedded on consolidated ground and the thinner tiles on a bed of sand. This hard finish could be laid reasonably level and could be cleaned by brushing and washing. Because of the nature of the material the floor would be both cold and noisy underfoot.

The two types of tile may be distinguished as floor quarries and clay floor tiles. The word quarry is derived from the French *carré* meaning square.

Floor quarries

Floor quarries have been manufactured in Staffordshire and Wales from natural plastic clays. The clay is ground and mixed with water and then moulded in hand operated presses. The moulded clay tile is then burned in a kiln. If the clay is of good quality and the tile is burned at the correct temperature the finished tiles will be very hard, dense and will wear extremely well. But as there is no precise examination of the clays used, nor accurate control of pressing or burning, the tiles produced vary considerably in quality, from very hard well burned quarries to soft underburned quarries unsuitable for any use in buildings.

The manufacturers grade the tiles according to their hardness, shape and colour. The first or best quality of these clay floor quarries are so hard and dense that they will suffer the hardest wear on floors for centuries without noticeably wearing. Because they are made from plastic clay, which readily absorbs moisture, quarries shrink appreciably when burned, and there is often a noticeable difference in the size of individual tiles in any batch. The usual colours are red, black, buff and heather brown.

Some common sizes are 100 × 100 × 12.5 mm thick, 150 × 150 × 12.5 mm thick and 229 × 229 × 32 mm thick.

Plain colours

Plain colours are manufactured from natural clays selected for their purity. The clay is ground to a fine dry powder and a small amount of water is added. The damp powder is heavily dust pressed into tile shape and the moulded tiles are burned. Because finely ground clay is used the finished tiles are very uniform in quality and because little water is used in the moulding, very little shrinkage occurs during burning. The finished tiles are uniform in shape and size and have smooth faces. The tiles are manufactured in red, buff, black, chocolate and fawn.

Because of their uniformity of shape these tiles provide a level surface, that is resistant to all but heavy wear, does not dust through abrasion, is easily cleaned with water and has a smooth, non-gloss finish which is reasonably non-slip when dry. They are used for kitchens, bathrooms and halls where durability and ease of cleaning are an advantage.

Some common sizes are 300 × 300 × 15 mm thick, 150 × 150 × 12 mm thick and 100 × 100 × 9 mm thick.

Vitreous floor tiles

The two types of vitreous (glass like) tiles are vitreous and fully vitreous. Vitreous tiles are made from clay and felspar which gives the tile a semi-gloss finish. Fully vitreous tiles contain a higher proportion of the vitrifying agent either in the tile itself or as a surface finish.

These tiles are made from felspar or other material which melts when the tile is burned and causes it to have a hard, smooth, glass-like surface which is impervious to water. By itself felspar would make the tile too brittle for use and it is mixed with both clay and flint. The materials are ground to a fine powder, a little water is added and the material is heavily pressed into tile shape and then burned.

The tiles are uniform in shape and size and have a very smooth semi-gloss or gloss surface which does not absorb water or other liquids and can be easily cleaned by mopping with water. Both vitreous and fully vitreous tiles may be moulded with a textured finish to provide a moderately non-slip surface.

A very wide range of both native and imported vitreous and fully vitreous floor tiles is available in the full range of colours possible. Sizes are generally similar to those of plain colour tiles.

The tiles are chosen in the main for the appearance of the semi- or fully gloss finish which enhances the colour and ease of cleaning. The gloss finish is impervious to most liquids, dust free and liable to be slippery, particularly when wet.

Laying clay floor tiles

The considerations that affect the choice of a method of laying floor tiles are

(1) provision of a material into, or onto, which the tile may be laid to take up variations in tile thickness to produce a reasonably level finish
(2) good adhesion to the base to provide solid support, particularly for thin tiles, to avoid cracking and
(3) to provide a means of accommodating relative structural, moisture and thermal movements between the base and the finish to prevent arching of the tile floor.

The following are the common methods of laying clay floor tiles.

Direct bedding method

The traditional method of laying tiles is to bed them on a layer of wet cement and sand spread over a screeded or level concrete floor. This direct bedding method is satisfactory for all but larger floor areas where it may be wise to form movement joints.

Quarry tiles are laid and bedded in sharp sand and cement, 1:3 or 1:4 mix, spread to a level thickness of 15 to 20 mm depending on the thickness of the tiles, on a fully dry concrete base. The cement and sand should be mixed with just sufficient water for workability and pressing the tiles into the bed. Too wet a mix will cause excess drying shrinkage. The main purpose of the bed is to accommodate the appreciable variations in thickness of the quarries to provide a reasonably level finish.

While there will be some little adhesion of the cement sand bed to the concrete, adhesion is not a prime consideration. The bed will provide a solid base for the heavy wear such surfaces are usually used for.

The joints between the quarries will be up to 15 mm wide, to allow for variations in shape, and filled with cement and sand and finished level with the floor surface, or just below the surface, to emphasise the individual tiles.

The direct bedding method of laying is used for plain colour clay tiles on a bed some 10 mm thick and with joints between 5 and 10 mm wide depending on variations in the size of tiles and the need to adjust tile width to that of a whole number of tiles with joints to suit a particular floor size.

Separating layer method

Some few instances of tiled floors 'arching' have been widely publicised and made much more of than is reasonable. The word arching is the effect of some tiles, usually in the centre of the floor, rising above their bed in the form of a shallow arch. Arching is caused by expansion of the tiles relative to their bed or contraction of the bed relative to the tiles.

Arching can be caused by shrinkage of the concrete base, or the bedding being too cement rich, wet mix shrinks on drying out. Other less usual causes are thermal shrinkage of the concrete base due to the greater thermal movement of concrete to that of clay, thermal expansion of tiles due to the use of hot water washing and creep (jelly like) deflection of concrete. Arching will be more pronounced with tightly edged butted tiles.

Where there is a realistic likelihood of arching the tiles may be laid on a bed spread over a separating layer so that movement of either the tiles or the base will not affect the floor finish.

A layer of polythene film, bitumen felt or building paper is spread with 100 mm lapped joints over the concrete floor. The tiles are then laid and bedded on a cement/sand mix spread and levelled to a thickness of from 15 to 25 mm, depending on the thickness of the tiles, and jointed in the same way as for direct bedding.

As an alternative to using a water impermeable separating layer a layer of dry sand may be used. A layer of dry sand or crushed stone, thoroughly sieved to remove large grains, is spread and raked level on the fully dried concrete base and the tiles are bedded and levelled directly on the dry sand.

This dry sand, separating layer method is suitable for thin tiles and is commonly used in southern European countries as a bed for thin stone slab floor surfaces, such as marble. The sand bed will accommodate relative movements and serve as a sound bed for thin tiles and slabs that are subject to all but the heaviest wear.

Thin bed adhesive method

The majority of the thin, vitreous tiles that are used today are bedded and laid on an adhesive that is principally used as a bond between the tiles and the base and to some small extent as a bed to allow for small variations in tile thickness.

The adhesives that are used are rubber latex cement, bitumen emulsion and sand and epoxy resins. These adhesives are spread on a level power floated concrete or a screed finish, to a thickness of from 3 to 5 mm, combed to assist bedding and the tiles are pressed and levelled in position.

Where the thin bed, epoxy resins are used as an adhesive for thin, vitreous tiles there should be no large protruding particles of aggregate or sand in the floor surface over which the brittle tile will crack under load.

Concrete tiles

Concrete tiles made of cement and sand, which is hydraulically pressed to shape as floor tiling, have been used as a substitute for quarry and plain colour clay tiles. The usual size of tiles is 300 × 300 × 25 mm, 225 × 225 × 19 mm and 150 × 150 × 16 mm. The material may be pigmented or finished to expose aggregate. The density and resistance to wear depend on quality control during manufacture and the nature of the materials used.

Because of the poor quality of colour possible by pigmentation, the necessarily coarse surface of the tile, which is not easy to clean, and the bad name given by poor quality tiles that have been produced, these tiles have lost favour.

They are laid on a level power floated concrete or screed surface and jointed in the same way as quarries and plain colours.

Stone slabs

The word tile is used in a general sense to describe square or rectangular units, thin relative to their length or width, of burned clay used as a floor or wall finish. The word slab is used to describe natural stone in units that are generally larger than tiles such as those used for outdoor paving which are also called paviours. A small slab could as well be described as a tile or a paviour when used for flooring.

A wide range of natural stone slabs is used as a floor finish, from the very hard slabs of granite to the less dense soft marbles. Stone is selected principally for the decorative colour, variations in colour, grain and polished finish that is possible and durability and freedom from dusting.

Because of their composition, all stones are hard and noisy underfoot and cold where the floor is not insulated.

The method of bedding natural stone slabs as an internal floor finish varies with the thickness, size, nature and anticipated wear on the surface.

Large, thick slabs of limestone or sandstone up to 50 mm thick are laid by the direct or separating layer method on cement and sand with cement and sand joints depending on the area to be covered and the anticipated, relative shrinkage of the bedding material. The bedding material may be of cement, or lime and sand.

Comparatively thick slabs of slate are bedded in the same material as thick slabs of limestone.

Thin slabs of granite and marble are laid by the thin bed adhesive method or the dry sand bed method, which is particularly used for marble.

Joints

The width of the joints between tiles and slabs as an internal floor finish is determined by the uniformity of shape of the material used. For quarries, joints of up to 12 mm may be necessary to allow for the considerable variations in size, and joints as little as 1 mm may be possible with very accurately cut and finished, thin slabs of granite and marble.

The disadvantage of wide joints is that the material used, such as cement and sand for quarries, will be more difficult to clean and will more readily stain than the floor material. Thin joints are used for highly polished, accurately cut granite slabs to provide the least obvious joint possible for appearance sake.

Ideally a jointing material should have roughly the same density, resistance to wear and ease of cleaning as the floor finish.

Movement joints

To an extent the joints between tiles and slabs will serve the purpose of accommodating some movement of the floor finish relative to the bedding and the concrete floor. Some small expansion or contraction of the floor finish will be taken up in the joints through slight cracks or crushing of the very many joints.

In any large structure it is practice to form movement joints (see Volume 4) to accommodate structural, moisture and thermal movement. These flexible joints should be continued through the rigid floor finishes as a flexible joint.

Because of a few failures of rigid floor finishes it has become practice to form movement joints whether they are reasonably necessary or not. A principal cause of the failure of rigid floor finishes is the use of cement rich mixes of bedding material or the concrete base. On drying cement shrinks fiercely. The richer the cement mix used, the greater the amount of water necessary in the mix and the greater the shrinkages.

By control of the mix of concrete, screeds and bedding to provide a workable mix using the least amount of cement and water, drying shrinkage may be minimised. There has been a recommendation to form movement joints around the perimeter of floors with an elastic sealant joint, others recommend dividing rigid floor finishes into bays of a variety of areas to be on the 'safe side'.

The considerable disadvantage of these joints is that the joint material is necessarily softer than the surrounding surface, difficult to keep clean and will encourage wear of the edges of the finish next to the joint. Good sense dictates the use of movement joints only where there is sound reason to anticipate gross relative movements.

Wood floor finishes

Natural wood floor finishes such as boards, strips and blocks are used for the advantage of the variety of colour, grain and texture of this natural material which is warm, resilient and comparatively quiet underfoot. The disadvantages of wood finishes is that they are difficult to clean and at the same time maintain their original attractive appearance. In those countries where wood is much used as a floor finish it is not uncommon for the visitor to be firmly invited to change outdoor shoes for slippers to avoid damage to polished wood floors.

Floor boards

Floor boards, described in more detail for upper floors, may be used as floor surface and finish for concrete floors. Either plain edge or tongued and grooved boards are used. The boards are nailed to wood battens set in a screed or to battens secured in floor clips. More usually wood strip flooring is used.

Wood strip flooring

The most pronounced drying shrinkage of wood is across the long grain so that wood boards shrink across their width. As the shrinkage is circumferential to the round section of the log from which the boards are cut, the boards will both shrink across their width and deform out of the flat section. Plainly the wider the board, the greater the loss of width and the greater the loss of shape. The purpose in cutting narrow strips of board is to minimise loss of width and shape that is due to the inevitable drying shrinkage.

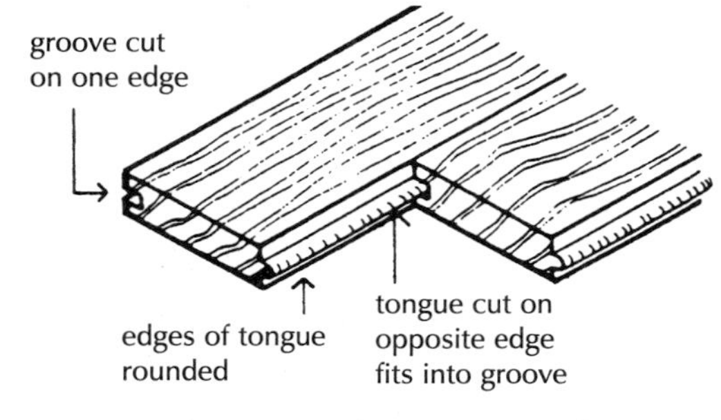

Fig. 135 Tongue and groove strip flooring.

Strips of hardwood or softwood of good quality, specially selected so as to be particularly free of knots, are prepared in widths of 90 mm or less and 19, 21 or 28 mm in thickness. The type of wood chosen is one which is thought to have an attractive natural colour and decorative grain. The edges of the strip are cut so that one edge is grooved and the other edge tongued, so that when they are put together the tongue on one fits tightly into the groove in its neighbour, as illustrated in Fig. 135. The strips are said to be tongued and grooved, usually abbreviated to T & G. The main purpose of the tongue and groove is to cause the strips to interlock so that any slight twisting of one strip is resisted by its neighbour.

There is always some tendency for wood strips to twist out of flat, due to the wood drying out, and to resist this the strips have to be securely nailed to wood battens which are secured to the concrete floor, either by means of galvanised metal floor clips, or in a cement and sand screed. The illustration of part of a concrete floor finished with wood strips nailed to battens, as in Fig. 136, will explain the arrangement of the parts.

The floor clips are of galvanised sheet steel which is cut and stamped to the shape shown in Fig. 136. These are usually set into the concrete or screed whilst it is still wet. They are placed in rows 350 to 450 mm apart and the clips in each row are spaced 450 to 750 mm apart so that when the concrete has hardened, the clips are firmly bedded in it. The strip flooring is usually laid towards the end of building operations and the 50 × 38 mm or 50 × 25 mm softwood battens are wedged up until they are level and the clips are then nailed to them. The strip flooring is then nailed to the battens.

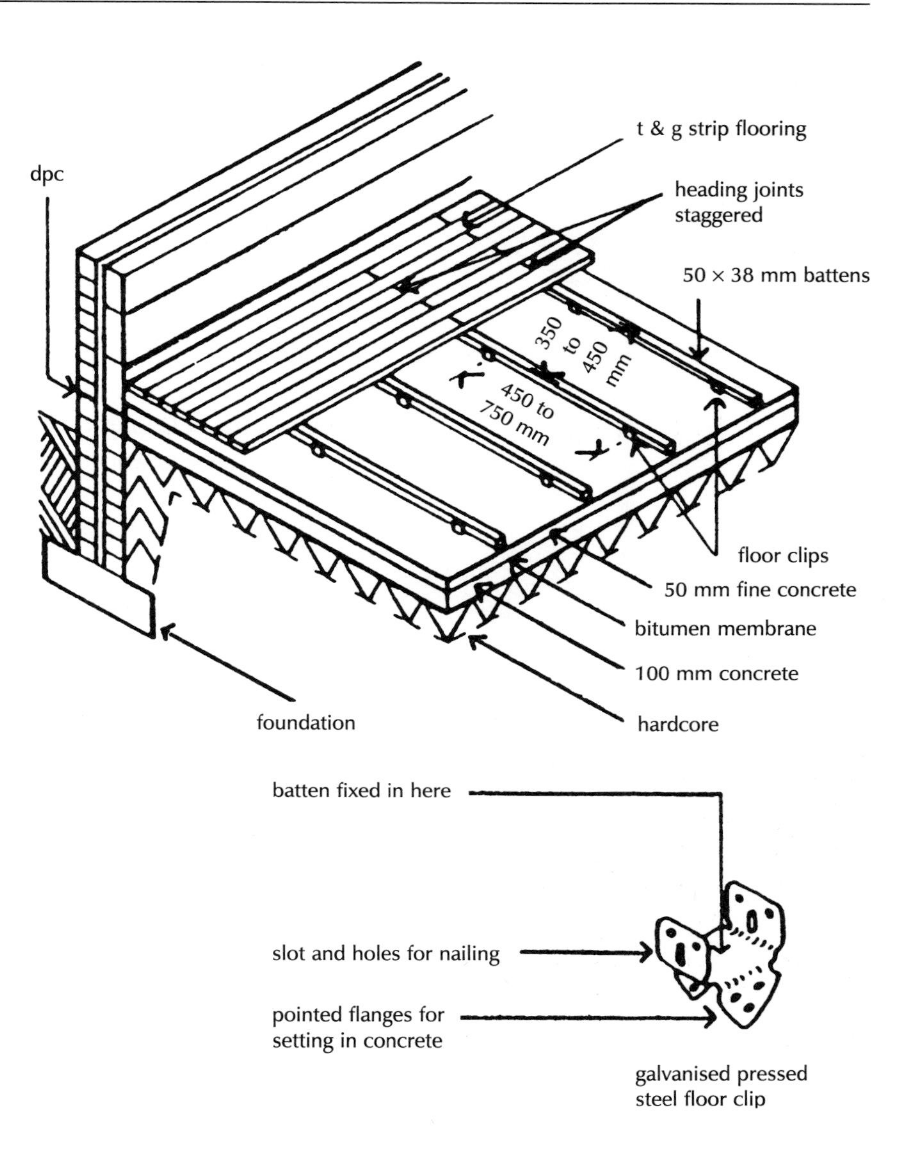

Fig. 136 Strip flooring fixed to battens and clips.

The advantage of the floor clips is that the battens may be wedged up to a true finished level and as the strips are fixed across battens there will be some resilience of the surface to provide the feeling of some springy softness underfoot.

Because wood strip flooring is an expensive, decorative, floor finish the strips of wood are nailed to the battens so that the heads of the nails do not show on the finished surface of the floor. This is termed secret nailing. If the strips have tongued and grooved edges the nails are driven obliquely through the tongues into the battens below so that the groove in the edge of the next board hides the nail. Even though the nails used have small heads they may split the narrow tongue off the edge of the strip and obstruct a close fit of tongues to grooves, so that there may be a poor fixing. To avoid this, the edges of

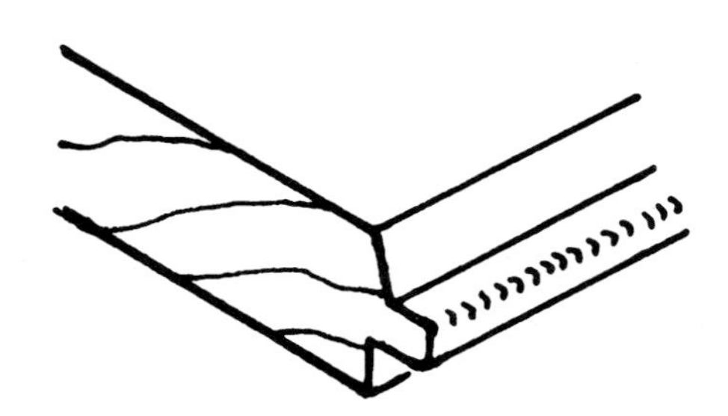

Fig. 137 Splayed tongued and grooved strip.

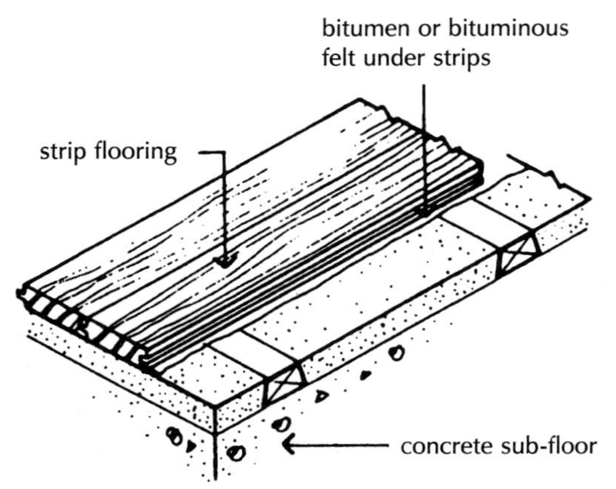

Fig. 138 Strip fixed to battens in screed.

the strips can be cut with splayed tongued and grooved joints as illustrated in Fig. 137.

An alternative method of securing battens to a concrete floor, that has been used, is to bed dovetailed wood battens in a sand and cement screed spread and levelled on a concrete base, as illustrated in Fig. 138. The disadvantages of this method of fixing are that it adds considerably to the labour of laying a wet screed. The moisture from the wet screed will cause the battens to expand and then shrink as the screed dries out. Even though the screed may make good adhesion to the concrete there is some possibility that deformation of the strips may lift a batten and the screed around it. This method of fixing has been largely abandoned.

Of recent years wood strip flooring has been fixed by the thin bed adhesive method. Comparatively short lengths of wood strip, 300 mm long, with tongued and grooved edges and joints, are used to minimise drying deformation. The strips are bedded on an epoxy resin adhesive spread over a true level screed on to which the strips are laid and pressed or rolled to make sound, adhesive contact. The strips are usually laid with staggered end joints.

This is a perfectly satisfactory method of laying wood strip flooring as the narrow width and short length of strip is unlikely to suffer drying deformation likely to tear strips away from the adhesive bed.

If timber is in contact with a damp surface it may rot and it is important to protect both the battens and the strip flooring from damp which may rise from or through the concrete subfloor. The battens should be impregnated with a preservative before they are fixed and either the surface of the concrete subfloor should be covered with a coat of bitumen or a waterproof membrane should be used in or under the concrete oversite.

Wood strip flooring is used as an expensive, decorative floor finish which is used where wear is light, as in households where the considerable care and labour required to maintain the colour and texture of the material is accepted.

After the finish is laid it is usually sanded to remove the thin top surface which is initially and subsequently polished with wax. Over the years a thin, hard wax finish is developed. This thin, non-gloss surface enhances the colour and grain of the wood. The surface is cleaned by dry mopping or dusting and polishing. To minimise labour it is not uncommon to apply a silicone seal to the surface which can then be mopped with a damp cloth or mop to remove dust. The seal provides a semi-gloss finish that obscures the colour and texture of the wood.

Wood block floor finish

Blocks of wood are used as a floor finish where resistance to heavy wear is required, as in halls, corridors and schools, to provide a surface which is moderately resilient, warm and quiet underfoot. An advantage of the comparatively thick blocks is that after wear the top surface may be sanded to reduce the block to a level surface. The word 'sanded' describes the operation of running a power driven sanding machine over the surface. The sanding machine has a rotating plate surfaced with carborundum or sand paper which removes the top surface.

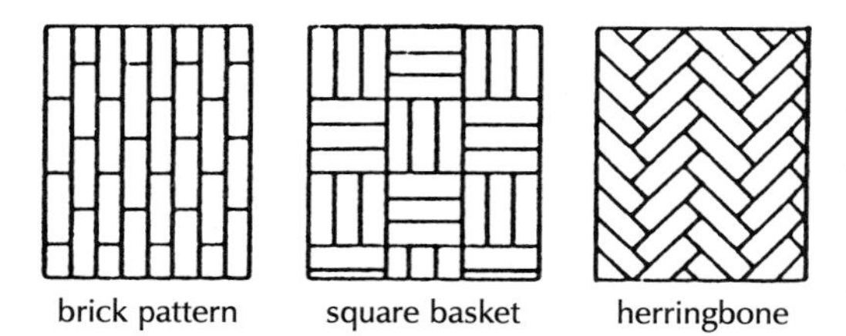

Fig. 139 Wood blocks patterns.

Blocks of some softwood or hardwood with good resistance to wear are cut. The blocks are usually 229 to 305 mm long by 75 mm wide by from 21 to 40 mm thick. The blocks are laid on the floor in a bonded, herringbone or basket weave pattern. The usual patterns are illustrated in Fig. 139. Moisture movement across the long grain of the blocks is balanced by alternating and cross laid blocks.

Wood blocks are laid on a thoroughly dry, clean, level cement and sand screeded surface which has been finished with a wood float to leave its surface rough textured. The traditional method of laying blocks is to spread a thin layer of hot bitumen over the surface of the screed into which the blocks are pressed. The lower edges of the blocks of wood are usually cut with a half dovetail incision so that when the blocks are pressed into the bitumen some bitumen squeezes up and fills these dovetail cuts and so assists in binding the blocks to the bitumen, as illustrated in Fig. 140.

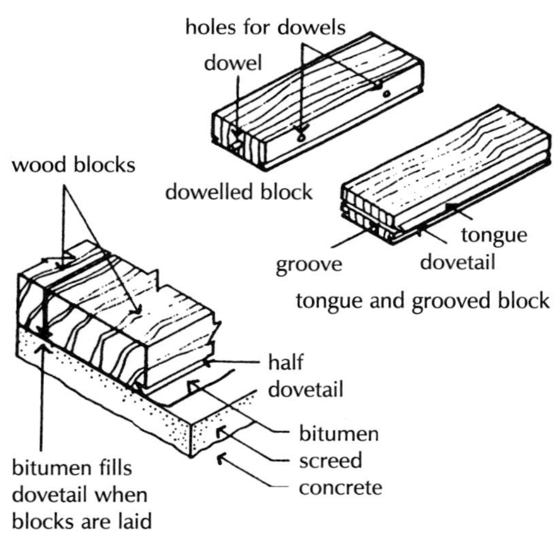

Fig. 140 Joints for wood blocks.

If the wood blocks have been thoroughly seasoned (dried) and they are firmly pressed into the bitumen they will usually be securely fixed to the floor. It is possible, however, that one or more blocks may not be firmly fixed and will come up. To prevent this happening good quality wood blocks 25 mm thick and over have either tongues and grooves cut on their edges or wood dowels to joint them, as illustrated in Fig. 140.

After the surface has been sanded to provide a level finish a seal is applied to provide an easily cleaned finish. Either a wax polish is used, which requires effort to maintain, or a polyurethane seal for ease of cleaning.

SUSPENDED TIMBER GROUND FLOORS

Many houses built in this country from about 1820 up to about 1939 were constructed with timber ground floor raised 300 mm or more above the packed earth, brick rubble or site concrete below. The purpose of raising the ground floor was to have the surface of the ground floor living rooms sufficiently above ground level to prevent them being cold and damp in winter. At that time imported softwood timber was cheap and this ground floor construction was both economical and satisfactory.

Since the end of the Second World War (1945), imported softwood timber has been expensive and for some years after the war its use was restricted by government regulations. Most ground floors today are formed directly off the site concrete and this is covered with one of the surface finishes described previously.

A suspended or raised timber ground floor is constructed as a timber platform of boards nailed across timber joists bearing on $\frac{1}{2}$B brick walls raised directly off the packed earth, brick rubble or site concrete, as illustrated in Fig. 141. The raised timber floor is formed inside the external walls and internal brickwork partitions, and is supported on brick sleeper walls.

Sleeper walls are $\frac{1}{2}$B thick and built directly off the site concrete up to 1.8 m apart. These sleeper walls are generally built at least three courses of bricks high and sometimes as much as 600 mm high. The walls are built honeycombed to allow free circulation of air below the floor, the holes in the wall being $\frac{1}{2}$B wide by 65 mm deep, as illustrated in Fig. 142.

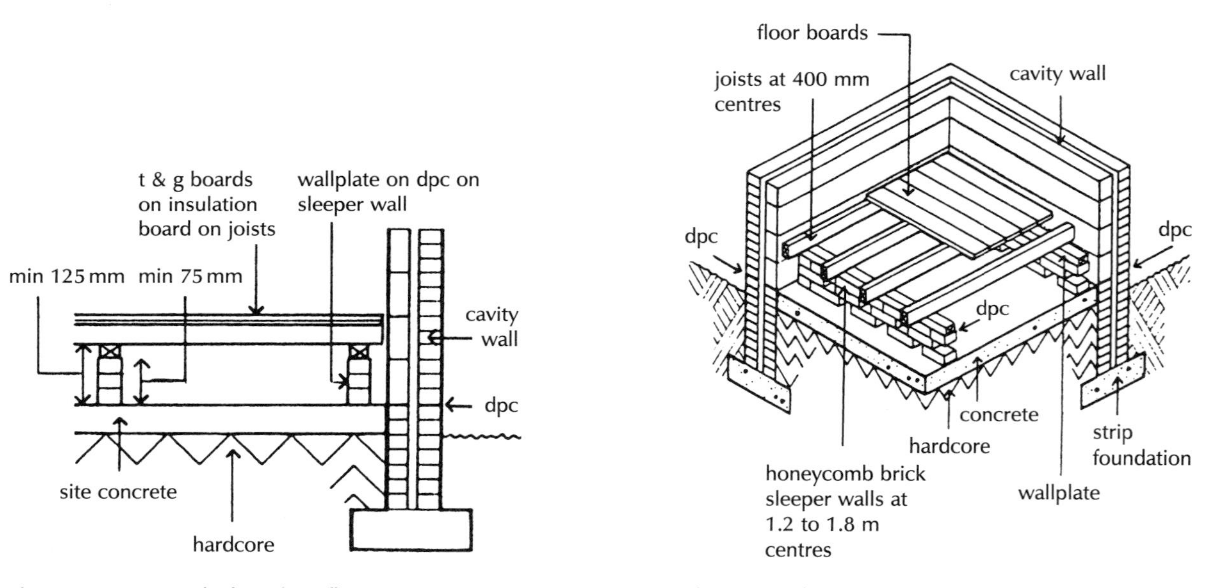

Fig. 141 Suspended timber floor.

Fig. 142 Sleeper walls, joists and boards.

The practical guidance in Approved Document C to the Building Regulations requires a space of at least 75 mm from the top of the concrete to the underside of a wall plate and at least 150 mm to the underside of the floor joists, concrete oversite at least 100 mm thick on a hardcore bed or concrete at least 50 mm thick on a dpm of 1200 gauge polythene on a bed which will not damage it. Underfloor ventilation should have a free path between opposite sides, with openings equivalent to 1500 mm^2 for each metre run of wall.

Wall plate

A wall plate is a continuous length of softwood timber which is bedded in mortar on a dpc. The wall plate is bedded so that its top surface is level along its length and it is also level with the top of wall plates on the other sleeper walls.

A wall plate is usually a 100 × 75 mm timber and is laid on one 100 mm face so that there is a 100 mm surface width on which the timber joists bear. The function of a wall plate for timber joists is two-fold. It forms a firm level surface on which the timber joists can bear and to which they can be nailed and it spreads the point load from joists uniformly along the length of the wall below.

Floor joists

Floor joists are rectangular sections of sawn softwood timber laid with their long sectional axis vertical and laid parallel spaced from 400 to 600 mm apart. Floor joists are from 38 to 75 mm thick and from 75 to 225 m deep. The span of a joist is the distance measured along its length between walls that support it. The sleeper walls built to support the joists are usually 1.8 m apart or less, and the span of the joists in this type of floor is therefore 1.8 m or less.

Timber, chipboard or plywood boards are laid across the joists and they are nailed to them to form a firm, level floor surface.

From a calculation of the dead and imposed loads on the floor the most economical size and spacing of joists can be selected from the tables in Approved Document A to the Building Regulations and from this the spacing of the sleeper walls to support the joists.

Similarly the thickness of the floor boards to be used will determine the spacing of the joists, the thicker the board the greater the spacing of the joists.

Floor boards

Any length of timber 100 mm or more wide and under 50 mm thick is called a board. Floor boards for timber floors are usually, 16, 19, 21 or 28 mm thick and 65, 90, 113 or 137 mm wide and in length up to about 5.0 m. The boards are cut from whitewood, which is moderately cheap, or from redwood which is more expensive but which provides better wear. The edges of the boards may be cut square, or plain edged, which is the cheapest way of cutting them. But as these boards shrink, ugly cracks may appear between them.

The usual way of cutting boards is with a projecting tongue on one edge and a groove on the opposite edge of each board, as in Fig. 135. The boards are then said to be tongued and grooved, abbreviated to T & G. The boards are laid across the floor joists and cramped together. Cramping describes the operation of forcing the edges of the boards tightly together so that the tongues fit firmly into the grooves and there are no open cracks between the boards. The boards, as they are cramped up, are nailed to the joists with two nails to each board bearing on each joist.

Heading joints

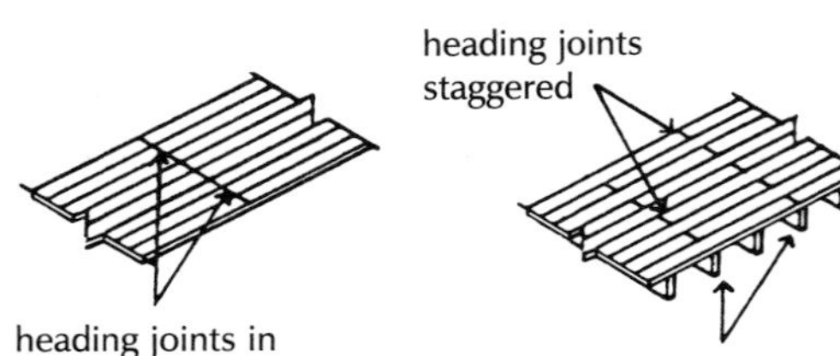

Fig. 143 Heading joints.

Floors of small rooms can often be covered with boards sufficiently long to run in one length from wall to wall, but in most rooms the ends of boards have to be cut to butt together. The joint between the end of one board and the end of another is described as the heading joint. The appearance of a boarded floor is spoiled if the heading joints run in a continuous line across the floor because the cut ends of the boards tend to be somewhat ragged and the continuous joint looks ragged and ugly (Fig. 143). The heading joints in floor boards should always be staggered in some regular manner. Obviously the heading joint ends of boards must be cut so that the ends of both boards rest on a joist to which the ends are nailed. A usual method of staggering heading joints is illustrated in Fig. 143.

End matched flooring

Hardwood strips are often prepared with tongues and grooves on the ends of the strips so that their ends firmly interlock and do not have to lie over a joist. The strips of flooring are said to be end matched. The end joint so formed provides a neater finish than a sawn end of ordinary boarding.

End support of floor joists

The floor joists of a raised timber ground floor bear on wall plates on sleeper walls and the best method of supporting the ends of the joists at external walls and at internal brick partitions is to build a honeycombed sleeper wall some 50 mm away from loadbearing walls to carry the ends of the joists , as illustrated in Fig. 142. The sleeper wall is built away from the main wall to allow air to circulate through the holes in the honeycomb of the sleeper wall. The ends of the joists are cut so that they are clear of the inside face of the wall by 50 mm.

Damp-proof course

A dpc should be spread and bedded on top of the sleeper walls under the wall plate to prevent any moisture rising through site concrete without a dpm and through sleeper walls to the timber floor. Any of the materials described in Chapter 1 may be used for this purpose.

Ventilation

The space below this type of floor is usually ventilated by forming ventilation gratings or bricks in the external walls below the floor so that air from outside the building can circulate at all times under the floor. The usual practice is to build air bricks into the external walls. An air brick is a special brick made of terra cotta (meaning earth burned) with several square or round holes in it and its size is 215 × 65, 215 × 140 or 215 × 215 mm (Fig. 144). These bricks are built into external and internal walls for each floor to provide 1500 mm of ventilation for each metre run of wall. The bricks are built in just above ground level and below the floor, as illustrated in Fig. 144.

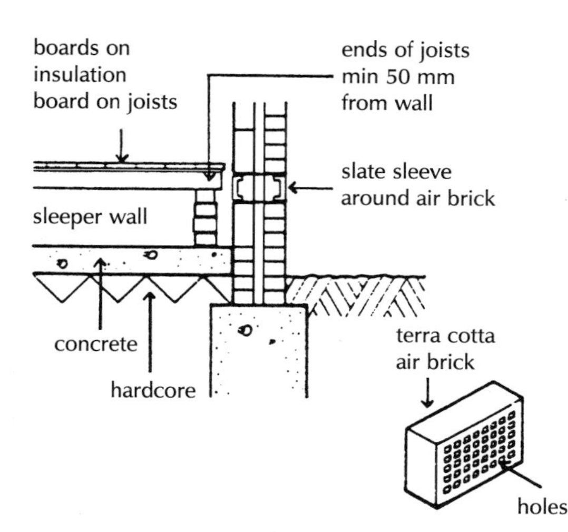

Fig. 144 Ventilation of raised timber floor.

The purpose of the air bricks is to cause air to circulate under the floor and so avoid stagnant damp air which might induce fungus to grow. The disadvantage of ventilating the space below this type of floor is that in winter the floor is liable to be cold. It is usual, therefore, to fix an insulating board or quilt under the floor boards to minimise transfer of heat to the ventilated spaces below the floor.

It is common practice today to prevent cold air from outside entering the cavity of a cavity wall to avoid the inner skin becoming cold. When ventilating air bricks are built into a cavity wall some means must be devised of preventing cold air getting into the cavity through the air brick. One common method of doing this is to build a roofing slate sleeve around the air bricks and across the cavity, as illustrated in Fig. 144, or a short length of pipe. A duct is made out of four pieces of slate which are built into the wall around the air bricks. Providing mortar droppings do not accumulate on top of the slate sleeve the slate will not convey water from the outer to the inner skin of the wall.

Thermal insulation

To meet the requirements of the Building Regulations for resistance to heat transfer through ground floors it may be necessary to insulate suspended timber ground floors that have a ventilated space below. The most practical way of insulating a suspended timber ground floor is to fix mineral wool roll, mat or quilt or semi-rigid slabs between the joists. Rolls or quilt of loosely felted glass fibre or rockwool are supported by a mesh of plastic that is draped over the joists and stapled in position to support the insulation. Semi-rigid slabs or batts of fibre glass or rockwool are supported between the joists by nails or battens of wood nailed to the sides of the joists.

Raised concrete ground floors

As an alternative to timber a raised ground floor may be constructed in concrete. It is not uncommon today to construct a raised concrete ground floor for the advantage of a solid floor surface raised above the ground level. This raised floor will not require protection against the possibility of damp causing rot in timber and provides the sense of being elevated above the surrounding ground, particularly in bungalows.

A concrete ground slab at least 100 mm thick is cast on the ground, after vegetable top soil has been stripped, with a damp proof membrane formed below the slab where the ground below the floor has been excavated below the surrounding ground level.

The space between the top of the concrete ground slab and the underside of the raised concrete is determined by the need to raise the floor level above the dpc in the walls. Where there is a risk of an accumulation of gas in the space, which might lead to an explosion, then the space between ground slab and underside of floor should be at least 150 mm and the space should be ventilated with openings at least 1500 mm^2 for each metre run of wall to provide cross ventilation.

Usual practice is to build a brick or concrete block sleeper walls on the ground slab to support the raised floor. The floor is constructed with one of the precast inverted 'T' beam and hollow concrete infill blocks described for use with upper floors. Spacing between sleeper walls is dictated by an economical span for the inverted 'T' beams. Concrete topping is spread and levelled over the precast concrete units.

UPPER FLOORS

Timber floors

Timber upper floors for houses and flats are about half the cost of comparable reinforced concrete floors. For upper floors of offices, factories and public buildings timber floors are not much used today because the resistance to fire of a timber floor, plastered on the underside, is not sufficient to comply with building regulations for all but small buildings. Concrete floors are used instead because of their better resistance to fire, and better resistance to sound transmission.

Strength and stability

Floor joists

A timber floor is framed, or carcassed, with sawn softwood timber joists which are usually 38 to 75 mm thick and 75 to 235 mm deep. The required depth of joists depends on the dead and imposed loads and the span. The spacing of the joists is usually 400, 450 or 600 mm measured from the centre of one joist to the next. Tables in Approved Document A to the Building Regulations set out the required size of timber joists for given spans and two strength classes, with given spacing of joists for various loads for single family dwellings of up to three storeys.

Where rigid plasterboard is used as the soffit (ceiling) of timber floors it is practice to use regularised joists. The depth of sawn softwood timbers may vary to the extent that where the joists are framed with their top surface level for floorboards the underside may be so out of level that plasterboard fixed to it will be noticeably out of level. The process of regularising sawn timber joists is carried out to produce joists of regular depth.

To economise in the use of timber, the floor joists of upper floors usually span (are laid across) the least width of rooms, from external walls to internal loadbearing partitions. The joists in each room span the least width. The maximum economical span for timber joists is between 3.6 and 4.0 m. For spans greater than 4.0 m it is economic to reduce the span of the joists by the use of steel beams.

Double floors

Where the span of a timber floor is greater than the commercially available length of timber and where, for example, joists span parallel to a cross wall, it is convenient and economic to use a steel beam or

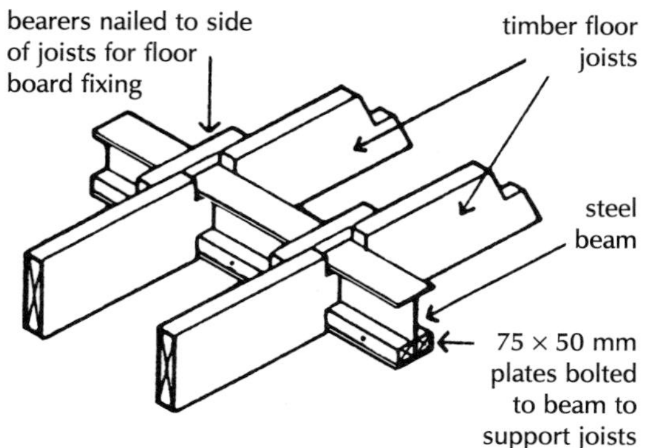

Fig. 145 Double floor.

timber beam to provide intermediate support for timber joists. This combination of beam and the joists is described as a double floor. Steel beams are generally used because of their small section.

The supporting steel beam may be fixed under the joists or wholly or partly hidden in the depth of the floor. To provide a fixing for the ends of the joists, timber plates are bolted to the bottom flange of the beam and the ends of the joists are scribed (shaped) to fit into the joist over the plates to which they are nailed. To provide a fixing for floor boards timber bearers are nailed to the sides of joists across the supporting steel beam, as illustrated in Fig. 145.

The ends of the supporting steel beam are built into loadbearing walls and bedded on a pad stone to spread the load along the wall.

Where the supporting steel beam projects below the ceiling it is cased in plasterboard as fire protection.

Strutting between joists

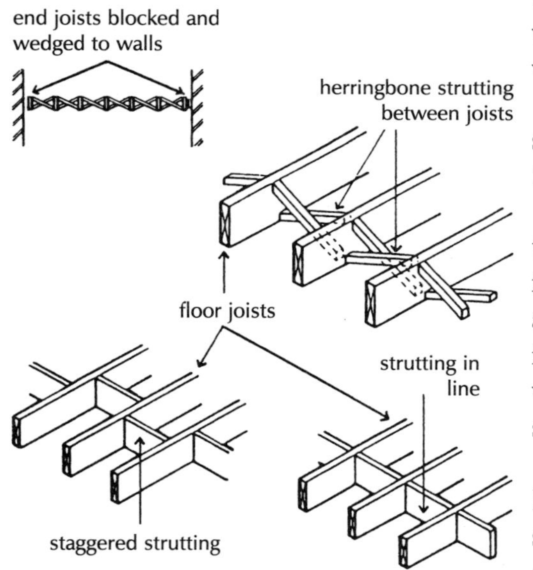

Fig. 146 Strutting between joists.

When timber is seasoned it shrinks, and timber such as in floor joists, which is not cut on the radius of the circle of the log, does not shrink uniformly. The shrinkage will tend to make the floor joists twist, or wind out of the vertical. To maintain joists in the vertical position in which they were initially fixed, timber strutting is used.

The type of strutting most used is that known as herringbone strutting. This consists of short lengths of softwood timber about 50 × 38 mm nailed between the joists, as illustrated in Fig. 146.

Each strut is cut with oblique faced ends to bear between the top and bottom edges of adjacent joists. A second system of struts is fixed across the first, as illustrated in Fig. 146. As the struts are nailed between the joists they tend to spread and secure the joists in an upright position. To provide rigid strutting between walls, wedges are fixed between the joists and walls at both ends of the strutting.

The recommendation in Advisory Document A to the Building Regulations is that joists which span less than 2.5 m do not require strutting, those that span from 2.5 to 4.5 m require one row of struts at mid-span and those more than 4.5 m span require two rows of struts spaced one-third of the span.

As an alternative to herringbone strutting a system of solid strutting may be used. This consists of short lengths of timber of the same section as the joist which are nailed between the joists either in line or staggered, as in Fig. 146. This is not usually so effective a system of strutting as the herringbone system, because unless the short solid lengths are cut very accurately to fit the sides of the joists they do not firmly strut between the joists.

As with herringbone strutting the end joists are blocked and wedged up to the surrounding walls.

End support for floor joists

For stability, the end of floor joists must have adequate support from walls or beams. If the floor is to be durable, timber joists should not be built into external walls where their ends may be persistently damp and suffer decay. Timber joists should not be built into or across separating or compartment walls where they may encourage spread of fire. Floor joists are, therefore, either built into internal and external walls or they are supported in hangers, or corbels projecting from the face of walls.

Timber floor joists that are built into walls may bear on a wall plate of timber or metal, which serves to spread the load from the floor along the length of the walls and as a level bed on which the joists bear.

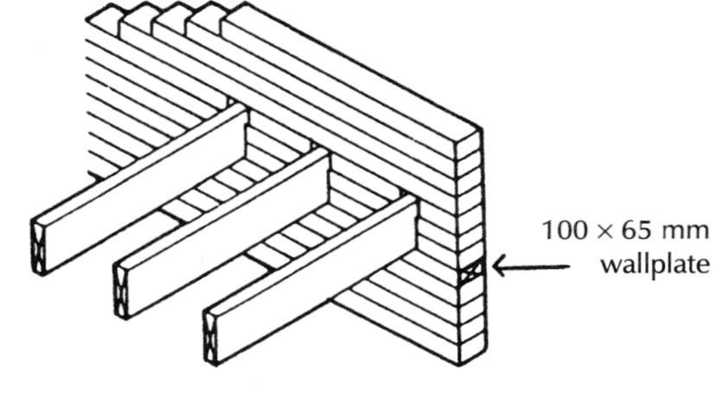

Fig. 147 Joists built into wall.

Timber wall plates are of sawn softwood, 100 × 65 mm, to course into brickwork, and laid with one 100 mm face horizontal. The wall plate is bedded level in mortar to take the ends of the joists which are nailed in position to the timber plate and the wall is then raised between and above the floor, as illustrated in Fig. 147, which illustrates joists built into an internal loadbearing brick wall.

A timber wall plate generally has sufficient compressive strength to support the loads from a wall above. For greater loads a steel wall plate may be used or the joist fixed to bear directly on the wall. Where joists both sides of an internal wall bear on the wall the joist ends of one side bear alongside those on the other side.

A disadvantage of building in the ends of floor joists is that there is a deal of wasteful cutting of bricks or blocks around joist ends, as illustrated in Fig. 148. Cutting around joist ends is more straightforward with the small units of brick than around larger blocks.

Timber joists built into the inner skin of a cavity wall must not project into the cavity and it may be wise to treat the ends of the joists with a preservative against the possibility of decay due to moisture penetration. The ends of joists built into a cavity wall may bear on a timber wall plate, which may also be treated with a preservative. The wall plate is bedded on the blockwork inner skin. As an alternative, a mild steel bar of 75 × 6 mm may be used. This metal wall plate is tarred and sanded and bedded level in mortar, and the joist ends bear on the plate, as illustrated in Fig. 148.

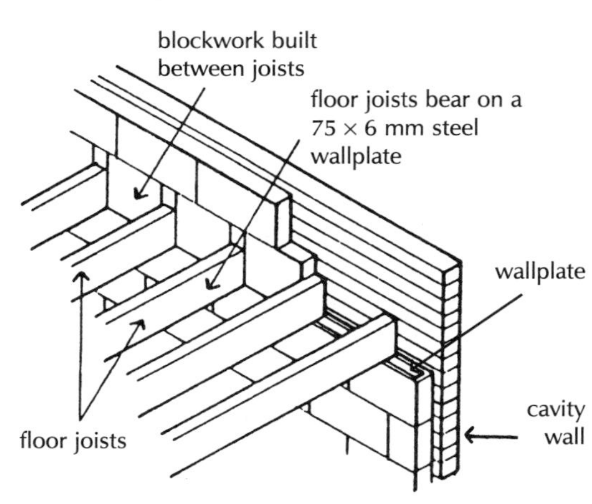

Fig. 148 Joist ends built into cavity wall.

Instead of using a timber or a metal wall plate the joists may bear directly on the brick or block wall with tile or slate slips in mortar packed under each joist end to level the joists. This is a somewhat laborious procedure and the slips may be displaced and the joists move out of level during subsequent building operations.

As an alternative to building in the ends of timber joists, joist hangers are used. Galvanised, pressed steel joist hangers are made with straps for building into horizontal courses and a stirrup to support a joist end, as illustrated in Fig. 149. The joist hangers are built into horizontal brick or blockwork as walls are raised and the joists fitted and levelled later or the joists, with the hangers nailed to their ends, are given temporary support as the brick or blockwork is raised and the hangers are built into horizontal courses.

The advantage of joist hangers is that joist ends are not exposed to possible damp, and there is no need for cutting brick or block to fit around joists.

Before the use of cavity external walls became common and today for buildings such as sheds and stores where a cavity wall is considered unnecessary, the ends of timber joists are supported by corbel brackets, or on brick courses corbelled out for the purpose as illustrated in Fig. 150. The purpose of the projecting corbel arrangement is to avoid building in the ends of timber joists to solid external walls where damp penetration might well cause rot.

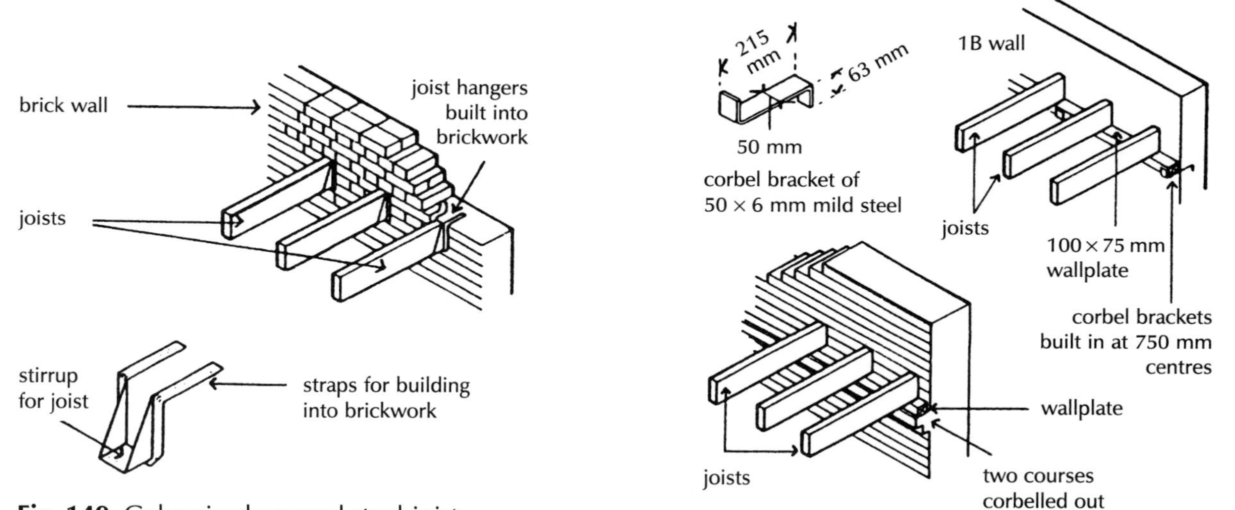

Fig. 149 Galvanised pressed steel joist hanger.

Fig. 150 Corbels to support joists.

The forged steel corbel brackets are usually coated in bitumen or tar and sanded and are built in at 750 mm centres as support for a timber wall plate to which the joists are nailed.

Two courses of brick are built to project $\frac{1}{4}$B from the face to provide a $\frac{1}{2}$B projecting corbel to support a wall plate to which the joists are nailed.

The projection of the corbel support is of no consequence in sheds and stores where walls are not decorated. When, in the past, corbels were used it was practice to disguise them with plaster cornices as part of the plaster finish to buildings.

Lateral restraint for walls

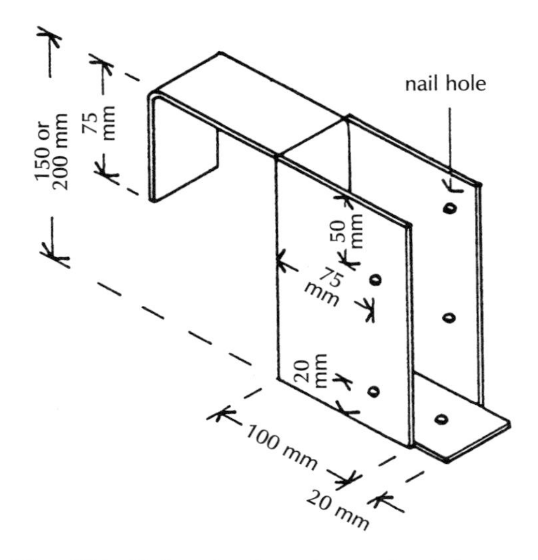

Fig. 151 Galvanised steel restraint joist hanger.

To provide lateral support to walls by floors Approved Document A to the Building Regulations recommends the use of straps or joist hangers to provide lateral support for walls at each storey floor level above ground to transfer lateral forces on walls, such as wind, to floors.

Lateral support is required to any external, compartment or separating wall longer than 3 m at every floor, roof and wall junction and any internal loadbearing wall, not being a compartment or separating wall, of any length at the top of each storey and roof.

Walls should be strapped to floors above ground level at intervals of not more than 2 m by the 30 × 5 mm straps illustrated in Fig. 67, shown in Chapter 2.

Straps are not required in the longitudinal direction of joists in houses of not more than two storeys if the joists are at no more than 1.2 m centres and where the joists are supported by restraint type joist hangers, illustrated in Fig. 151, at not more than 2 m centres.

Notches and holes

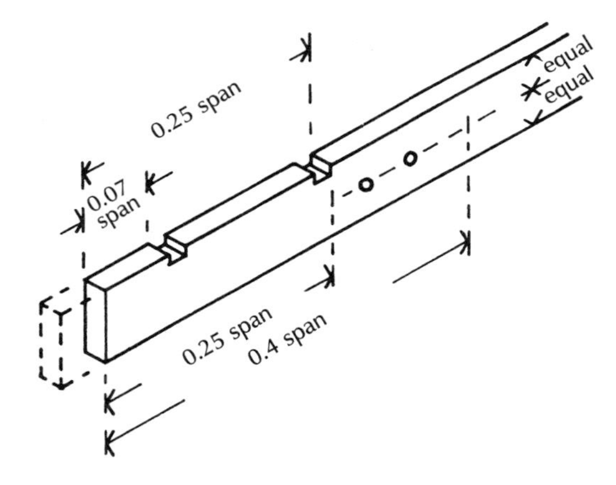

Fig. 152 Notches and holes in timber joists.

So that notches and holes cut in timber joists for electric cables and water and gas service pipes run in the floor do not seriously weaken the strength of the floor, limitations of size are given as practical guidance to meeting the requirements of the Building Regulations for small domestic buildings.

Notches should be no deeper than 0.125 times the depth of a joist and cut no closer to the support than 0.07 of the span nor further away than 0.25 times the span.

Holes should be of no greater diameter than 0.25 the depth of joist, drilled on a neutral axis and not less than 2 diameters apart (centre to centre) and located between 0.25 and 0.4 times the span from the support, as illustrated in Fig. 152.

Floor boards

The surface of a timber framed upper floor is formed by nailing tongued and grooved (T & G) softwood boards across the joists as a finished floor surface or, more usually, as a base for carpet or one of the plastic sheet or tile finishes. The boards are cramped, nailed and laid with staggered heading joints as described for raised timber ground floors.

The recommendation in Approved Document A to the Building Regulations is T & G boards nailed to joists spaced at up to 500 mm should be at least 16 mm finished thickness and at wider spacings up to 600 mm, 19 mm finished thickness.

Manmade boards of compressed wood chips, chipboard, are commonly used today as a substitute for T & G boards as a base for carpets or one of the sheet or tile finishes. The use of large tongued and grooved chipboards minimises joints. The boards are nailed or screwed to joists.

Fire safety

Structural floors of dwelling houses of two or three storeys are required to have a minimum period of fire resistance of half an hour. Timber floors with tongued and grooved boards or sheets of plywood or chipboard at least 15 mm thick, joists at least 37 mm wide and a ceiling of 12.5 mm plasterboard with 5 mm neat gypsum plaster finish or at least 21 mm thick tongued and grooved boards or sheets of plywood or chipboard on joists at least 37 mm wide with a ceiling of 12.5 mm plasterboard with joints taped and filled, will both have a resistance to fire of half an hour.

Resistance to the passage of heat

Timber upper floors that are exposed to outside air, such as floors over car ports, have to be insulated against heat transfer to meet the requirements of the Building Regulations. The maximum U value of exposed floors has to be 0.45 W/m^2K, the same as external walls. Insulation between joists in the form of low density glass fibre or rockwool rolls, mats or quilts can be laid on to ceiling finish or semi-rigid batts, slabs or boards of fibre glass or rockwool which are friction fitted between joists and supported by nails or wood battens nailed to the sides of joists. To avoid cold bridges the insulation must extend right across the floor in both directions up to surrounding walls.

Resistance to the passage of sound

A boarded timber floor with a rigid plasterboard ceiling affords poor resistance to the transmission of sound.

Airborne sound

Sound is transmitted through a floor by vibrations of air from the source of sound, such as a loudspeaker, which spread out and set up vibrations in the floor which in turn set up vibrations in the air on the opposite side of the floor. This is sometimes described as transmission of 'airborne sound'.

Impact sound

The other source of sound is caused by impact on a hard surface, such as footsteps on a boarded floor. The footsteps cause the floor structure to vibrate. Vibration of the ceiling below in turn causes vibration of air in the form of what is sometimes called 'impact sound'.

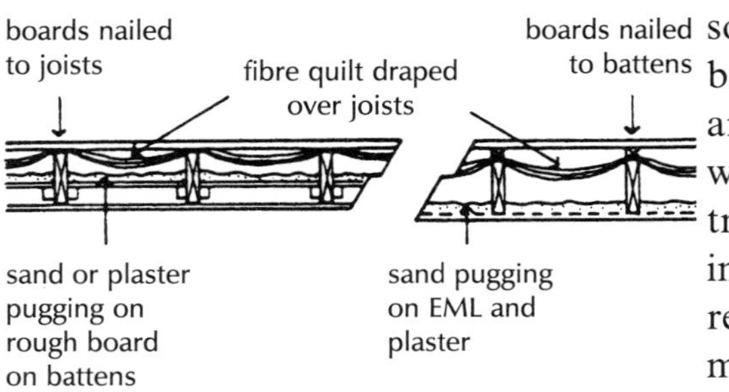

Fig. 153 Sound insulation of timber floor.

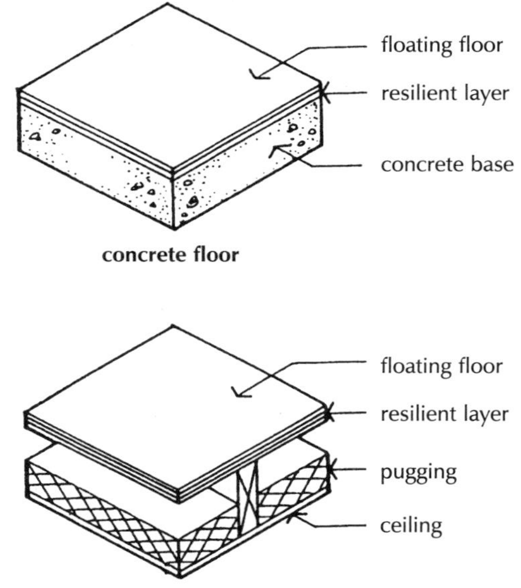

Fig. 154 Sound insulation of floors.

To reduce the transmission of airborne sound it is necessary to increase the mass of a floor to restrict the flow of energy through it. To reduce the transmission of impact sound it is necessary to provide some soft material, such as a carpet, between the cause of the impact and the hard surface.

The traditional method of insulating timber floors against sound was to spread a layer of plaster or sand on rough boarding fixed between the joists or sand on expanded metal lath and plaster, as illustrated in Fig. 153. The layer of plaster or sand was termed pugging. Pugging is effective in reducing the transmission of airborne sound but has little effect in deadening impact sound. To deaden impact sound it is necessary to lay some resilient material on the floor surface or between the floor surface material and the structural floor timbers. The combination of a resilient layer under the floor surface and pugging between joists will effect appreciable reduction of impact and airborne sound. Where the floor boards are nailed directly to joists, through the resilient layer, as illustrated in Fig. 153, much of the impact sound deadening effect is lost.

Approved Document E, giving practical guidance to the requirements of the Building Regulations for resistance to the passage of sound, includes specifications for the construction of concrete and timber floors.

For concrete floors the resistance to airborne sound depends mainly on the mass of the concrete. Resistance to impact sound may be provided by a soft covering of carpet. As an alternative the resistance to airborne sound is provided mainly by the mass of the concrete floor and partly by a floating top layer. The top layer consists of a platform of T and G boards or T & G chipboard nailed to timber battens that are laid on a 13 mm thickness of resilient material as indicated in Fig. 154, with the edges of the resilient layer turned up around the edges of the floating top layer. This floating layer is laid loose on the concrete base as a surface for one of the plastic sheet or tile finishes.

For timber floors pugging is placed or fixed between the joists for resistance to airborne sound and a floating floor as resistance to impact sound and to some extent resistance to airborne sound, as indicated in Fig. 155.

This platform floor is constructed as a floor of 18 mm thick T & G boards or chipboard with all joints glued. The boarded platform is spot bonded to a base of 19 mm thick plasterboard. The boarded platform is laid on a 25 mm thick layer of resilient mineral fibre on a floor base of 12 mm thick boarding or chipboard nailed to the joints, as illustrated in Fig. 155.

The ceiling is two layers of plasterboard, with joints staggered, to a

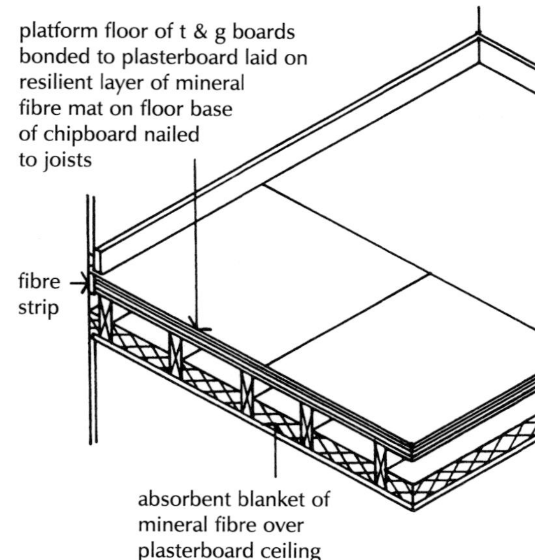

Fig. 155 Platform floor.

finished thickness of 30 mm on which 100 mm of absorbent mineral fibre is laid.

To provide a level floor surface for platform floors it is necessary to fix the floor boards by nailing to battens or by fixing the boards to a firm level base of plasterboard or similar material, so that the boards can be cramped together. The boards are bonded to the plasterboard base with strips or pads of adhesive to keep them flat, as the tongues in the edges of the boards are cramped up into the grooves of adjacent boards to produce a level floor finish. In this operation it is plainly simpler to cramp up the joints between large chipboards than the much less wide timber boards, which are only used when they are to be the finished floor surface.

To minimise flanking transmission of sound from the floor surface to the surrounding walls, a strip of resilient fibre material is fixed between the edges of platform and ribbed floors and surrounding solid walls, and a gap of at least 3 mm is left between skirtings and floating floors. To limit the transmission of airborne sound through gaps in the construction it is important to seal all gaps at junctions of wall and ceiling finishes and to avoid or seal breaks in the floor around service pipes.

The materials used as resilient layer in platform floors and as strips under ribbed floors are fibre glass or rockwool in rolls or mat and the materials used as absorbent pugging between joists are fibre glass or rockwool in the form of rolls or semi-rigid slabs.

The construction of the platform and ribbed floors detailed in Approved Document E is so laborious and costly that good sense would dictate the use of a reinforced concrete floor instead of timber. The concrete floor would be little if any more expensive than timber and have the added advantage of better resistance to fire.

REINFORCED CONCRETE UPPER FLOORS

Reinforced concrete floors have a better resistance to damage by fire and can safely support greater superimposed loads than timber floors of similar depth. The resistance to fire, required by building regulations for most offices, large blocks of flats, factories and public buildings, is greater than can be obtained with a timber upper floor so some form of reinforced concrete floor has to be used.

The types of reinforced concrete floor that are used for small buildings are self-centering 'T' beams and infill blocks, hollow beams and monolithic in situ cast floors. The word centering is used to describe the temporary platform on which in situ cast concrete floors are constructed and supported until the concrete has sufficient strength to be self-supporting. The term self-centering is used to define those precast concrete floor systems that require no temporary support.

Precast 'T' beam and infill block floor

This type of reinforced concrete floor is much used for comparatively small spans and loads. The great advantage of this floor system is that the small units can be handled by two men without the need for lifting gear.

Solid reinforced concrete beams usually shaped like an inverted T in section are precast in the manufacturer's yard to the required length. The depth of the beams is from 130 to 250 mm and 20 mm wide at the bottom. The beams are made in lengths of up to 6 m. The 'T' beams are reinforced with mild steel reinforcing bars to provide adequate support for the dead weight of the floor and anticipated dead and live loads.

Hollow precast lightweight concrete infill blocks are made to fit between and bear on the 'T' beams. These blocks are hollow and made for lightness in handling and to minimise the weight of the floor.

The beams are placed at 270 mm centres with their ends built into walls or bearing on beams of at least 90 mm. The blocks are placed in position and the floor completed with a layer of constructional concrete topping, 50 mm thick spread and levelled ready for a screed or power floated finish as illustrated in Fig. 156. The purpose of the

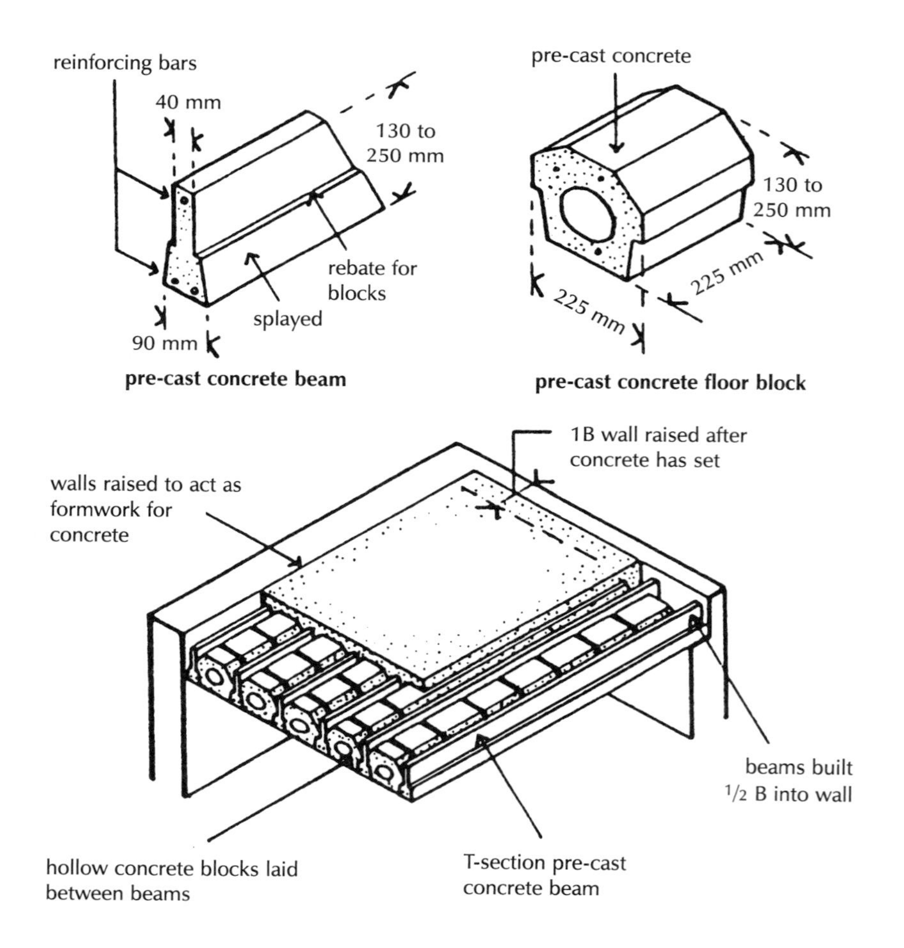

Fig. 156 Pre-cast concrete beam and block floor.

constructional concrete topping is to spread the loads on the floor over the blocks and beams.

The underside or soffit of the floor is covered with plaster or will provide support for a suspended ceiling.

This comparatively cheap floor system provides reasonable resistance to airborne sound and resistance to fire. This floor system is particularly suited for use as a raised (suspended) ground floor.

Hollow beam floor units

Hollow, reinforced concrete beams are precast around inflatable formers to produce the hollow cross section. The beams are rectangular in section with the steel reinforcement cast in the lower angles of the beam. The sides of the beams are indented to provide a key for the concrete topping, as illustrated in Fig. 157.

These beams are usually 355 mm wide, from 130 to 205 mm deep and up to 6 m long. The depth of the beam depends on the superimposed loads and the span.

Because of their length and weight, lifting gear is necessary to raise and lower the beams into place.

The beams are placed side by side with their ends bearing $\frac{1}{2}$B on or into brick loadbearing walls or on to steel beams. If the ends of the beams are built into walls the ends should be solidly filled with concrete as the hollow beam is not strong enough to bear the weight of heavy brickwork.

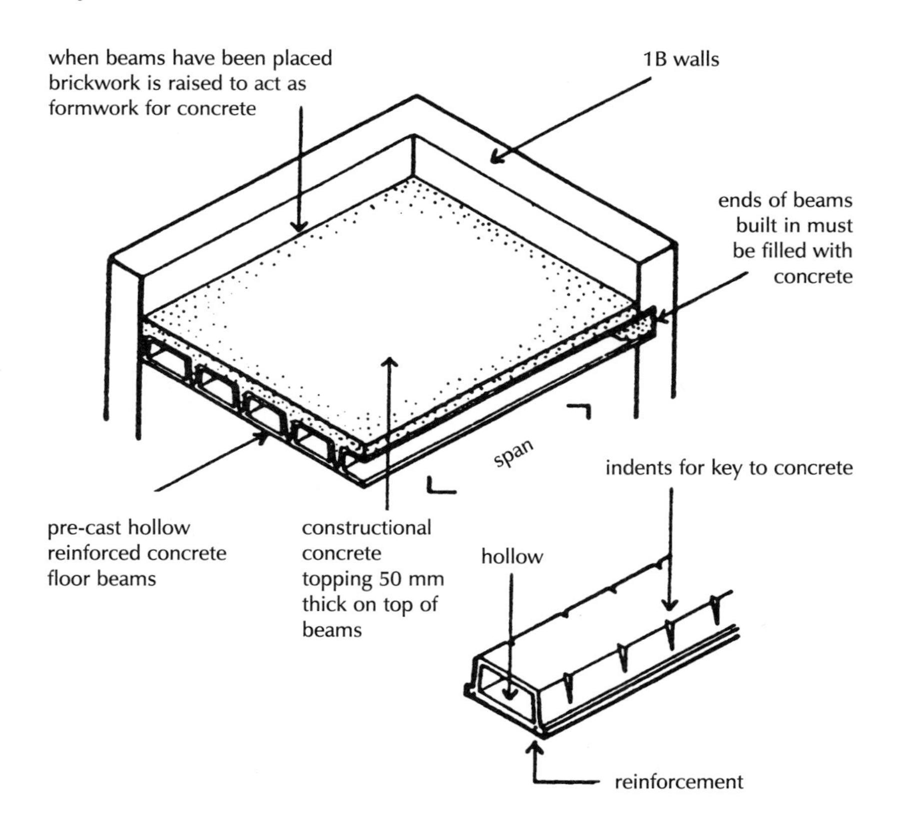

Fig. 157 Hollow concrete beam floor.

The walls of the beams are made thin so that they are light in weight for transporting and hoisting into position. The thin walls of the beams are not strong enough to carry the direct weight of say furniture, and over them is spread a layer of concrete usually 50 mm thick which serves to spread point loads. The concrete is termed constructional concrete topping, and it is an integral part of this floor system. The concrete is mixed on the building site and is spread and levelled on top of the beams as illustrated in Fig. 157.

The hollow beam floor system is particularly suited to multi-storey buildings where lifting equipment such as the tower crane is used. This floor system was used with structural steel frame construction with the beams bearing on steel beams.

Reinforced concrete and clay block floor

The resistance to damage by fire of a reinforced concrete floor depends on the protection, or cover, of concrete underneath the steel reinforcement. Under the action of heat, concrete is liable to expand and come away from its reinforcement. If, instead of concrete, pieces of burned clay tile are cast into the floor beneath the reinforcing bars the floor has a better resistance to fire than it would have with a similar thickness of concrete.

The particular advantage of this type of floor is its good resistance to damage by fire, and it is sometimes termed 'fire-resisting reinforced concrete floor'.

To keep the dead weight of the floor as low as possible, compatible with strength, it is constructed of in situ reinforced concrete beams with hollow terra cotta infilling blocks cast in between the beams. The words terra cotta mean 'earth burned'. The words terra cotta are used in the building industry to describe selected plastic clays which contain in their natural state some vitrifying material. After burning, the clay has a smooth hard surface which does not readily absorb water. The blocks are made hollow so that they will be light in weight and the smooth faces of the blocks are indented with grooves during moulding, to give a good 'key' for plaster and concrete.

A typical TC (terra cotta) block is shown in Fig. 158. This type of floor has to be given temporary support with timber or steel centering. The TC blocks and the reinforcement are set out on the centering, and pieces (slips) of clay tile are placed underneath the reinforcing bars. Concrete is then placed and compacted between the TC blocks and spread 50 mm thick over the top of the blocks. Figure 158 is an illustration of part of one of these floors.

The floor is built into walls $\frac{1}{2}$B thick as shown. This type of floor can span up to 5.0 m and the depth of the blocks, the depth of the finished floor and the size and number of reinforcing bars depend on the superimposed loads and span. This type of floor, which is much less used today in this country than it was because of the considerable

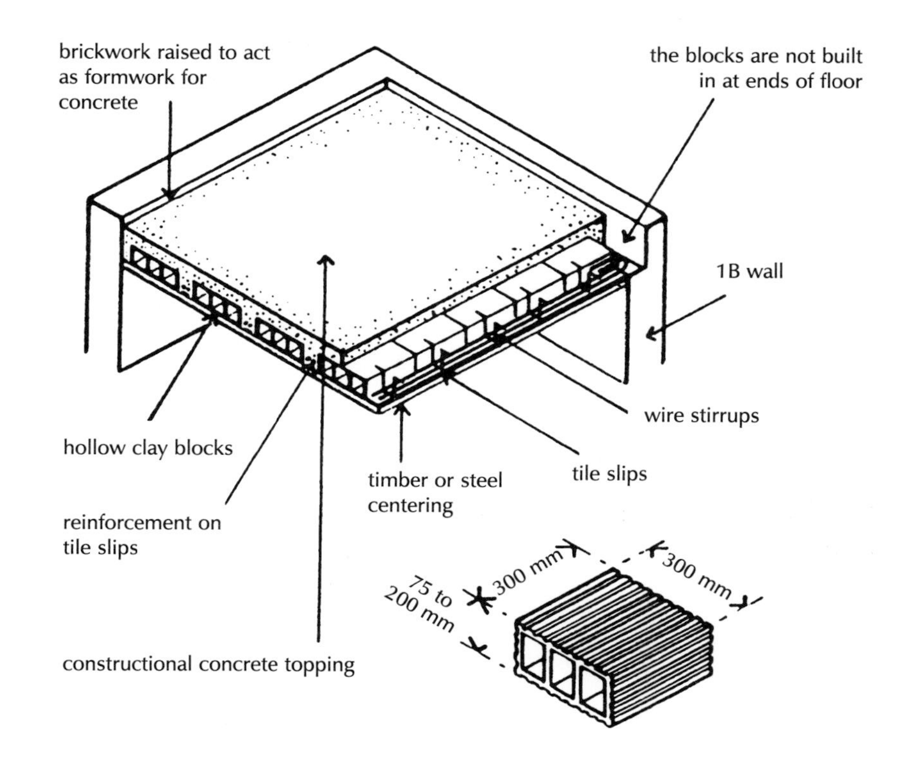

Fig. 158 Terra cotta block floor.

labour in placing the hollow TC pots and reinforcement and need for temporary support, is suited to those countries where hollow clay blocks are extensively used for infill walls to reinforced concrete frame buildings.

Monolithic reinforced concrete floor

The word monolithic is used to describe one unbroken mass of any material. A monolithic reinforced concrete floor is one unbroken solid mass, between 100 and 300 mm thick, of in situ cast concrete, reinforced with mild steel reinforcing bars. To support the concrete while it is still wet and plastic, and for 7 days after it has been placed, temporary centering has to be used. This takes the form of rough timber boarding, plywood, block board or steel sheets, supported on timber or steel beams and posts. The steel reinforcement is laid out on top of the centering and raised 20 mm or more above the centering by means of small blocks of fine concrete which are tied to the reinforcing bars with wire or by plastic spacers (see Volume 4). The wet concrete is then placed and spread on the centering, and it is compacted and levelled off.

It is usual to design the floor so that it can safely span the least width of rooms and two opposite sides of the concrete are built into walls and brick partitions $\frac{1}{2}$ B each end or where the floor gives lateral support to walls it may be built in parallel to its span. Figure 159 illustrates a single monolithic concrete floor with part of the concrete taken away to show reinforcement and timber centering.

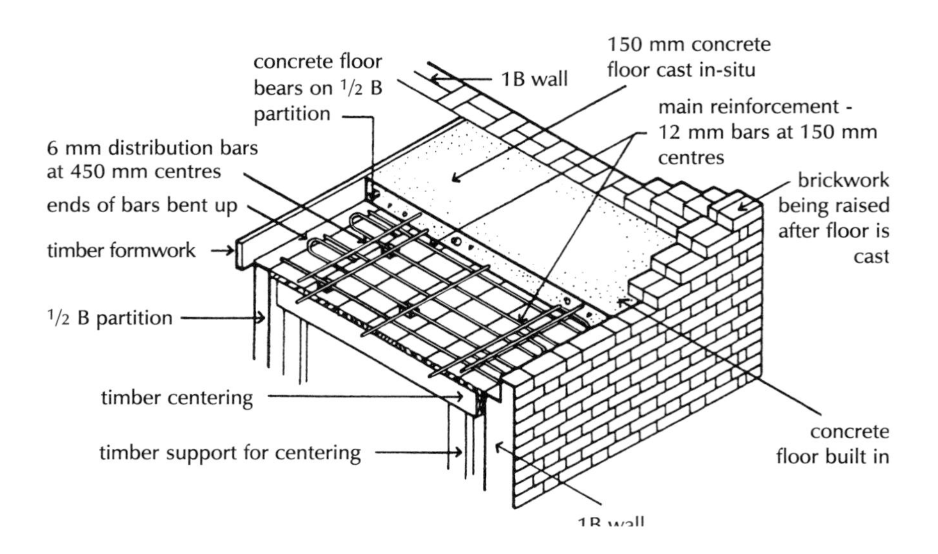

Fig. 159 Monolithic reinforced concrete floor.

Centering

The temporary timber, plywood, blockboard or sheet steel support for monolithic concrete floors or roofs is termed centering. The word centering was originally used for the timber formwork on which brick and stone arches and vaults were formed, but today it is used to include the temporary support for concrete floors even though there is no curvature to the underside of the floor.

Reinforcement of concrete

A concrete floor has to carry loads, just as a concrete lintel does, and when loaded tends to bend in the same way. The steel reinforcing bars are cast into the underside of the floor with 20 mm or more of concrete cover below them to prevent the steel rusting and to give it protection in case of fire. The thicker the concrete cover to reinforcement the greater the resistance of the floor to fire.

When the engineer designs a reinforced concrete floor he usually calculates the amount of steel reinforcement required for an imaginary strip of floor 300 mm wide spanning between walls, as though the floor were made up of 300 mm wide concrete beams placed side by side. The engineer will first calculate the combined superimposed and dead load that the floor has to support. The superimposed load is determined just as it is for timber floors and the dead load will include the actual weight of the concrete, the floor finish and the plaster on the soffit.

From the loads and the span the required thickness of concrete will be determined and then the cross-sectional area of steel reinforcement for every 300 mm width of floor calculated. A rough method of determining the thickness of concrete will be decided and then the cross-sectional area of steel reinforcement for every 300 mm width of floor calculated. A rough method of determining the thickness of concrete required for floors of houses and flats is to allow 15 mm thickness of concrete for every 300 mm of span. The main reinforcement consists usually of 12 mm diameter mild steel rods spaced from 150 to 225 mm apart, and these span across the floor between walls supporting the floor.

The 6 mm diameter mild steel rods wired across the main reinforcement are spaced at 450 to 900 mm apart and are called distribution rods or bars. These rods are tied to the main reinforcement with wire and keep the main reinforcing rods correctly spaced whilst the concrete is being placed and their main purpose is to assist in distributing point loads on the floor uniformly over the mass of the concrete.

In designing a reinforced concrete floor, as though it consisted of 300 mm wide beams, it is presumed that it bends in one direction only when loaded. In fact a monolithic concrete floor bends just as the skin of a drum does, when it is pressed in the middle. In presuming that the floor acts like a series of 300 mm wide beams the engineer can quite simply design it. But as no allowance is made for bending across the span, the floor as designed will be heavier and more expensive than it need be to safely carry its loads. The work involved in allowing for the bending that actually occurs in monolithic floors is considerably more than that required if it is presumed that the floor is a series of beams 300 mm wide.

Of recent years several firms have specialised in designing and constructing reinforced concrete and in order to be competitive their engineers make the more complicated calculation so as to economise in concrete and reinforcement.

Because the centering required to give temporary support to a monolithic concrete floor tends to obstruct and delay building operations 'self-centering' concrete floors are largely used today for multi-storey buildings with monolithic concrete floors used for heavily loaded and specially designed construction and for stairs, ramps and small spans.

Cold rolled steel deck and concrete floor

Of recent years profiled cold rolled steel decking, as permanent formwork and the whole or a part of the reinforcement to concrete, has become one of the principal floor systems for multi-storey framed buildings, as described in Volume 4.

Fire safety

The resistance to fire of a reinforced concrete floor depends on the thickness of concrete cover to steel reinforcement, as the expansion of the steel under heat will tend to cause the floor to crack and ultimately give way. The practical guidance given in Approved Document B to the Building Regulations specifies least cover of concrete for reinforcement to give a specific notional period of fire resistance.

Resistance to the passage of heat

A reinforced concrete upper floor that is exposed to outside air and one that separates a heated from an unheated space has to be insulated against excessive transfer of heat by a layer of some insulating material that is usually laid on the top of the floor under a screed or boarded platform floor surface.

Resistance to the passage of sound

The mass of a concrete floor will provide some appreciable resistance to the transfer of airborne sound. Where it is necessary to provide resistance to impact sound a form of floating floor surface may be necessary.

4: Roofs

HISTORY

Pitched roofs

Before the development of the railway system in the nineteenth century the form of roof common to most buildings in this country was dictated by the availability of local materials used as roof coverings.

Thatch

In lowland areas, such as Norfolk and Suffolk, thatch was common. Long straight stalks of water and marsh plants, reeds, were cut, dried, bound together and laid up the slopes of pitched (sloping) roofs as thatch. Thatch efficiently drains rainwater, excludes wind and acts as an effective insulator against transfer of heat, a combination of advantages that no other roof covering offers.

The disadvantage of thatch is that the dry material readily ignites and burns vigorously and the thick layer of thatch is an ideal home for small birds, rodents and insects.

Tiles

Extensive beds of clay in midland and southern England, suited to pressing and burning to the shape of roof tiles, provided a ready supply of material for making the small, thin slabs of burned clay used as a roof covering to the traditional pitched roof form of most buildings in the area.

The advantage of the small, flat units of tile is that laid overlapping up the slopes of pitched roofs they effectively drain rainwater, accommodate the structural, moisture and thermal movements common to roofs and are durable for the life of most buildings. A disadvantage is that the very many joints between tiles do not exclude wind and tiles do not serve as an efficient insulator against transfer of heat by themselves.

Slates

In rocky upland areas such as Cornwall, Wales, northern England and parts of Scotland, beds of rock that can be split into comparatively smooth plates were used as a roof covering.

Welsh slates can be split into thin, uniform thickness slabs of rock suited to cutting into standard sizes. Most Cornish, Westmorland, northern England and Scottish rocks can at best be split into comparatively thick, uneven surfaced slabs which are more laborious and wasteful to cut to standard sizes. These slates, which are often used in random sizes, are sometimes described as stone tiles.

Laid overlapping up the slopes of pitched roofs slates effectively and rapidly drain rainwater, accommodate structural, moisture and thermal movement and are durable for the life of most buildings. Like tiles, slates do not exclude wind or serve as an effective insulator against heat transfer by themselves.

With the development of an extensive railway system in the nineteenth century, the use of the common roofing materials, tile and slate, was no longer confined to the original local areas of use.

Flat roofs

The least slope or pitch necessary for tile and slate coverings had for centuries determined the symmetrical pitch form of roof common to the majority of buildings.

Bitumen felt

Asphalt

During the first half of the twentieth century the fashion for flat roofs led to the use of materials such as bitumen felt and asphalt that had previously been used as a temporary roof covering and for paving roads, respectively.

Initially two or three layers of bitumen felt on timber roofs and asphalt on concrete roofs were used with a modest degree of success providing they were renewed every 20 to 25 years. At the time there was no requirement to insulate buildings against transfer of heat to outside.

With the introduction of statutory requirements to minimise transfer of heat it became necessary to introduce a layer of some insulating material to roofs. A layer of insulating material was formed under flat roof coverings. The very considerable temperature fluctuations that a roof covering suffers between hot days and cold nights were no longer borne by both the coverings and the roof below to the detriment of felt and asphalt coverings that rapidly deteriorated and failed.

With the introduction of more exacting requirements for the conservation of energy and increased insulation the flat roof form has lost favour to the extent that many flat roofs have been rebuilt as pitched roofs to minimise repair and replacement costs.

Low pitch roofs

At about the same time that the flat roof form was adopted it became fashionable to use low pitch roofs, particularly for single storey buildings, as a revolt against the traditional pitched roof forms. The roofs of comparatively wide span, low rise buildings were pitched at from 5° to 10° to the horizontal, a slope too shallow for the traditional tile or slate coverings.

Instead of using felt or asphalt flat roof coverings it became fashionable to employ sheets of non-ferrous metals such as copper and aluminium joined with standing seams down the slope and welted seams across the slope. This form of covering was also used as a weather surface to arched and curved reinforced concrete roof forms. The appearance and durability of this sheet metal covering justified the cost.

Today the majority of new houses are built with traditional pitched roof structures covered with tile or slate to satisfy the inbred prejudice against the modern in favour of the sense of the majority of house buyers, that a house should look like a house.

FUNCTIONAL REQUIREMENTS

The functional requirements of a roof are:

Strength and stability
Resistance to weather
Durability and freedom from maintenance
Fire safety
Resistance to the passage of heat
Resistance to the passage of sound

Strength and stability

The strength and stability of a roof depend on the characteristics of the materials from which it is constructed and the way in which the materials are formed as a horizontal platform or as a triangular framework.

Flat roof

A roof may be constructed as a flat roof, that is a timber, metal or concrete platform which is usually horizontal or inclined at up to 5° to the horizontal.

The strength and stability of a flat roof depend on adequate support from walls or beams and sufficient depth or thickness of timber joists or concrete relative to span to avoid gross deflection under the dead load of the roof itself and the load of snow and wind pressure or uplift that it may suffer.

Sloping roof

A sloping or sloped roof is inclined at between 5° and 10° to the horizontal, either as a sloping platform or as a shallow frame, as illustrated in Fig. 160. Both the monopitch roof and the butterfly roof are constructed with shallow timber or metal trussed rafters designed to support the dead load of the roof and imposed loads of snow and wind.

The butterfly roof is in effect two monopitch roofs which depend for support on a central beam which is carried on internal columns or end walls.

Both the monopitch and the butterfly roofs depend for strength and stability on the depth of the trussed rafters.

The monopitch roof with sloping soffit is constructed as a flat roof inclined out of horizontal, with timber or metal rafters providing strength and stability. Because of the shallow slope, this roof does impose some small lateral pressure on the wall under the lowest edge of the roof, which is designed to support both the lateral and horizontal pressure from the roof.

Pitched roof

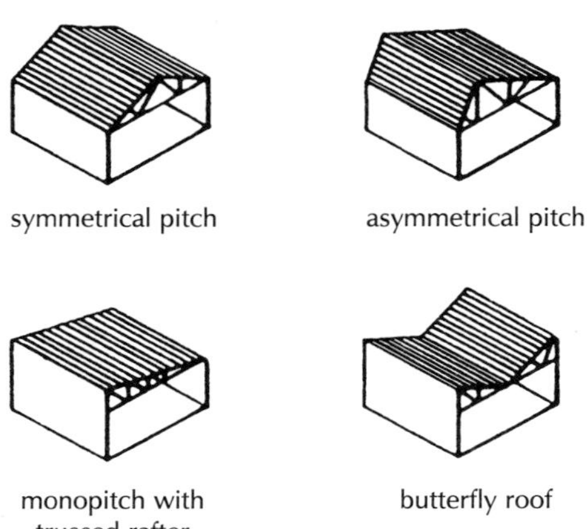

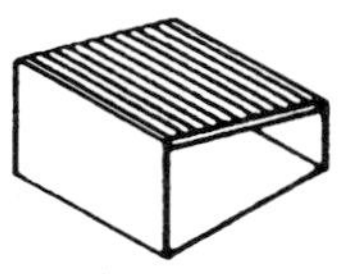

Fig. 160 Sloping roofs.

The use of the word pitch derives from the word describing the form of a tent, in which slender lengths of wood are pitched, thrown up to a central point as a framework. The word pitch is used to describe the angle of inclination to horizontal of the rafters that are pitched up at an angle of more than 10° to a ridge.

The traditional pitched roof has rafters at equal slopes rising to a central ridge in the form of a symmetrical pitch roof, as illustrated in Fig. 160. A variant of this form is the asymmetrical roof illustrated in Fig. 160, where the slopes are different.

The usual construction of this roof is as triangular frames of sloping rafters tied, trussed, together with horizontal ceiling joists, usually with a system of struts and ties for economy in the use of timber and rigidity.

The strength and stability of this form of roof depends on the depth of the triangular frames at mid-span. There is an inherent instability across the slopes of this roof, parallel to the ridge, to the extent that wind pressure may cause the frames to rack or fall over like a stack of books on a shelf. To resist racking the frames are braced by gable end walls, hipped ends or by cross-bracing of diagonal roof boarding or braces across slopes.

The lean-to roof illustrated in Fig. 160 is a monopitch roof supported by an external wall and walls enclosing an annexe or addition, with a pitch to suit the roof covering used.

Resistance to weather

A roof excludes rain through the material with which it is covered, varying from the continuous impermeable layer of asphalt covering that can be laid horizontal to exclude rain, to the small units of clay tiles that are laid overlapping down slopes so that rain runs down to the eaves.

In general, the smaller the unit of roof covering, such as tile or slate, the greater must be the pitch or slope of the roof to exclude rain that runs down in the joints between the tiles or slates on to the back of another tile or slate lapped under and so on down the roof. Larger units such as profiled sheets (see Volume 3) can be laid at a lower pitch than that required for tiles. Impermeable materials such as asphalt and bitumen that are laid without joints can be laid flat and sheet metals such as lead and copper that are overlapped or joined with welts can be laid with a very shallow fall.

The small open jointed units of tile and slate which provide little resistance to the penetration of wind into the roof require a continuous layer of felt or paper to exclude wind.

A roof structure will be subject to movements due to variations in loading by wind pressure or suction, snow loads and movements due to temperature and moisture changes. The great advantage of the traditional roofing materials slate and tile, is that as the small units are hung, overlapping down the slope of roofs, the very many open joints between the tiles or slates can accommodate movements in the roof structure without breaking slates or tiles or letting in rainwater, whereas large, unit size materials and continuous roof coverings may fail if there is inadequate provision of movement joints.

Durability and freedom from maintenance

The durability of a roof depends on the ability of the roof covering to exclude rain, snow and the destructive action of frost and temperature fluctuations. Persistent penetration of water into the roof structure may cause or encourage decay of timber, corrosion of steel or disintegration of concrete.

The traditional materials, slates and tiles, when laid at an adequate pitch (slope) and properly double lapped to exclude rain, will, if undisturbed, have a useful life of very many years and will require little if any maintenance and may survive the anticipated life of buildings. Because of the variations in shape, colour and texture of natural slates and handmade clay tiles it is generally accepted that these traditional materials are often an initial and continuing attractive feature of buildings. The uniformity of shape, colour and texture of machine made slates and tiles have, by common acceptance, a less pleasing appearance.

The non-ferrous sheet metal coverings, lead, copper, zinc and aluminium, overlapped and jointed and fixed to accommodate movement and with a slope adequate for rainwater to run off, will have a useful life of many years and should require little if any maintenance during the life of most buildings. As these durable roof coverings are not a familiar part of the appearance of the majority of buildings in this country, there is no broad consensus of opinion as to their effect on the appearance of buildings.

The flat roof materials asphalt and bitumen felt are, by their nature, short to medium term life materials due to the gradual oxidisation and hardening of the material which has a useful life of some 20 to 30 years at most. While the covering to a flat roof is not generally one of the more visible features of buildings, it is generally accepted that these roofing materials do not have an attractive appearance.

Fire safety

The requirements for fire safety in the Building Regulations are concerned for the safe escape of occupants to the outside of buildings. The regulations require adequate means of escape, and limitation to internal and external fire spread. On the assumption that the structure of a roof will ignite after the occupants have escaped there is no requirement for resistance to fire of most roofs.

The requirements do limit the external spread of fire across the surface of some roof coverings to adjacent buildings by limits to the proximity of buildings. Bitumen felt roof coverings, by themselves, do to an extent encourage spread of flame unless they are covered with stone chippings at least 12.5 mm deep, non-combustible tiles, sand and cement screed or macadam.

Resistance to the passage of heat

The materials of roof structures and roof coverings are generally poor insulators against the transfer of heat and it is usually necessary to use some material which is a good insulator, such as lightweight boards, mat or loose fill to provide insulation against excessive loss or gain of heat. The requirement of the Building Regulations for the insulation of roofs of dwellings is a maximum U value of 0.25 W/m^2K.

The most economical method of insulating a pitched roof is to lay or fix some insulating material between or across the ceiling joists, the area of which is less than that of the roof slope or slopes. This insulating layer will at once act to reduce loss of heat from the building to the roof space, and reduce gain of heat from the roof space to the building. As the roof space is not insulated against loss or gain of heat it is necessary to insulate water storage cisterns and pipes in the roof against possible damage by freezing. Where the space inside a pitched roof is used for storage or as part of the building it is usual to insulate the roof slopes.

With insulation at ceiling level in a pitched roof it may be necessary to ventilate the roof space so that moisture vapour that might condense on cold surfaces does not adversely affect the performance of insulation or the structure. In warmer climates than the United Kingdom it is often practice to provide ventilation to the roof to reduce the temperature inside the roof.

With the insulation at ceiling level the roof will be roughly at outside air temperature. This is a form of cold construction, or a cold roof. With the insulation fixed across the roof rafters, under the roof covering, the roof is insulated against changes in outside air temperature and is a form of warm construction or a warm roof. Obviously the roof space below insulation across roof rafters need not be ventilated to outside air.

Insulating materials may be applied to the underside or top of flat roofs or between the joists of timber flat roofs. Rigid structural materials, such as wood wool slabs, that serve as a roof deck, and rigid insulation boards are laid on the top of the roof and non-rigid materials either between joists or on top of the roof below some form of decking.

The position for insulation will be affected by the type of flat roof structure, the nature of the insulation material, the most convenient place to fix it and the material of the roof finish. The most practical place to fix the insulation is on top of the timber or concrete roof deck or structure, under the roof covering.

With insulation on top of the roof deck the structure will be insulated from outside air and maintained at roughly the inside air temperature. This arrangement is sometimes described as warm construction, as the roof structure is as warm as the inside of the building, or the roof is said to be a warm roof. The advantage of the warm roof, particularly with timber roofs, is that there is no necessity to ventilate the roof itself against condensation moisture.

The very considerable disadvantage of the warm roof is that as the insulation is directly under the roof covering, the material of the covering will suffer very considerable temperature fluctuations between hot sunny days and cold nights, as the roof structure will absorb little if any of the heat or cold that the roof covering is subjected to. These very considerable temperature fluctuations will cause tar and bitumen coverings, such as asphalt and felt, to oxidise more rapidly, become brittle and fail and cause severe mechanical strain in other coverings. An inverted or upside down warm roof, with the insulation on top of the roof covering, will protect the roof covering from severe temperature fluctuations.

With the insulation below, the roof structure is subject to the fluctuations of temperature between hot sunny days and cold nights, and the construction is sometimes referred to as cold construction or cold roof. The disadvantage of cold construction or cold roof is that the moisture vapour pressure of warm inside air may cause vapour to penetrate the insulation and condense to water on the cold side of the insulation, where it may adversely affect the performance of the insulant. Cold roofs are often protected by a vapour check on the warm side of the insulation and roofs may be ventilated against build-up of moisture.

Vapour check

To control the movement of warm moist air from inside a building to the cold side of insulation, it is practice to fix a layer of some impermeable material, such as polythene sheeting, to the underside, the warm side, of insulation that is permeable to moisture vapour, as a check to the movement of moisture vapour. By definition a vapour check is any material that is sufficiently impermeable to check the movement of moisture vapour without being an impenetrable barrier.

The term vapour barrier is sometimes unwisely used in lieu of the term vapour check. The use of the word barrier suggests that the material serves as a complete check to the movement of vapour. The practical difficulties of forming an unbroken barrier between sheets of polythene at the junction of walls and ceiling and around roof lights and pipes suggest the term vapour check is more appropriate.

Some insulating materials, such as the organic, closed cell boards, for example extruded polystyrene, are substantially impermeable to moisture vapour. When these boards can be close butted together or provided with rebated or tongued and grooved edges and cut and close fitted to junctions with walls and around pipes, they will serve as a vapour check and there is no need for an additional layer of vapour check material.

Resistance to the passage of sound

The resistance of a roof to the penetration of airborne sound is not generally considered unless the building is close to a busy airport. The mass of the materials of a roof is the main consideration in the reduction of airborne sound. A solid concrete roof will more effectively reduce airborne sound than a similar timber roof. The introduction of mineral fibre slabs, batts or boards to a timber roof will have some effect in reducing intrusive, airborne sound.

PITCHED ROOFS

Strength and stability

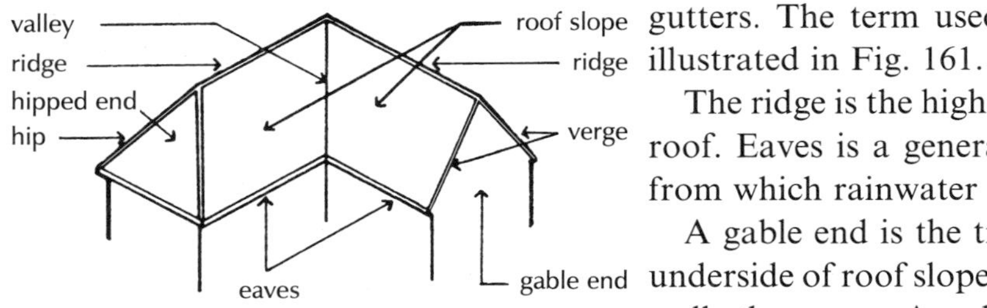

Fig. 161 Pitched roof.

The majority of pitched roofs are constructed as symmetrical pitch roofs with equal slopes pitched to a central ridge. The least slope of a pitched roof is determined by the minimum slope necessary for the roof covering to exclude and drain rainwater to eaves or valley gutters. The term used to describe the parts of a pitched roof are illustrated in Fig. 161.

The ridge is the highest, usually central horizontal part of a pitched roof. Eaves is a general term to describe the lowest part of a slope from which rainwater drains to a gutter or to ground.

A gable end is the triangular part of a wall that is built up to the underside of roof slopes, and the junction of slopes at right angles to a wall, the verge. A valley is the intersection of two slopes at right angles. A hipped end is formed by the intersection of two, generally similar slopes, at right angles.

The traditional pitched roof is constructed with slopes pitched at least 20° to the horizontal for slates and 40° to 60° for tiles.

Couple roof

The simplest form of pitched roof structure consists of timber rafters pitched up from supporting walls to a central ridge. This form of pitched roof is termed a couple roof as each pair of rafters acts like two arms pinned at the top and the mechanical term for such an arrangement is a couple. Figure 162 is an illustration of a couple roof.

Pairs of rafters are nailed each side of a central ridge board. The lower part or foot of each rafter bears on, and is nailed to, a timber wall plate which is bedded on the walls.

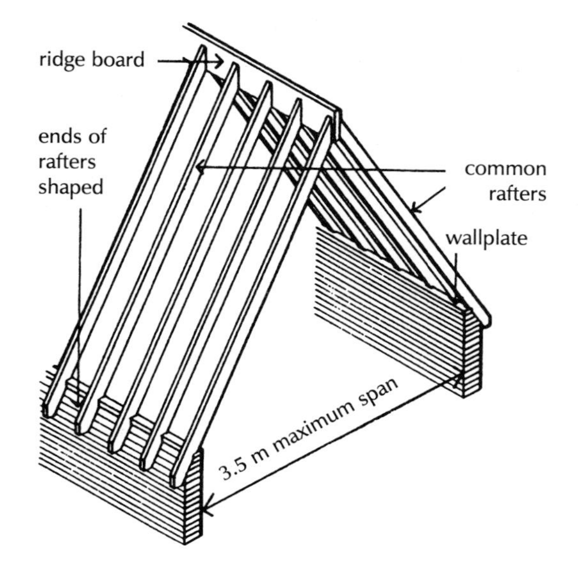

Fig. 162 Couple roof.

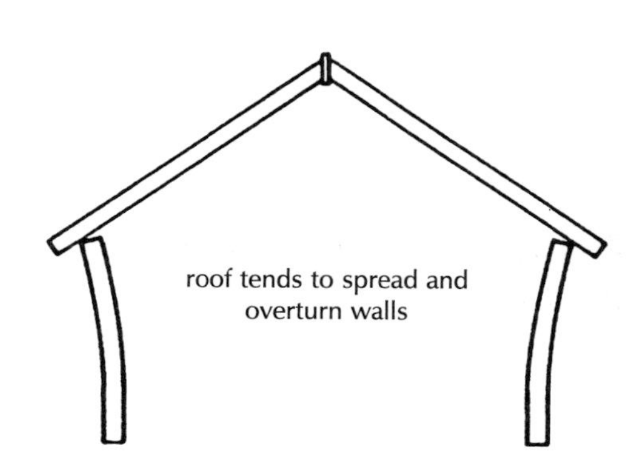

Fig. 163 Couple roof tends to spread and overturn walls.

When this form of roof is covered with slates or tiles and subject to wind pressure there is a positive tendency for the foot of the rafters to spread and overturn the walls on which they bear, as illustrated in Fig. 163. Spreading of rafters is only weakly resisted by the nailed connection of rafters to ridge board which does not act as an effective tie. The maximum span of this roof is generally limited to 3.5 m.

Couple roofs have been used to provide shelter for farm buildings and stores with the ends of such buildings formed with gable ends built in brick or stone or timber framed and rough boarded.

The most straightforward way of providing a tie to resist the spread of the foot of rafters is to fix a metal tie rod between the foot of pairs of rafters. A round or flat section of wrought iron was bolted between the foot of the fourth to sixth pair of rafters. By this device working spans of up to 5 m are practical.

Rafters are spaced at from 400 to 600 mm apart to provide support for tiling or slating battens. Tiles are hung on and nailed to battens and slates nailed to battens.

The necessary size of rafter depends in part on the spacing of rafters and mainly on the clear span, measured up the length of the rafter from the support on the wall plate to the ridge. Sawn softwood rafters for a typical couple roof would be 100 mm deep by 38 mm thick to provide reasonable support for the roof, its covering and wind and snow loads.

Ridge board

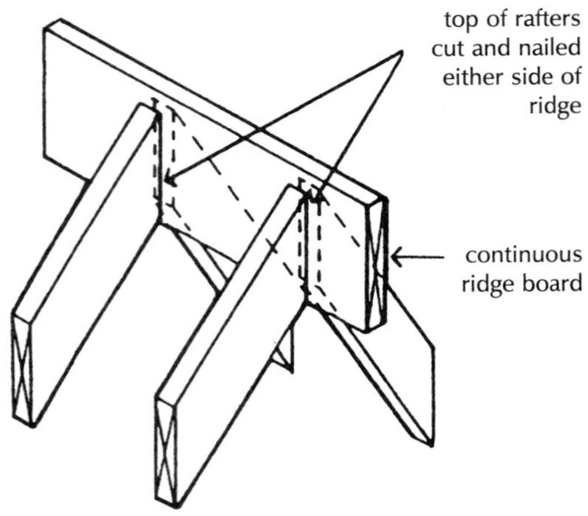

Fig. 164 Ridge board.

The purpose of the ridge board is to provide a means of fixing the top of pairs of rafters. A softwood board is fixed with its long axis vertical and its length horizontal. The top of rafters is cut on the splay so that pairs of rafters fit closely to opposite sides of the ridge board to which they are nailed, as illustrated in Fig. 164.

The ridge board is one continuous length of softwood usually 32 mm thick. The depth of the ridge board is determined by the depth of the splay cut ends of rafters that must bear fully each side of the board. Obviously the necessary depth of ridge board is set by the pitch of the roof and the depth of the rafters. A steeply pitched roof and deep rafters will need a deeper ridge board than a shallow pitched roof.

The ridge board is usually some 50 mm deeper than required as bearing for rafter ends, to provide fixing for battens.

Wall plate

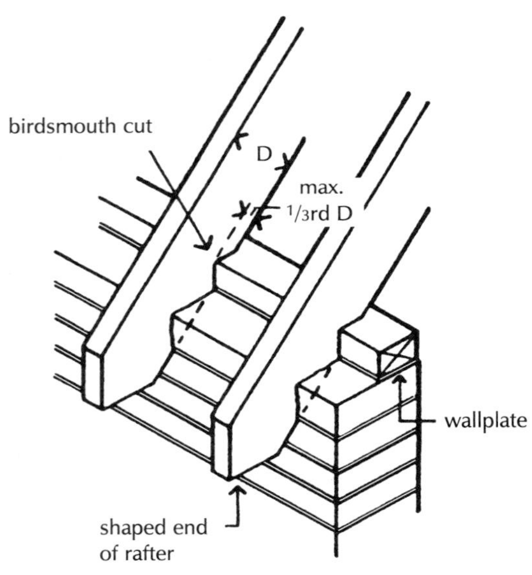

Fig. 165 Wall plate.

A continuous timber wall plate is bedded in mortar on walls as a means of providing support and fixing for the foot of rafters. The sawn softwood plate, usually 100 × 75 mm in section, which is laid on its 100 mm face, serves to spread the load from rafters along the length of the wall.

A birdsmouth cut is made in the top of each rafter to fit closely round the wall plate, as illustrated in Fig. 165. Rafters are nailed to the wall plate. The birdsmouth cut should not be greater than one-third of the depth of a rafter to limit loss of strength in rafters that often extend beyond the wall face as a projecting eaves.

The lower or eaves ends of rafters are cut to form a vertical face as fixing for a gutter board to support gutters and a lower horizontal face to provide fixing for a soffit board to form a closed eaves to exclude wind. As an alternative the soffit board is omitted to form open eaves for farm buildings and sheds.

The bearing of the rafter ends on the wall plate does not effectively resist the tendency of a couple roof to spread under load, to the extent that the plate may be moved by the spreading action of the roof.

Close couple roof

Pitched roofs to small buildings such as houses and bungalows are framed with rafters pitched to a central ridge board with horizontal ceiling joists nailed to the side of the foot of each pair of rafters, as illustrated in Fig. 166. The ceiling joists serve the dual purpose of ties to resist the natural tendency of rafters to spread and as support for ceiling finishes. Because the ceiling joists act as ties to the couple of

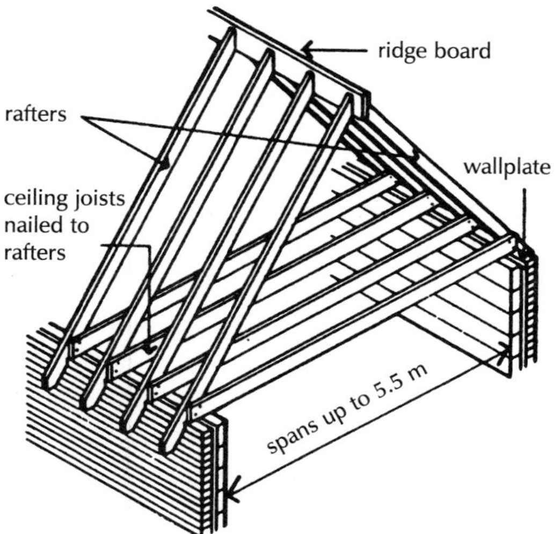

Fig. 166 Close couple roof.

pairs of rafters, this form of roof construction is a close couple or closed couple roof.

Ceiling joists are usually 38 or 50 mm thick and 97 to 220 mm deep sawn softwood, the size of the joists depending on their spacing and span between supports. The maximum span between supporting walls for the close couple roof illustrated in Fig. 166 is 5.5 m. For this span, ceiling joists with no intermediate support from internal walls would be 220 mm deep by 50 mm thick to support ceiling finishes without undue deflection.

The advantage of the triangular space inside the roof above the ceiling joists is that it will to some extent provide insulation, provide a convenient space for water storage cisterns and provide the storage space that is lacking in most modern house designs.

The disadvantage of the close couple roof structure by itself is that the considerable clear spans of rafters and ceiling joists require substantial timbers as compared to similar roofs where there is intermediate support to rafters from purlins and to ceiling joists from binders.

Collar roof

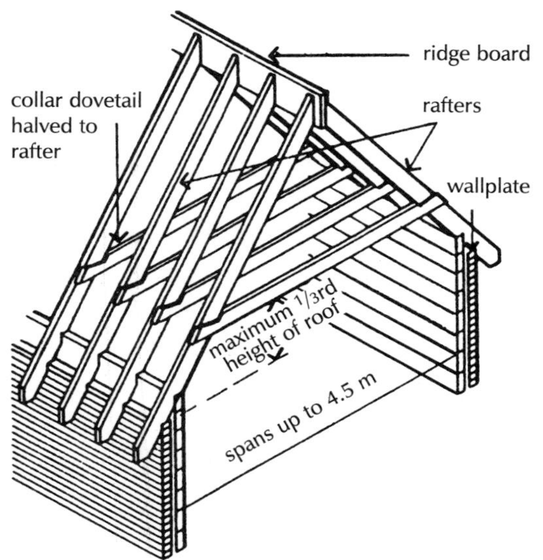

Fig. 167 Collar roof.

Another form of tied couple roof is framed with collars joined across pairs of rafters, at most one-third up the height of the roof, as illustrated in Fig. 167. The purpose of this arrangement is to extend first floor rooms into the roof space and so limit the largely unused roof space. A disadvantage of this arrangement is that the head of windows formed in a wall will be some distance below ceiling and give less penetration of light. To provide normal height windows a form of half dormer window is often used with the window partly built into the wall and partly as a dormer window in the roof.

A collar, fixed at up to a third up the height of a roof, does not so effectively tie pairs of rafters as does a ceiling joist fixed to the foot of rafters. To provide a secure joint between the ends of collars and rafters, a dovetail half depth joint is formed. The ends of collars are cut to half their width in the shape of a half dovetail to fit into a similar half depth housing in rafters and the two nailed together, as illustrated in Fig. 167.

Because a collar is a less effective tie than a ceiling joist the maximum span of this roof is limited to 4.5 m. To provide solid framing the rafters are usually 125 × 44 mm and the collars 125 × 44 mm.

A collar roof provides substantially less storage space and limited head room for access to water storage cisterns. For insulation, the roof space insulation has to be extended down the lower part of roof slopes, between rafters.

Purlin or double roofs

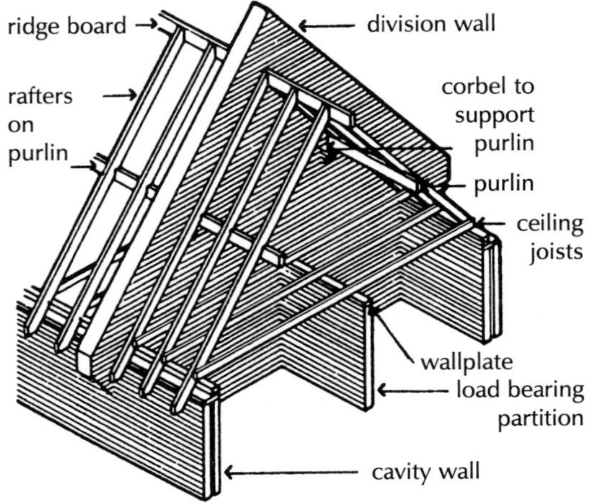

Fig. 168 Purlin roof.

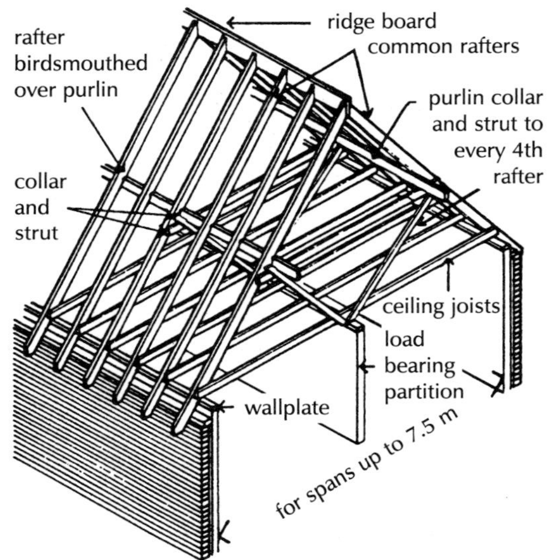

Fig. 169 Double roof.

To economise in the section of roof rafters it has been practice to provide intermediate support up the slope of roofs by the use of purlins. Purlins are horizontal timbers supported by end walls or struts to internal loadbearing walls to support rafters usually half way up slopes. By the use of a comparatively substantial timber purlin an appreciable saving in timber rafter size can be effected.

The most economic form of housing is as a terrace with solid brick or block division walls carried up to or above roof level. To limit spread of fire between houses no combustible material such as wood may be built into division or separating walls.

Timber purlins to provide support for roof rafters are supported on brick, stone or concrete corbels built to project from separating walls or on metal joist hangers built into walls. The word corbel describes any solid material solidly built into and projecting out some 100 mm to provide support. The purlins are fixed with their long axis vertical to span between separating walls, as illustrated in Fig. 168.

The roof illustrated in Fig. 168 is framed with purlins for support for rafters and an internal loadbearing wall to provide intermediate support for ceiling joists to economise in timber sizes. The purlins illustrated are 175 × 75 mm, the rafters 125 × 50 mm and the ceiling joists 125 × 50 mm sawn softwood.

Where there are no separating or gable end walls to provide support for purlins an internal loadbearing wall can be used to support timber struts fixed between the wall and rafters, as illustrated in Fig. 169.

The purlins are fixed with their long axis vertical with rafters notched over and nailed to purlins. Horizontal collars are nailed to the side of every fourth pair of rafters under purlins. Pairs of struts are notched around a wall plate bedded on the internal wall, notched under and around purlins and nailed to the side of collars. Ceiling joists nailed to the foot of rafters and bearing on the internal wall serve as ties to the close couple roof and as ceiling support.

Struts 75 × 75 mm support 150 × 50 mm purlins with 125 × 50 mm collars, 150 × 50 mm rafters and 125 × 50 mm sawn softwood ceiling joists.

An advantage of this somewhat complicated pitched roof, sometimes referred to as a double roof, is that it is singularly suited to hipped end roofs. The system of struts, collars and purlins can more readily be returned under the sloping hipped end than other forms of roof framing. A disadvantage is that the struts and collars impede head room and access for storage.

Hipped ends of roof

To limit the expanse of roof to detached buildings such as houses, for appearance sake, the ends of a pitched roof are sometimes formed as slopes described as hipped ends. The hipped, sloping ends of pitched roofs are usually framed at the same slope as the main roof. To provide fixing and support for the short length of rafters which are pitched up to the intersection of roofs, a hip rafter is used as illustrated in Fig. 170.

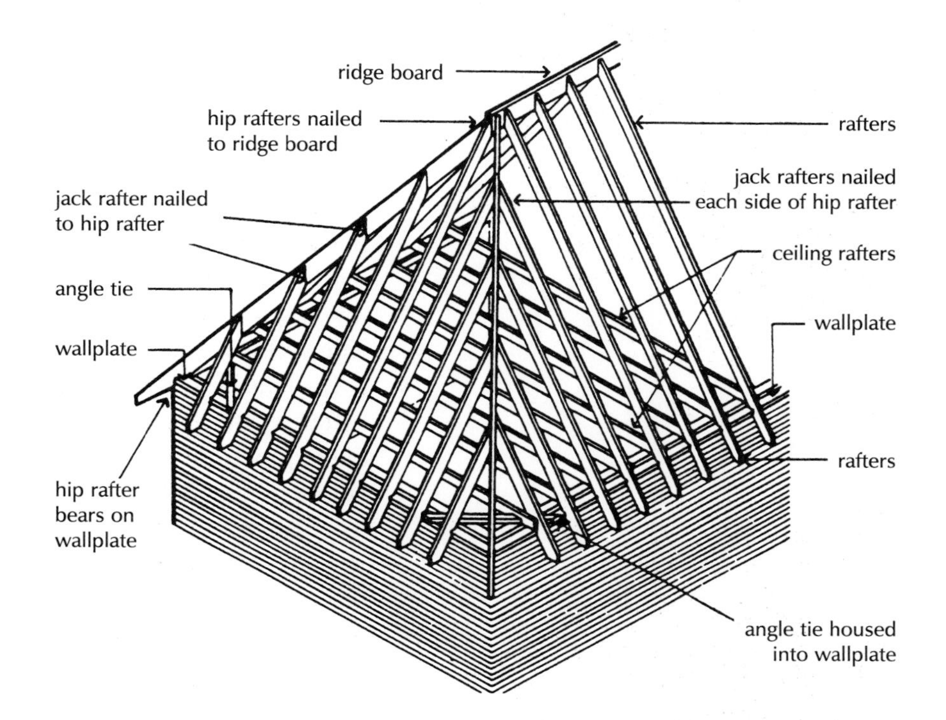

Fig. 170 Hipped end.

To provide adequate bearing for the splay cut ends of rafters pitched up to it, the hip rafter is usually 200 to 250 mm deep and 38 mm thick. The short lengths of rafter pitched to the hip rafter are the same size as the rafters of the main roof.

Jack rafters

These shorter lengths of rafter, commonly termed jack rafters, are nailed to the hip rafter, and finished with shaped ends for gutter and soffit boards.

The hip rafters are oblique, splay cut to bear each side of the ridge board to which they are nailed. The foot of hip rafters is notched to fit around the wall plate to which it is nailed.

A hip rafter provides bearing and support for the ends of jack rafters which support the roof covering and wind and snow loads. Because of the considerable load that a hip rafter carries it will tend to spread and displace the junctions of the wall plates on which it bears and overturn the walls which support it. To maintain the right angles junction of wall plates and wall against the spread of a hip rafter it is necessary to form a tie across the right angle junction.

Angle tie

Angle ties are cut from 100 × 75 mm sawn softwood timbers. These ties are either bolted to the wall plate or dovetail housed into the wall plates some 600 mm from the angle across which they are fixed, as illustrated in Fig. 170.

The hip end illustrated in Fig. 170 is shown as a close couple roof. More usually this form of roof is framed as a purlin or double roof, as previously described and illustrated.

An advantage of a hip end roof is that the hip end acts to provide stability to the main roof against the tendency of a pitched roof to rack and overturn parallel to its ridge. The disadvantage of a hip end roof is the very considerable extra cost in the wasteful cutting of timber and roof covering at the hips.

Valleys

The valley formed at the internal angle junction of two pitched roofs is framed in much the same way, in reverse, as a hip end. A valley rafter is pitched up from the junction of the wall plates at the internal angle junction of walls to the intersection of ridge boards, splay end cut to fit to ridge and notched and housed over wall plate and nailed in position.

The 38 or 50 mm thick valley rafter is of the depth required to provide a bearing for the full depth of the splay cut ends of the rafters it supports and finish level with the tops of these rafters. The cut, jack, rafters are nailed to the side of the ridge board and valley rafter respectively.

The valley rafter is fixed in position with its top edge level with adjoining rafters so that it does not obstruct battens run down into the valley for valley tiles or valley boards for a non-ferous metal valley gutter.

Trussed roof construction

The span of a close couple roof is limited to 5.5 m, which is less than the width of most buildings. A purlin roof which can span up to 7.5 m depends for support on loadbearing walls conveniently placed and these partitions often restrict freedom in planning the rooms of buildings.

A traditional method of constructing pitched roofs, with spans adequate to the width of most buildings, that do not need intermediate support from internal partitions was to form timber trusses. The word truss means tied together and a timber roof truss is a triangular frame of timbers securely tied together. The traditional timber roof truss was constructed with large section timbers that were cut and jointed with conventional mortice and tenon joints that were strapped with iron straps screwed or bolted to the truss. The traditional large section timber king post and queen post trusses supporting timber purlins and rafters were designed to be used without intermediate support over barns, halls and other comparatively wide span buildings.

The combination of shortage of timber that followed the end of the Second World War (1945) and the need for greater freedom in planning internal partitions prompted the development of the economical timber truss designed by the Timber Development Association. The timbers of the truss were bolted together with galvanised iron timber connectors, illustrated in Fig. 171, bolted between timbers at connections. The strength of the truss was mainly in the rigidity of the connection.

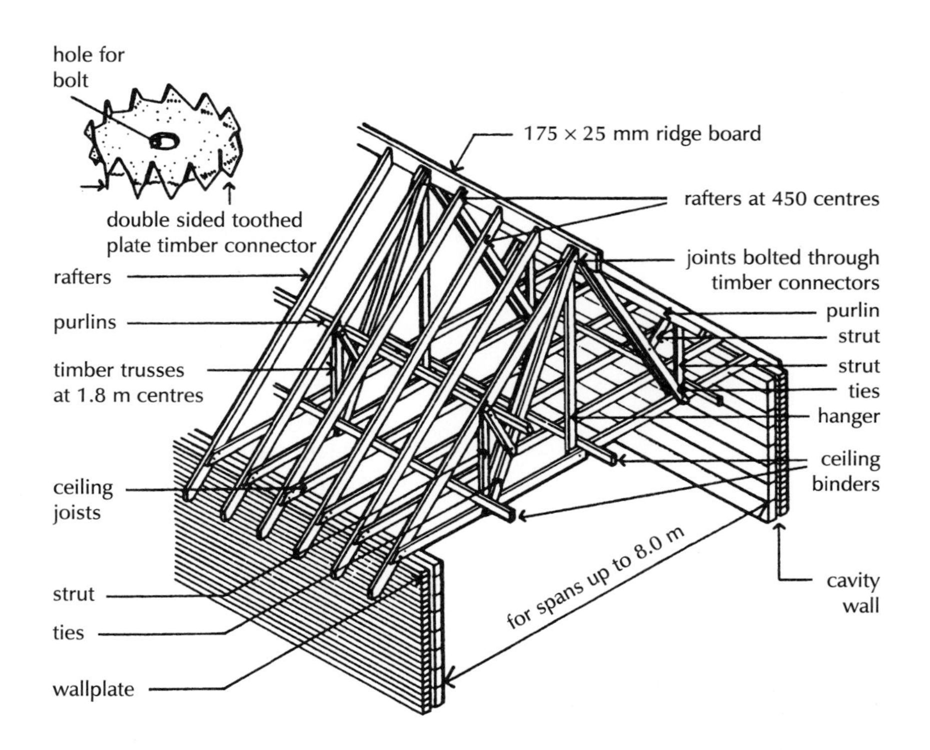

Fig. 171 Trussed roof.

These timber trusses were designed for spans up to 8 m. Each truss was framed with timbers of the same section as the rafters and ceiling joists they provided intermediate support to, through the purlins and binders they supported. The prefabricated timber trusses were fixed in position at 1.8 m centres bearing on and nailed to wall plates. The 175 × 25 mm ridge board was fixed and nailed to trusses. Purlins, 150 × 50 mm, were placed in position supported by struts and nailed to struts and rafters. Ceiling binders 125 mm deep by 50 mm wide were placed in position on 100 × 38 mm ceiling joists.

To complete the roof framing 100 × 38 mm rafters were cut to bear on the ridge board, notched over purlins and wall plate and nailed in position. Ceiling joists were nailed to the foot of rafters and the underside of ceiling binders.

These trusses were designed to use the same roof rafter and ceiling joist sections for continuity of level for fixing of roof covering and ceiling finish and trussed to provide strength in supporting purlins and binders.

The advantage of this trussed roof construction is that the continuity of the ridge board, purlins and binders along the length of the roof together with roofing battens provided adequate stability against racking.

Trussed roof construction has been largely replaced by trussed rafter roofs that require somewhat less timber and can be and are very often fixed by unskilled labour to effect a comparatively small cost saving relative to the total cost of a building. The consequence has been the failure of trussed rafter roofs due to inherent instability parallel to the ridge and poor workmanship.

Trussed rafter roof

For the maximum economy in site labour and timber the trussed rafter roof form was first used in this country in 1964. Each pair of rafters and ceiling joist to a pitched roof was formed as a truss, as illustrated in Fig. 172.

Trussed rafters are fabricated from light section, stress graded timbers that are accurately cut to shape, assembled and joined with galvanised steel connector plates. Much of the preparation and fabrication of these trussed rafters is mechanised, resulting in accurately cut and finished rafters that are delivered to site ready to

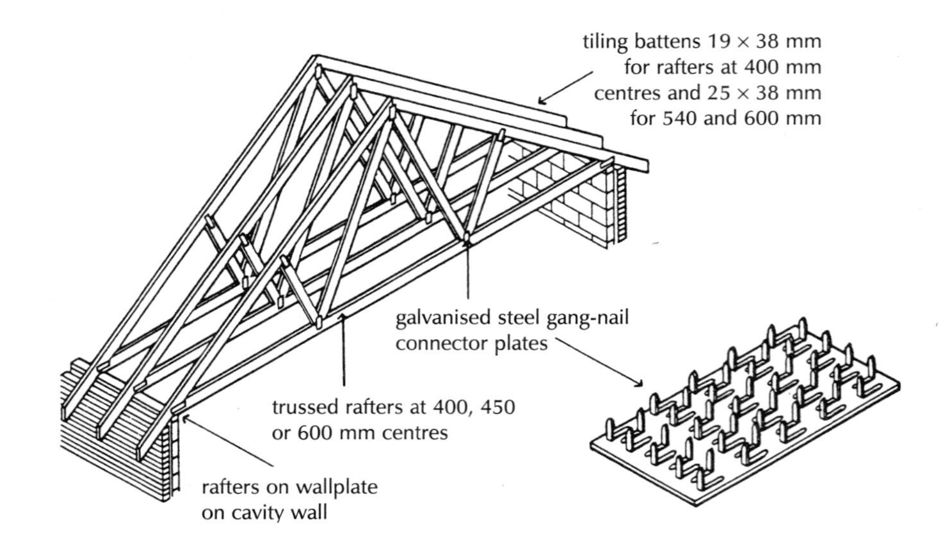

Fig. 172 Trussed rafter roof.

be lifted and fixed as a roof frame with the minimum of site labour.

The members of the truss are joined with steel connector plates with protruding teeth that are pressed into timbers at connections to make a rigid joint. Trussed rafters, that serve as rafters and ceiling joists, are fixed at from 400 to 600 mm centres, as illustrated in Fig. 172 for spans up to 12 m for roofs pitched at 15° to 40°. The trussed rafters bear on and are nailed to wall plates. As the rafters are trussed there is no need for a ridge board to provide a bearing and fixing for rafters.

Since they were first introduced into this country, trussed rafters have been very extensively used for houses, particularly on housing estates of similar buildings where repetitive production of similar trusses provided the greatest economy.

A consequence of the pressure for economy by the use of slender timber sections for trussed rafters spaced at maximum centres and careless workmanship, has been the failure of some trussed rafter roofs due to the inherent instability of a pitched roof parallel to the ridge and buckling of the slender section trussed rafters.

Approved Document A, giving practical guidance to meeting the requirements of the Building Regulations for small buildings, states that for stability, trussed rafter roofs should be braced to the recommendations in BS 5268: Part 3.

Because the members of a trussed rafter roof are delivered to site as framed, triangulated units, it has been common to employ unskilled labour in the erection of these roofs. The result has been that trusses, which were often not accurately spaced nor erected and fixed true vertical, have buckled under load. To take account of practical factors regarding the use of trussed rafters on site and the limit of experience of their behaviour under test and in service, there are recommendations in BS 5268: Part 3 for their use, erection and stability bracing. For the roofs of domestic buildings the members of trussed rafters with spans up to 11 m should be not less than 35 mm thick and those with spans of up to 15 m, 47 mm thick.

These slender section trusses are liable to damage in storage and handling. They should be stored on site either horizontal on a firm level base or in a vertical position with adequate props to avoid distortion. In handling into position each truss should be supported at eaves rather than mid-span to avoid distortion. The trussed rafters should be fixed vertical on level wall plates at regular centres and maintained in position with temporary longitudinal battens and raking braces. The rafters should be fixed to wall plates with galvanised steel truss clips that are nailed to the sides of trusses and wall plates. A system of stability bracing should then be permanently nailed to the trussed rafters. The bracing that is designed to maintain the rafters in position and to reduce buckling under load is illustrated in Fig. 173. The bracing members should be 25 × 100 mm and nailed with two 3.35 × 75 mm galvanised, round wire nails at each cross-over.

The longitudinal brace at the apex of trusses acts in much the same way as a ridge board and those halfway down slopes like purlins to maintain the vertical stability of trusses. The longitudinal braces, also termed binders, at ceiling level serve to resist buckling of individual trussed rafters. The diagonal, under rafter braces and the diagonal web braces serve to stiffen the whole roof system of trussed rafters by acting as deep timber girders in the roof slopes and in the webs.

The bracing shown in Fig. 173 is for stability. Depending on the position of exposure of the building, additional bracing may be necessary against wind pressures.

The implication of the recommendations in BS 5268: Part 3 is that some control and better workmanship is required in the erection of trussed rafter roofs than has been common. There is thus less saving in site labour and materials than previously and the economic advantage of the trussed rafter roof over, for example, the TDA truss roof is considerably less than it was.

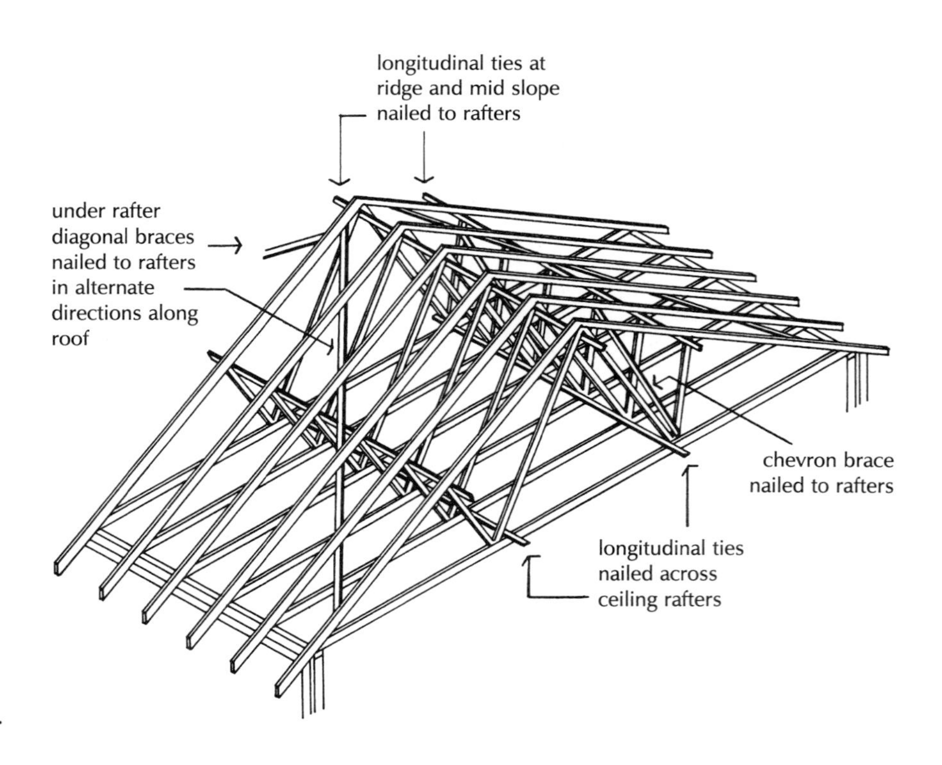

Fig. 173 Stability bracing to trussed rafters.

Eaves

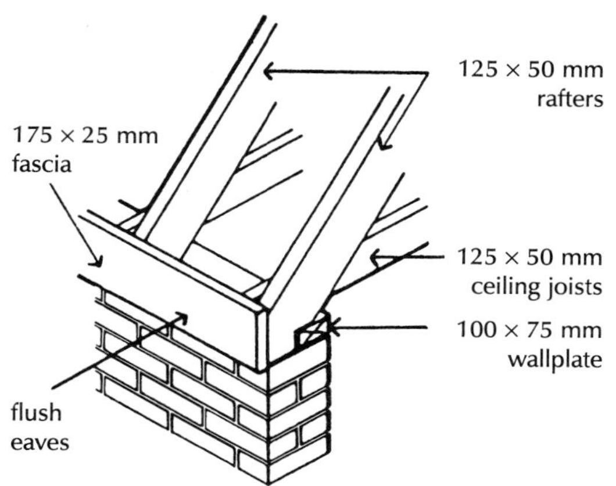

Fig. 174 Flush eaves.

Eaves is a general term used to describe the lowest courses of slates or tiles and the timber supporting them. The eaves of most pitched roofs are made to project some 150 to 300 mm beyond the external face of walls. In this way the roof gives some protection to the walls, and enhances the appearance of a building.

The eaves of the roof of sheds, outhouses and other outbuildings are sometimes finished flush with the face of the external wall of the building. The purpose of this is to economise in timber and roof covering. The construction of flush eaves is illustrated in Fig. 174. The ends of the rafters and ceiling joists are cut flush with the outside face of the wall below. A fascia or gutter board 25 mm thick is nailed to rafters and ceiling joist ends.

The necessary depth of the gutter board is such that it covers the cut ends of rafters and projects some 25 to 30 mm above the top of rafters to act as a bearing for the lowest courses of tiles or slates which project over it some 25 mm to discharge rainwater into a gutter fixed to the gutter board. Because the gutter board covers the ends of rafters this form of eaves is effectively a type of closed eaves.

Projecting eaves

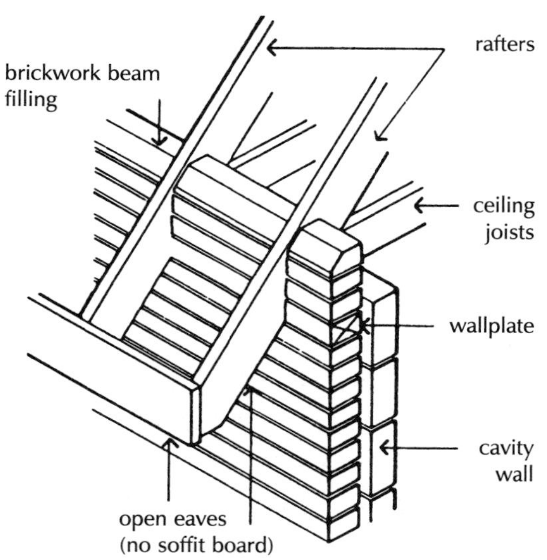

Fig. 175 Open eaves.

Projecting eaves are constructed as either open eaves or closed eaves.

Open eaves project with the ends of roof rafters exposed or open beyond the face of the wall below, as illustrated in Fig. 175. Open eaves may be constructed where the eaves project some distance beyond the face of the wall to provide protection for the wall below and rainwater runs off the edge of slopes to the ground or paved surface below.

An advantage of this arrangement is that there are no gutters and down pipes to keep clear of debris and maintain by periodic painting. The disadvantage is that the appreciable projection may obstruct penetration of light through the head of windows formed close to eaves level. Where the ground, such as sand or gravel, readily absorbs water and where paved areas fall to drains there will be little danger of rainwater falling off roofs causing adverse damp in walls.

The ends of roof rafters may be cut to provide a fixing for a fascia board for appearance sake or the rafter ends may be left exposed. Rafter ends should be stained or painted for protection and appearance sake.

To exclude wind and nesting birds brick or stonework is built between the rafters as beam filling, as illustrated in Fig. 175.

The conventional finish to projecting eaves to the majority of small buildings in this country is with a fascia or gutter board and a horizontal soffit, as illustrated in Fig. 176.

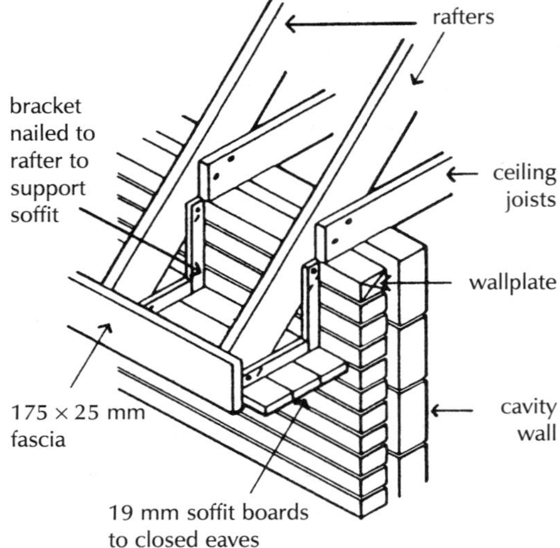

Fig. 176 Closed eaves.

The ends of rafters are shaped to provide a vertical face for fixing the gutter board and a horizontal face for fixing the soffit. The top edge of the gutter board finishes some 25 mm or more above the top of rafters as bearing for eaves tiles or slates and some 25 mm or more below the rafter ends to cover the soffit board.

The closed (enclosed) soffit to the rafter ends is fixed to a system of 50 × 25 mm softwood brackets nailed to the side of each rafter end, as illustrated in Fig. 176. Either tongued and grooved softwood boards or one of the external quality plywood or wood particle boards is nailed to the soffit brackets.

Both the gutter board and the soffit are primed and painted for protection and appearance sake.

The disadvantage of this boarded, closed eaves is that it requires comparatively frequent painting for protection and appearance.

Dormer windows

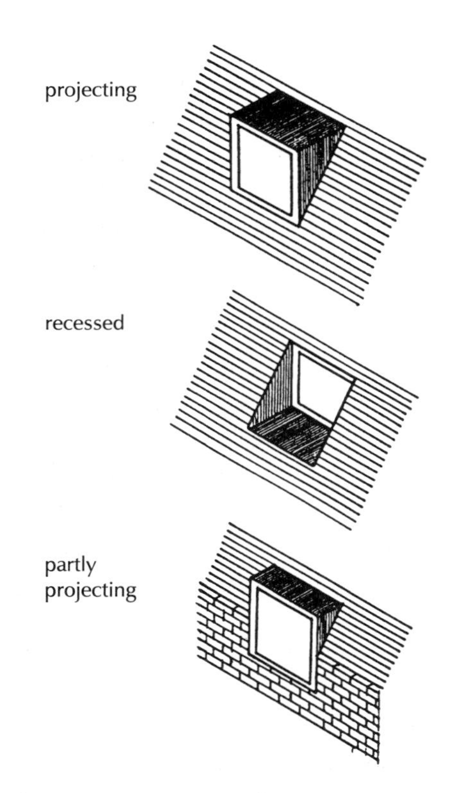

Fig. 177 Dormer windows.

Dormer windows are framed in a slope of pitched roofs as a vertical window for daylight to rooms inside the roof space whereas roof lights are formed as a glazed opening in the slope of the roof.

Dormer windows may be framed as a projection from the roof slope, recessed behind the slope or partly projecting from the roof and partly in the face of the wall below, as illustrated in Fig. 177.

A projecting dormer window will give limited penetration of daylight into rooms, because of the projection of the window head, some limitation of diffusion of light because of the dormer sides or cheeks and a wide angle of view out. A recessed dormer will give better penetration and diffusion of light into rooms than a projecting dormer, but a restricted angle of view out because of the recessed cheeks.

Projecting and partly projecting dormer windows are the traditional means of providing daylight to rooms formed wholly or partly inside the space in steeply pitched roofs. Dormer windows to roofs pitched at the shallow slopes suited to slates are little use because of the restricted useful head room inside such roofs.

To prevent an appreciable volume of rainwater running down the face of a projecting dormer window it is necessary to drain the roof of the dormer to run down the sides or cheeks. A disadvantage of the partly projecting dormer is that the lower part of the window, set in the wall below, will interrupt eaves gutters and require additional and unsightly rainwater pipes.

A recessed dormer window is particularly suited to a Mansard form of roof. Mansard roofs have been used in buildings in northern European cities where notional angles of light were imposed to provide some penetration of daylight to ground floor windows of

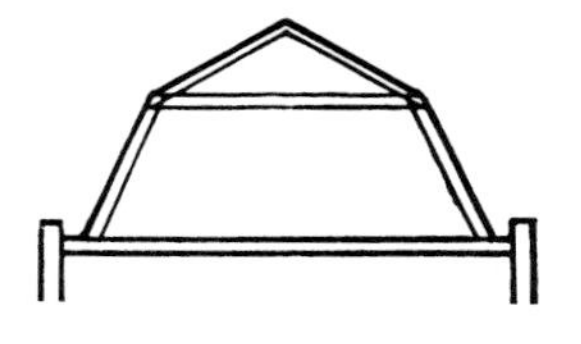

Fig. 178 Mansard roof.

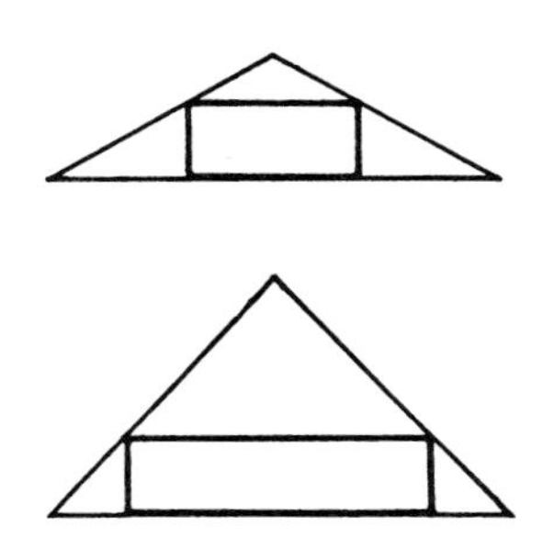

Fig. 179 Space in roofs.

buildings facing streets. It was a requirement that no part of a building project beyond a line drawn at 60° to the horizontal, drawn from the junction of opposite buildings and the street. To provide the maximum number of floors it was practice to form one or several floors inside a roof sloping at 60° with recessed dormer windows to provide daylight.

The advantage of the Mansard slope form of pitched roof, illustrated in Fig. 178, is that it provides maximum useful floor area. Where there are no restricting angles of light a Mansard roof space may be formed with projecting dormer windows.

The very limited clear headroom and restricted useful floor area inside a shallow slope roof pitched at 30°, suited to slate covering, does not justify the use of dormer windows. The more extensive head room and useful floor area inside a roof pitched at 45° has the disadvantage of an appreciable volume and area of roof above that does not provide useful space, as illustrated in Fig. 179.

The most advantageous roof form for dormer windows is a Mansard type roof with the rooms and dormer windows in the steeply pitched roof slopes and a shallow pitch or flat roof above.

Dormer window framing

The opening in roof rafters for dormer windows is framed between trimming rafters and head and cill trimmers which provide support for the trimmed rafters pitched up from the top trimmer and bearing on the bottom trimmer, as illustrated in Fig. 180. For narrow width dormers a normal rafter will serve and for wider windows the trimming rafter thickness is some 25 mm more than that of the main rafters.

The lower or cill trimmer is usually fixed vertically between trimming rafters as a base for the cill of the window. The depth of the cill trimmer is such that it will provide bearing and fixing for the trimmed roof rafters and will extend above the top level of rafters to accommodate a flashing of lead dressed down over the roof covering. The top or head trimmer may be fixed normal to rafters or vertical depending on convenience in constructing the dormer roof. Both top and bottom timber trimmers are through tennoned to trimming rafters.

To gain the maximum penetration of daylight the top, head trimmer should be close to the ceiling of rooms. Whether framed in the slope of a steeply pitched roof or in a Mansard slope it is convenient that a purlin of timber or steel coincides with the head of a dormer window so that it may be designed to carry the weight of the roof above the dormer and such framing to the dormer roof as is necessary. In a Mansard roof the head trimmer should ideally support the shallow pitch or flat roof above the Mansard slope.

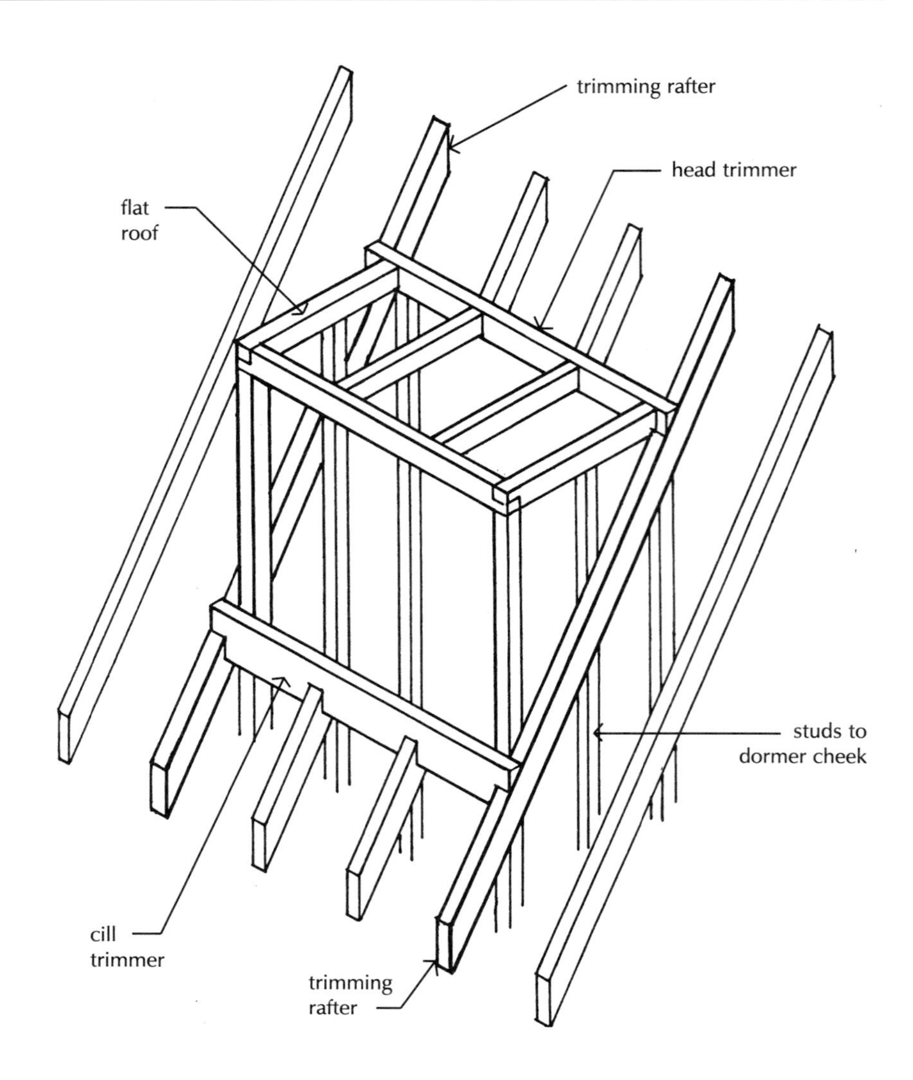

Fig. 180 Dormer window framing.

For the least timber framing and wasteful cutting of timber and roof covering a flat roof dormer is selected. The construction of a small pitched roof over a dormer with a main roof involves a deal of labour in forming valleys and cutting of slates or tiles.

The cheeks, sides, of a dormer are framed with 75 × 50 mm vertical timber studs, at centres suited to fixing internal and external finishes. The ends of the studs are either oblique cut and nailed to the trimming rafters with a corner stud as fixing for the window, or the studs are carried down to the floor.

The flat roof of the dormer should be framed to fall towards a gutter at the junction of the dormer roof and that of the main roof or to fall towards the sides of the dormer to avoid an unsightly gutter over the head of the dormer window. The 75 × 50 mm rafters of the flat roof span towards the main roof or side to side as fixing for tapered, timber firring pieces to provide the necessary fall for the non-ferrous metal roof covering. The rafters are supported by the head of the window or a head and the top trimmer or by a head to dormer cheek framing depending on the fall, slope, of the flat roof covering.

The roof and cheeks of the dormer are boarded for lead or copper sheet. The junction of the main roof covering with the sheet metal covering of the roof and cheeks of the dormer is weathered with lead or copper flashings and gutter.

PITCHED ROOF COVERING

The traditional covering for pitched roofs in England and northern European countries has been clay tile and natural slate. These traditional roof coverings are still extensively used for new houses and other small buildings for their appearance, durability and freedom from maintenance.

TILES

From the earliest human settlements clay tiles have been used as a common form of roof covering for the majority of permanent buildings. Readily available local clay was hand pressed to shape and burned with crude equipment in the making of roofing tiles.

Flat, plain tiles were used in the wet climates of northern Europe and half round and flat unders with round overs in the drier climates of southern Europe and the Mediterranean basin. These roughly moulded and crudely burnt tiles varied noticeably in both shape and colour.

For the last hundred years many clay tiles have been made with selected, fine clay that is machine pressed to shape and burned in controlled kilns. Machine pressed tiles, which are uniform in shape, can be pressed to form tiles that combine flat unders with round overs in one tile, called a pantile. Pantiles can be accurately moulded with grooves in their edges that interlock to exclude wind and rain.

Plain tiles

Plain tiles continue to be one of the chosen roof coverings for small buildings, such as houses, in this country for their durability and appearance as a familiar roof covering.

A plain tile is a rectangular roofing unit of burned clay. Plain tiles are made in the standard size which was set by 'An Act for the making of Tile' passed in 1477 in the reign of Edward IV at $10\frac{1}{2}$ inches long, $6\frac{1}{4}$ inches wide and $\frac{1}{2}$ to $\frac{1}{8}$ inches thick (265 × 165 mm), to facilitate the replacement of missing or broken tiles by the imposition of a standard size.

Plain tiles are made with a small upward camber, so that when laid overlapping, the tail of a tile bears directly on the back of a tile below.

Early plain tiles were made with two holes for oak pegs on which the tiles were hung on battens. Plain tiles are now made with two nibs for hanging to battens, as illustrated in Fig. 181. The two holes near the head of the tile are for nailing tiles to battens.

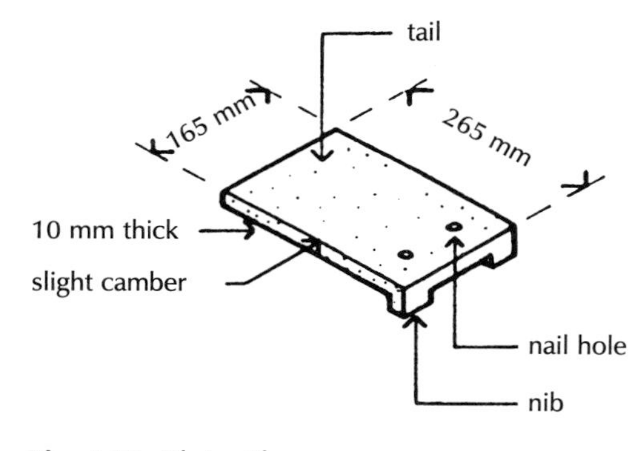

Fig. 181 Plain tile.

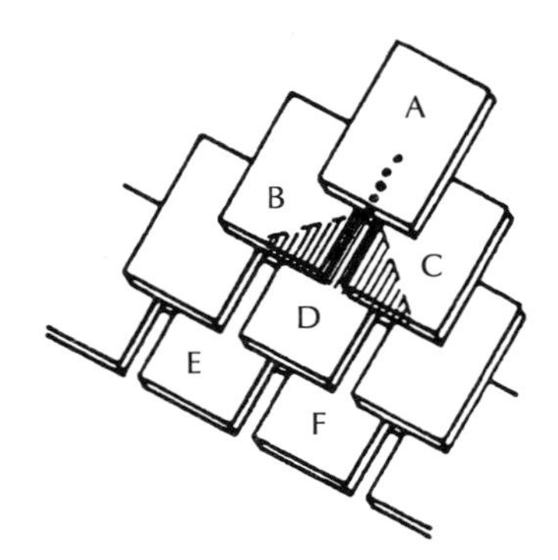

Fig. 182

Plain tiles are hung on battens in overlapping horizontal courses with the butt, side joints between tiles bonded up the roof so that rain runs down the slope from tile to tile to the eaves.

There are at least two thicknesses of tile at any point on the roof and also a 65 mm overlap of the tail of each tile over the head of tiles two courses below. The reason for this extra overlap is that, were the tiles laid in a straightforward overlap rain might penetrate the roof.

Rain running off tile A in Fig. 182 will run into the gap between tiles B and C, and spread over the back of tile D, as indicated by hatching, and probably run over the head of tiles E and F into the roof. It is against this possibility that a double end lap is made.

Lap, gauge and margin

With a 65 mm double end lap, plain tiles are laid to overlap 100 mm $\left(\frac{265-65}{2}\right)$ up the roof slope. The softwood battens on which the tiles are hung must therefore be fixed at a gauge (measurement) of 100 mm centres. The tail end of each tile shows 100 mm of the length, which is described as the margin, as illustrated in Fig. 183.

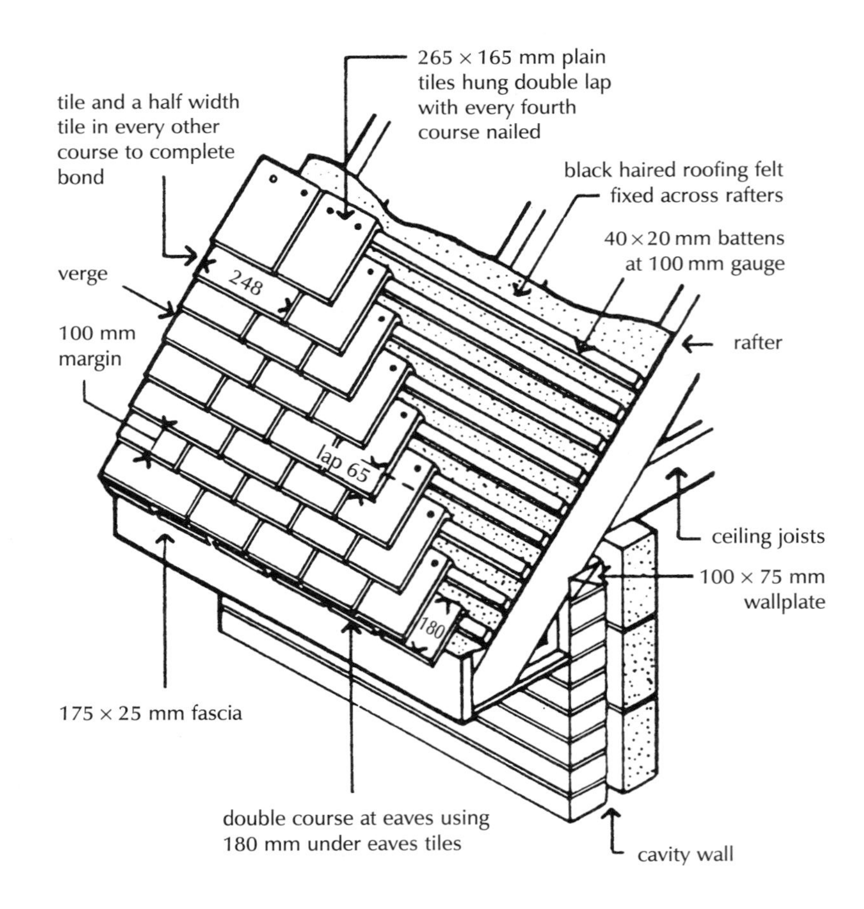

Fig. 183 Plain tiling.

Minimum slope of roof

The minimum roof slope for plain tiles depends on the density of the burned tile. Hand made tiles which fairly readily absorb water should not be laid on a roof slope pitched at less than 45° and the more dense machine pressed tiles at not less than 35°. Because of the thickness and double lap of these tiles the actual slope of a tile on a roof will be a few degrees less than that of the roof slope.

Roofing felt

The open butt, side joints of tiles and inexact fit of tile over tile would allow wind to blow into the roof space. To exclude wind it is practice to cover roof slopes with roofing felt. Rolls of bitumen impregnated roofing felt 813 mm wide are laid across roof rafters from the eaves upward with widths of felt lapped 75 to 150 mm up to the ridge. The felt covering is secured by timber battens, as illustrated in Fig. 183.

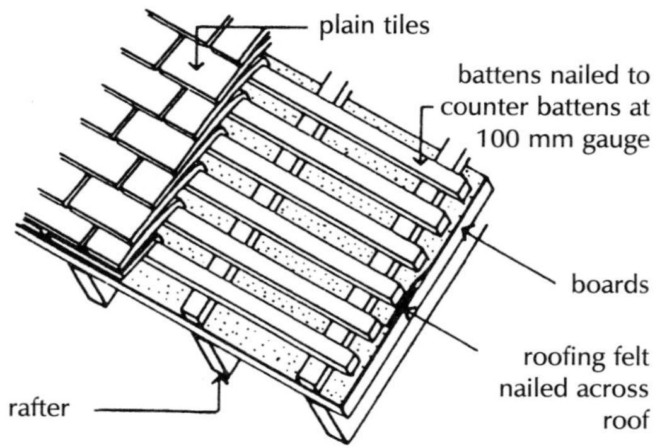

Fig. 184 Roof boarding and counter battens.

When it is planned to use a roof space for dry storage it is usual to cover roof slopes with either plain edge or tongued and grooved boarding laid across and nailed to rafters to prevent wind and dust entering more effectively. Rather than nail tiling battens directly to roof boarding it is usual to fix counter battens up the slopes, over roofing felt as illustrated in Fig. 184. The roofing felt will then conduct any water that penetrates the tiles down to eaves.

An advantage of roof boarding, sometimes called sarking, is that it acts to brace a pitched roof against its inherent instability across slopes.

Tiling battens

Plain tiles are hung on 38 × 19 mm or 50 × 25 mm sawn softwood battens which are nailed across rafters at 100 mm centres (gauge). As protection against the possibility of rain penetrating tiles and causing damage to battens by rot, the battens should be impregnated with a preservative. So that nails do not rust and perish they should be galvanised or made of a non-ferrous alloy.

The tiles in every fourth course are nailed to battens as a precaution against high wind lifting and dislodging tiles. In exposed positions every tile should be nailed.

Eaves

So that there are two thicknesses of tile at the eaves a course of eaves tile is used. These special length tiles are 190 mm long so that when hung to battens their tails lie directly under the tail of the course of full tiles above, as illustrated in Fig. 183. The tails of both the eaves tiles and full tiles above bear on the gutter board and project beyond the face of the board to discharge run off water into the gutter.

At the ridge a top course of tiles is used to overlap the course of full length tiles below to maintain the necessary two thicknesses of tile. These top course tiles are 190 mm long and hung on 40 × 40 mm battens which are thicker than normal battens, so that top course tiles ride over the tiles below. The ridge tiles are bedded on the back of the top course tiles so that the usual margin of tile is showing, as illustrated in Fig. 185.

The traditional and still commonly used method of covering the ridge of plain tile covered roofs is by one of the sections of clay ridge tiles made for the purpose. Half round and segmental ridge tiles are made for use in lower pitch roofs and the angle and hog back ridge tiles for more steeply pitched roofs. Figure 186 is an illustration of these tiles.

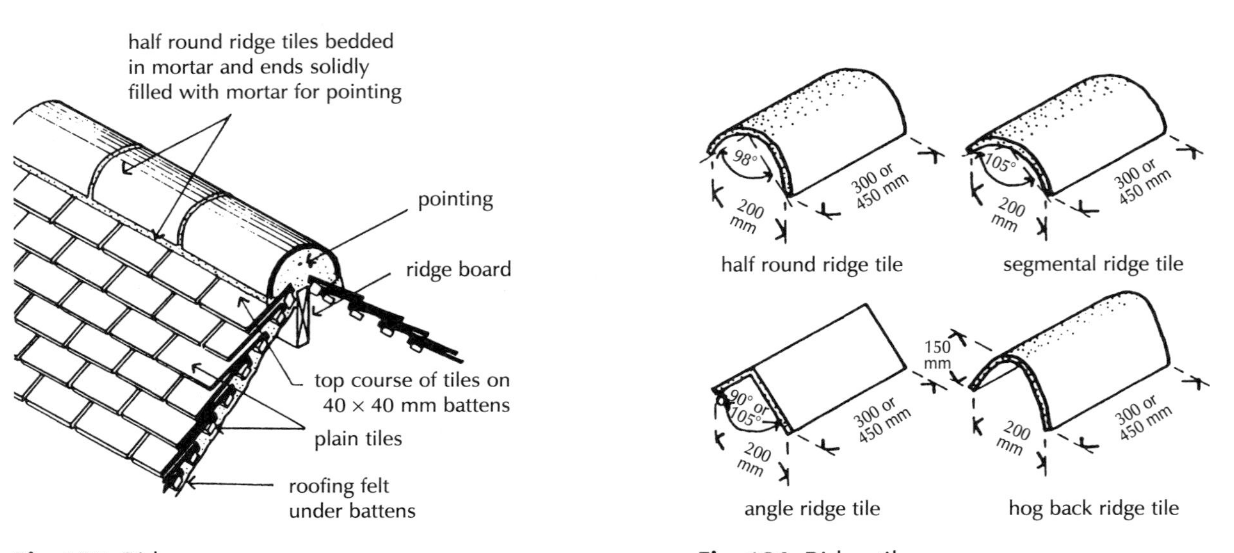

Fig. 185 Ridge.

Fig. 186 Ridge tiles.

The ridge tiles are bedded in cement and sharp sand mortar on to the back of the top course of tiles. The ends of the ridge tiles are solidly filled with mortar as a backing for the mortar pointing which is run in the butt end joints between ridge tiles, as illustrated in Fig. 185.

Which of the four sections of ridge tile is used depends in part on the slope of the roof and also on appearance. In addition to the ridge tiles illustrated a range of 'specials' is produced for decorative purposes.

Verges

The bonding of tiles at verges of plain tile roofing at the junction of a roof with a gable end and the junction of square abutments of slopes to parapet walls is completed with tile and a half width tiles

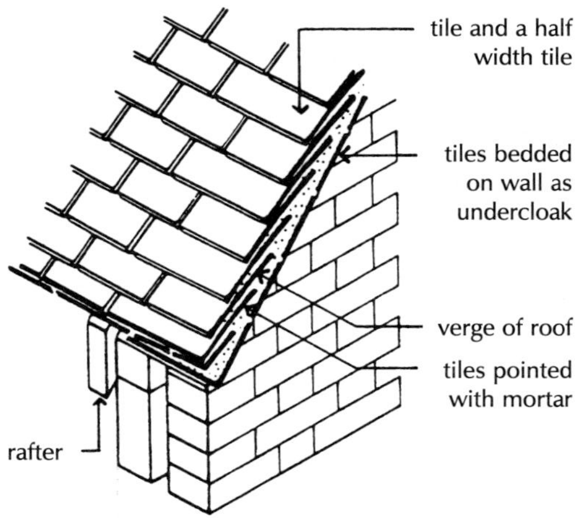

Fig. 187 Verge.

248 mm wide in every other course. The use of tile and a half width tiles at the verge of a tiled roof and a gable end is illustrated in Fig. 187. For details of the abutment and weathering to chimney stacks see Volume 2.

Tile and a half width tiles should be used instead of half width tiles, which might be dislodged by subsequent building work or by a heavy wind. In the gable end verge illustrated in Fig. 187, the verge tiling is hung to overhang the gable end wall by some 25 mm and the tiles are tilted slightly towards the roof slope to encourage rain to run down the roof slope rather than down the gable end wall.

As a bed for cement mortar pointing a layer of tiles without nibs is bedded in mortar on the gable end wall. On this layer of tiles cement mortar is spread as pointing between the verge tiles as weathering and for appearance sake.

Hip ends

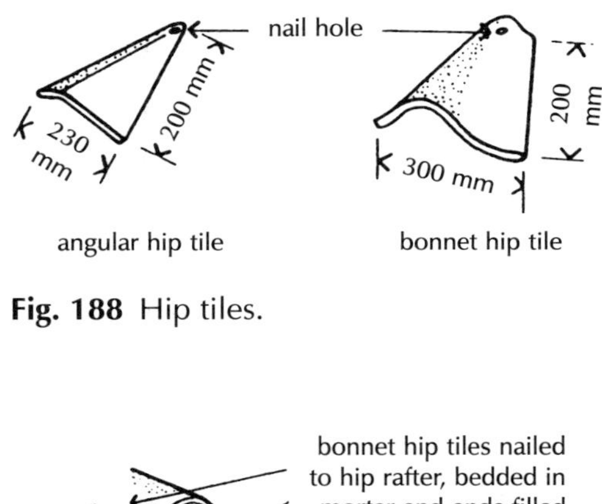

Fig. 188 Hip tiles.

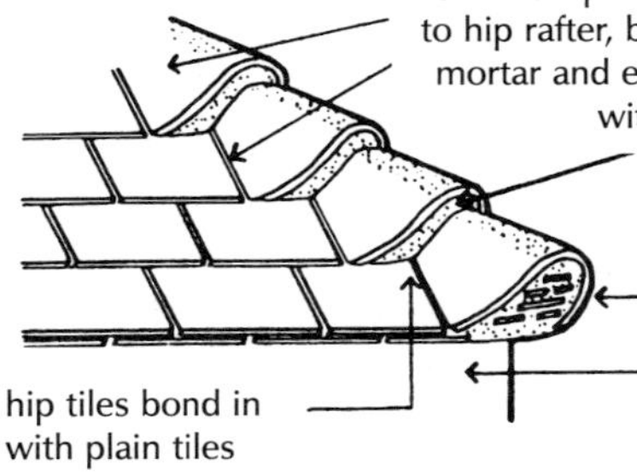

Fig. 189 Bonnet hip.

The hip ends of plain tile roofing slopes may be covered with one of the ridge tiles illustrated in Fig. 186. The plain tile courses each side of the hip are cut to fit the angle of the junction of the main slope and the hip end slope, using tile and a half width tiles as necessary, and the ridge tiles are bedded in and pointed with cement mortar as are ridge tiles on ridges. To prevent these tiles slipping down the slope of the hip a galvanised iron or wrought iron hip iron is fixed to the hip end or fascia to support the end hip tile.

Hip tiles are manufactured to bond in with the courses of tile in the adjacent slopes to provide a more pleasing appearance to the roof. Two sections of hip tile are produced. The angular and the bonnet hip tile, illustrated in Fig. 188, have holes for nailing the tiles to the hip rafter.

The hip tiles are nailed to the hip rafter up the slope of the hip and overlapping so that the tail of each tile courses in with a course of plain tiles. The hip tiles are bedded in mortar on the back of plain tiles which are cut to fit close to the hip and the end of hip tiles is filled with cement mortar as illustrated in Fig. 189. For the sake of appearance the end hip tile is filled with mortar with slips of cut tile bedded in the mortar.

Of the two hip tiles used the bonnet hip tile provides the most satisfactory appearance of continuing the course of tiles from a main roof slope to the hip end slope by the sweep of the bonnet hip end.

Valleys

At the junction of plain tile covered roof slopes in a valley, formed by an internal angle of walling, either a lead gutter is formed or special valley tiles are used.

The more usual valley is formed as a lead lined gutter dressed into a gutter formed by timber valley boards. The valley gutter should be

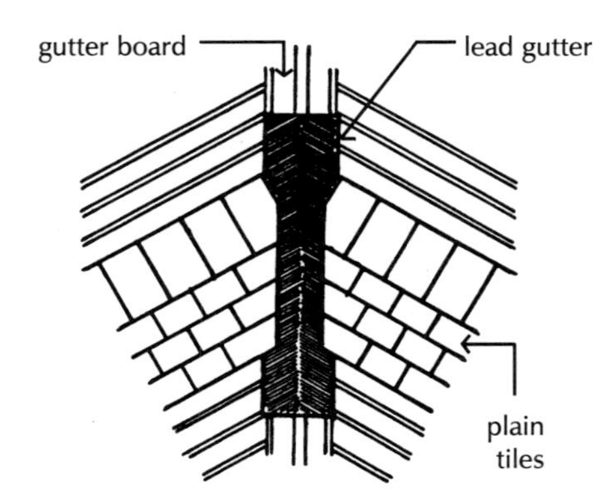

Fig. 190 Lead valley gutter.

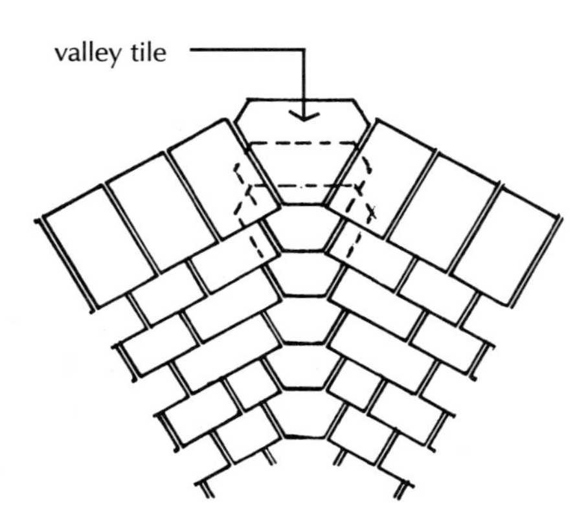

Fig. 191 Valley tiles.

wide enough to allow such debris as dust and leaves to be washed down to eaves without being obstructed and so blocking the gutter. A clear width of at least 125 mm is usual.

Gutter boards 25 or 38 mm thick and some 200 mm wide are nailed to the top of rafters, each side of the valley, with a triangular fillet of wood in the bed of the gutter as illustrated in Fig. 190. Tiling battens are continued down over the edges of the gutter boards. Roofing felt is laid under battens and over a double depth batten nailed at the cut ends of roofing battens.

A sheet or lapped end sheets of Code No 4 or 5 lead is laid on sheathing felt in the bed of the gutter, dressed up and nailed to the double depth battens. Plain tiles and tile and a half width tiles are cut to the rake of the valley and hung to battens as an under tile, over which full tiles are hung.

As an alternative to a valley gutter special valley tiles may be used to course in with the tile slopes each side of the valley.

Some of the specialist manufacturers of tiles provide valley tiles of the same material as their plain tiles. These valley tiles are shaped so that they bond in with tiles on main slopes pitched at specific angles. The length of the tile provides for the normal end lap. The dished tile tapers towards its tail, as illustrated in Fig. 191. The valley tiles are nailed to tiling battens continued down to the valley with plain tiling hung to side butt to valley tiles using tile and a half width tiles as necessary.

The advantage of this arrangement is that the tiled slope on one side is swept down into and around the internal angle of the valley to the adjoining slope for the sake of appearance. Because of the limitation of slope to suit standard valley tiles they are used principally for roofs that are a feature of the design where the roof is a clearly visible feature.

Concrete plain tiles

Concrete plain tiles were for a time used as a substitute for clay tiles. The tiles are made from a mixture of carefully graded sand, Portland cement and water which is compressed in a mould. A thin top dressing of sand, cement and colouring matter is pressed into the top surface of the tile. The moulded tiles are then left under cover for some days to allow them to harden.

Concrete tiles are uniform in shape, texture and colour. The colours in which these tiles have been made are poor when compared to natural clay tiles. The colouring of these tiles is only in the top surface and in course of time the colour bleaches in the sun and tends to be washed out, leaving anaemic looking tiles. For this reason concrete plain tiles have lost favour.

Single lap tiling

Single lap tiling is so called because each tile is laid to end lap the tile below down the slope of a roof and to side lap over and under the tiles next to it, in every course of tiles. The two types of single lap tile that have, for very many centuries, been used throughout the world are the rounded unders (channel) and overs (cover) and the flat unders (channel) and round overs (cover) tiles illustrated in Fig. 192. These traditional tiles are called Spanish and Italian tiles respectively in this country, even though they are not specific to those countries.

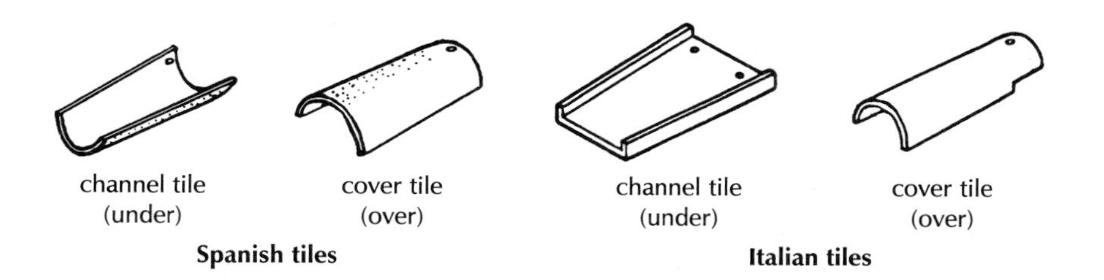

Fig. 192 Single lap tiles.

Spanish tiles

The traditional, rounded, tapering unders and overs of these tiles have been hand pressed or turned from local clay and burned in crude kilns to the shapes illustrated in Fig. 192 in a variety of sizes. Typical sizes are 350 mm length for both unders and overs, 180 and 130 mm width for unders, 200 and 155 mm width for overs and 70 and 100 mm depth for unders and overs respectively. After moulding, tiles may be perforated with nail holes in the bed of the head of unders and the sides of overs for fixing.

In hot, dry climates the tiles have been laid between and over rough roof rafters. Unders are laid between rafters, spaced for the purpose, with the wider head of the unders nailed to the sides of rafters and the narrower tail of the unders overlapping the lower under by some 75 mm. The narrower head of the overs is nailed to a rafter to overlap unders and end lap over the back of the over below.

The crudely shaped overs and unders were laid at slopes of as little as 20° to provide cover against rain with the open joints between tiles unfilled to encourage circulation of air into and out of the roof space to minimise build-up of heat.

Used in colder, wetter climates these tiles have more recently been laid as covering to roofs pitched at 20° to 30°. Plain edge roof boarding covered with roofing felt is nailed across the rafters. Battens 75 × 50 mm are nailed with the long axis vertical at centres to suit the unders. The unders are nailed to the side of battens and the overs to the top of battens with 75 mm end laps, as illustrated in Fig. 193.

Ridge tiles are bedded in mortar on the back of tiles and butt end joints pointed. At eaves the ends of unders and overs overhang the

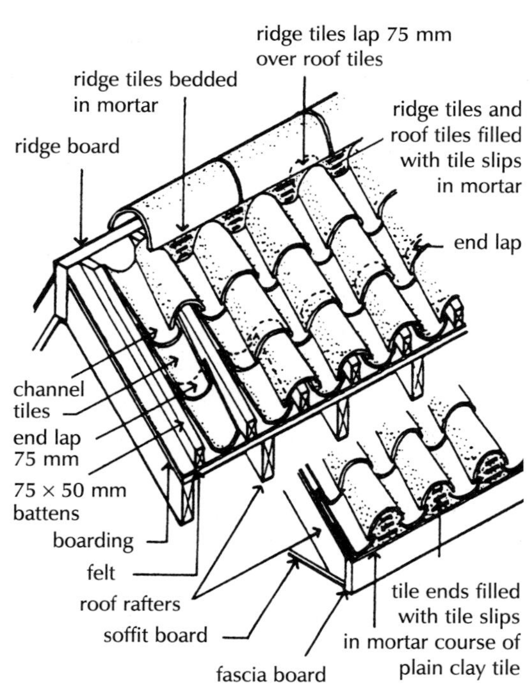

Fig. 193 Spanish tiles.

gutter board to shed water and the open ends of overs are filled with mortar and tile slips on the back of plain tiles.

The roof boarding and felt is used to exclude wind from the roof space as a more sophisticated wind check than the bedding of the overlap of tiles in some form of mortar which has been used in the past.

Italian tiles

The traditional flat channel (under) tile and rounded cover (over) tile of Italian tiles, illustrated in Fig. 194, were laid as roof covering in the same way as Spanish tiles with the head of unders nailed to roof boarding and the over to battens with an end lap of 75 mm. Ridge tiles are bedded in mortar and the open ends of overs at eaves filled with mortar, as illustrated in Fig. 195.

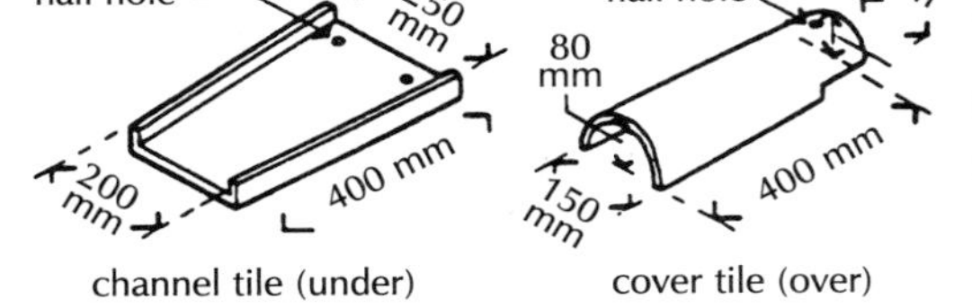

Fig. 194 Italian tiles.

Spanish and Italian tiles which are little used in this country have been replaced by pantiles in which the rounded unders and overs are combined as one pantile and the flat unders and rounded overs of Italian tiles are combined as single and double Roman tiles, illustrated in Fig. 196.

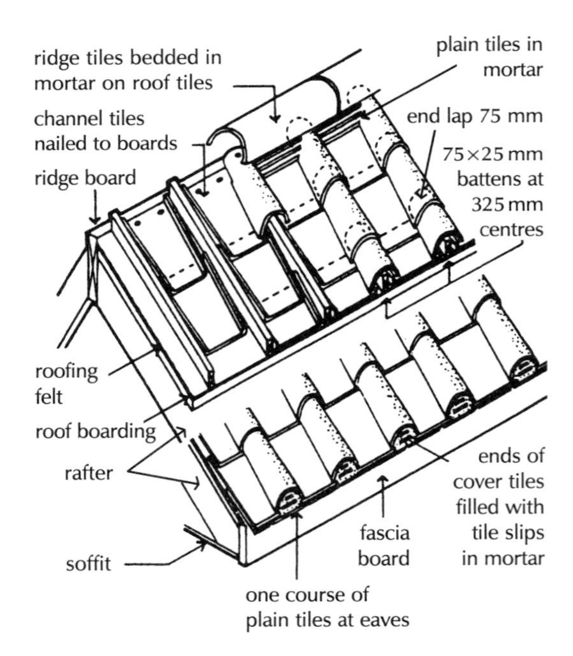

Fig. 195 Italian tiles.

Pantiles

A pantile incorporating the rounded unders and overs of Spanish tiles in one tile is illustrated in Fig. 196. By the use of selected clays or cement and sand and machine pressing this larger tile can be produced in standard sizes of uniform shape. The advantage of this tile is that it can be hung to be comparatively close fitting to exclude wind and rain.

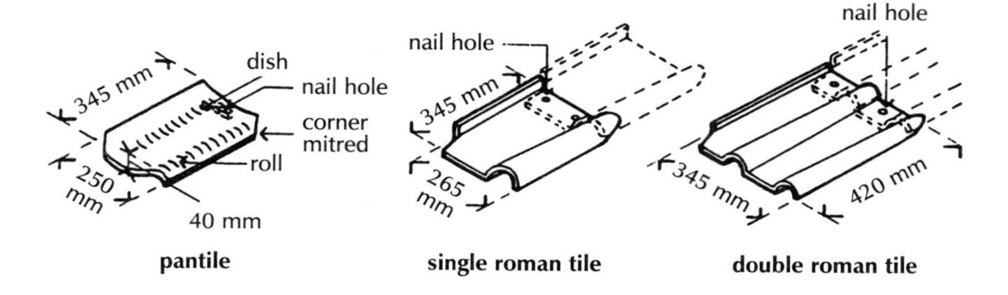

Fig. 196 Pantiles.

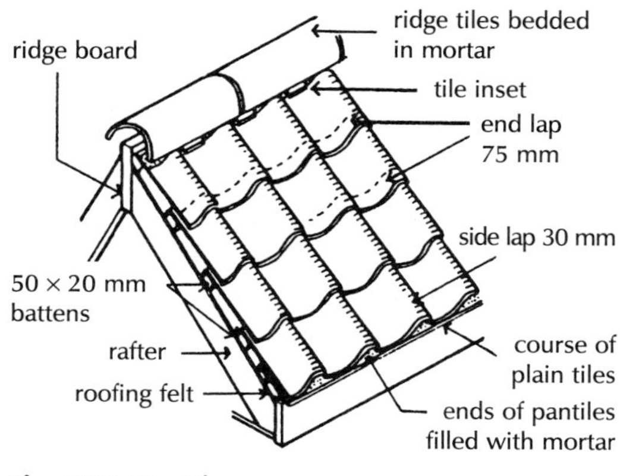

Fig. 197 Pantiles.

Both clay and concrete pantiles were fairly extensively used before the advent of interlocking tiles. Clay pantiles have been made with a fired on glazed finish to all exposed faces and edges, in a range of vivid colours, for the dramatic effect of this roof covering.

Like plain tiles, pantiles are made with a nib for hanging on to battens and a nail hole for securing tiles by nailing to battens against wind uplift. The tiles are hung on to 50 × 20 mm sawn softwood battens, nailed to rafters over roofing felt. The tiles are hung with an end lap of 75 mm at a gauge of 270 mm. The purpose of the mitred corners of pantiles is to facilitate fixing. But for the mitred corners there would be four thicknesses of tile at the junction of horizontal and vertical joints which would make it impossible to bed tiles properly.

Ridge tiles are bedded over the backs of pantiles and pointed at butt joints in cement and sand. As a bed for mortar filling, a course of plain tiles is hung at eaves on which mortar and tile slips are used to end fill pantiles. The pitch of roof for pantiles may be as low as 20° to 35°, depending on exposure. Figure 197 is an illustration of a roof covered with pantiles.

Interlocking single lap tiles

By careful selection of clay or sand and cement and machine pressing, a range of clay and concrete interlocking tiles is produced. Grooves in the vertical long edges of tiles are designed to interlock under and over the edges of adjacent tiles to exclude wind and rain. A range of profiles is made from the interlocking double pantile to the flat pantile illustrated in Fig. 198. There is a wider range of concrete than clay pantiles because of the difficulty of controlling shrinkage of clay during firing to produce a satisfactory interlock at edges.

There is no accepted standard size of interlocking tile although they are of similar size of about 420 mm long by 330 mm wide for concrete and about 320 mm long by 210 mm wide for clay.

These comparatively thick tiles are made with nibs and hung on 38 by 25 mm sawn softwood battens on a roofing felt underlay on rafters at a pitch of from 15° to $22\frac{1}{2}°$ to horizontal. A single end lap of 75 mm is usual with the side butt lap dictated by the system of interlock, as illustrated in Fig. 199.

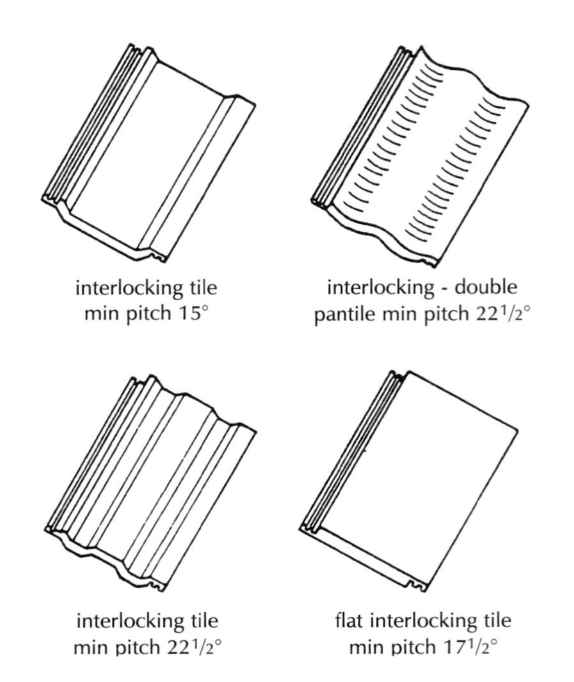

Fig. 198 Interlocking tiles.

These tiles are not usually made with holes for nails, instead a system of aluminium clips is used. The aluminium clips hook over an upstand side edge at the head of a tile, under the end lap, and are nailed or screwed to the back edge of battens. Clips are used in every course in severe exposure or every other course or more for less exposed conditions.

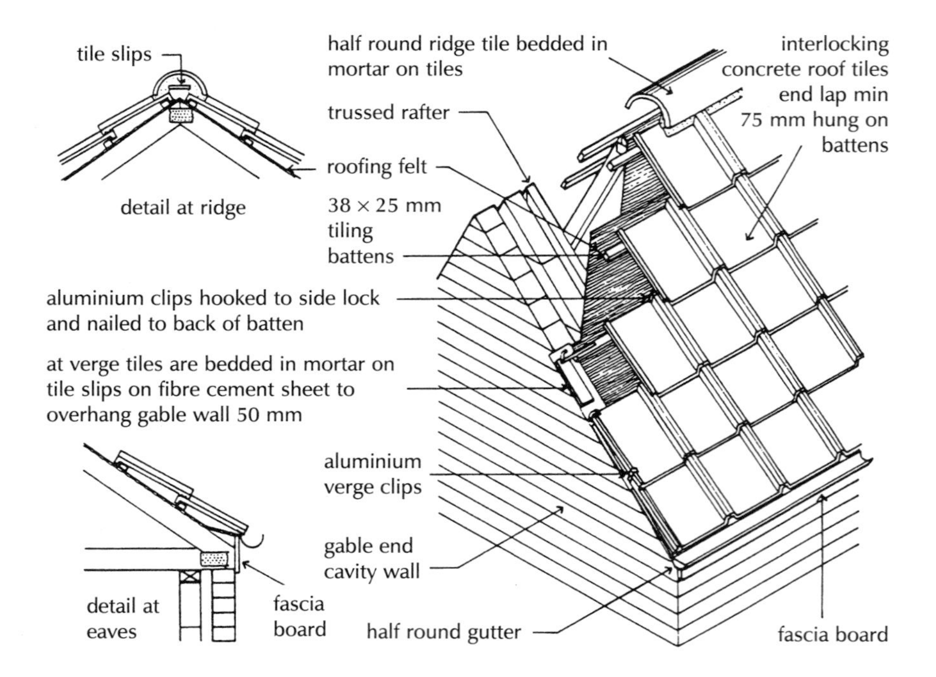

Fig. 199 Interlocking tiles.

At the ridge a half round ridge tile is bedded in mortar with end joints pointed. At the eaves a course of plain tiles provides a bed for mortar filling to the open ends of profiled tiles. Because of their size these tiles are best suited to the larger expanses of roof slopes.

Hips and valleys

At hips and valleys profiled tiles have to be cut to fit either under hip tiles or to a lead lined gutter to valleys. Because of the thickness of the tiles they have to be cut mechanically to provide a neat cut edge. No matter how carefully cut, the junction of profiled tiles with hips and valleys provides a ragged finish.

SLATES

Thin slabs of natural stone have for many centuries been a traditional roof covering material in areas where natural stone can be split and shaped as a slate, sometimes called stone tiles. The thickness, size and durability of natural stone slate depends on the nature of the stone and its ease of splitting into a useable size. Natural stone slates vary from the smooth thin Welsh slate, to the thicker rough Westmorland slate and the thick stone slabs used in the highlands of Scotland.

Welsh slates

Welsh slates were extensively quarried in the mountains of Wales up to the middle of the twentieth century, and continue, to a limited extent, to be today. Many buildings covered with these slates survive today throughout this country.

The stone from which they are split is hard and dense and varies in colour from light grey to purple. In the quarry the stone consists of thin layers or laminae of hard slate with a very thin, somewhat softer, layer of slate between the layers. By driving a wedge of steel into the stone between the layers it can be quickly split into fairly large, quite thin slates of thicknesses varying from 4 to 10 mm. The splitting results in slates of varying thickness and the thicker slates are classified as best, strong best and medium, and the thinner slates as seconds and thirds.

Traditional sizes of slate were 24 × 12, 20 × 10, 18 × 9 and 16 × 8 old Imperial inches, the equivalent metric sizes being 600 × 300, 500 × 250, 450 × 225 and 400 × 200 mm.

Good quality Welsh slates are hard and dense and do not readily absorb water. They are not affected by frost or the dilute acids in industrial atmospheres and will have a useful life as a roof covering of very many years.

Reclaimed slates

Because of the considerable cost of new Welsh slates it has been practice for some years to use reclaimed slates recovered from demolition work. The disadvantage of these reclaimed slates is that they may vary in quality, be damaged by being stripped from old roofs and there is little likelihood of recovering the slate and a half width slates necessary to complete the bond at verges and abutments.

Westmorland slates

Westmorland slates are quarried from stone in the mountain region of the Lake District. The stone varies in colour from blue to green with flecks and streaks of browns and greys. The stone is very hard and dense, and consists of irregular layers or laminae of hard stone, separated by very slightly softer stone. The laminae do not run in regular flat planes as in Welsh slate and the stone is more difficult to split than the Welsh. As a consequence it is not so economic to split slates and cut them to uniform size and Westmorland slates are very hard and so dense that they absorb practically no water no matter how long they are immersed in water. These slates are practically indestructible. Because of the considerable labour required to cut the stone and fix slates in random courses these slates are expensive and less used than they were.

Spanish slates

For some years now natural stone Spanish slates have been imported. These slates, which are considerably cheaper than new Welsh slates, are of a uniform dark grey colour, comparatively thin and smooth faced and cut to the traditional English slate size. Some of these slates have flecks of what is called fools' gold in them, which may affect their density and durability. These natural looking slates which have been in use for some years now appear to be a satisfactory substitute for Welsh slates. Their long term durability in this climate has yet to be tested.

Chinese slates

More recently, Chinese slates have been imported in traditional English sizes. They have much the same appearance as Spanish slates. It is too early to judge their durability.

Fibre cement slates

From the middle of the twentieth century slates made from asbestos fibres, cement and water were used as a cheap substitute for Welsh slates. Because of the hazard to the health of those making these slates, by the use of asbestos, they are now made with natural and synthetic fibres.

These slates are made from pigmented Portland cement and water, reinforced with natural and synthetic fibres. The wet mix is compressed to slates which are cured to control the set and hardening of the material. The standard slate, known by the trade name 'Eternit', is 4 mm thick, with a matt coating of acrylic finish and a protective seal on the underside against efflorescence and algal growth. The standard slate is rectangular and holed for nail and rivet fixing. These slates are made in three colours, blue/black, brown and rose and with either square tails or curved, angled or chamfered tails for decoration.

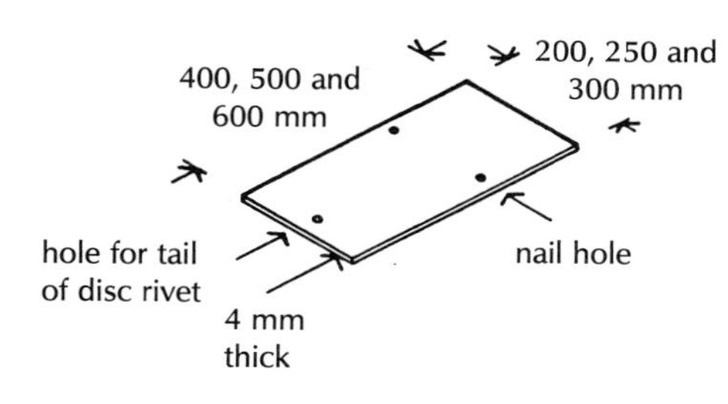

Fig. 200 Fibre cement slate.

The standard fibre cement slate is uniform in shape, colour and texture. Fibre cement slates have a useful life of about 30 years in normal circumstances as compared to a useful life of 100 years or more for sound natural Welsh slates.

Standard fibre cement slates are made in sizes of 600 × 300, 500 × 250 and 400 × 200 mm, as illustrated in Fig. 200, with a comprehensive range of fittings for ridge, eaves, verges and valleys.

Two somewhat more expensive forms of fibre cement slates are made to simulate the appearance of natural slate in colour and texture and the other with a random brick coloured surface.

Fixing natural slates

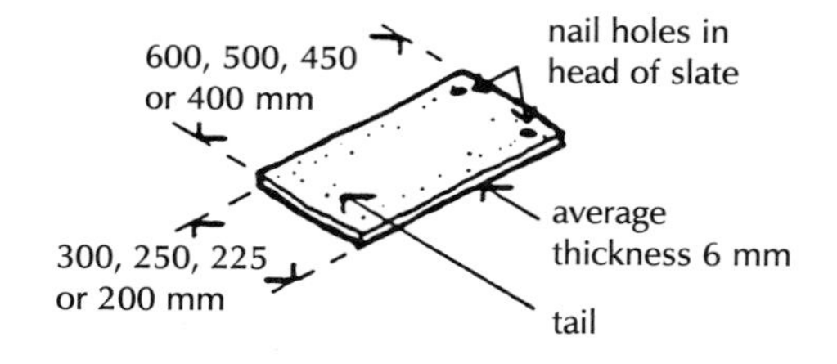

Fig. 201 Natural slate.

Slates are nailed to 50 × 25 mm sawn softwood battens with copper composition or aluminium nails driven through holes which are punched in the head of each slate. Galvanised steel nails should not be used as they will in time rust and allow the slates to slip out of position. Two holes are punched in each slate some 25 mm from the head of the slate and about 40 mm in from the side of the slate, as illustrated in Fig. 201.

The battens are nailed across the roof rafters over roofing felt and the slates nailed to them so that at every point on the roof there are at least two thicknesses of slate and so that the tail of each slate laps 75 mm over the head of the slate two courses below. This is similar to the arrangement of plain tiles and is done for the same reason. Because the length of slates varies and the end lap is usually constant it is necessary to calculate the spacing or gauge of the battens. The formula for this calculation is

$$\text{gauge} = \frac{\text{length of slate} - (\text{lap} + 25)}{2}$$

For example the gauge of the spacing of the battens for 500 mm long slates is:

$$\frac{500 - (75 + 25)}{2} = 200$$

The 25 that is added to the lap represents the 25 mm that the nail holes are punched below the head of the slate so that the 75 mm end lap is measured from the nail hole.

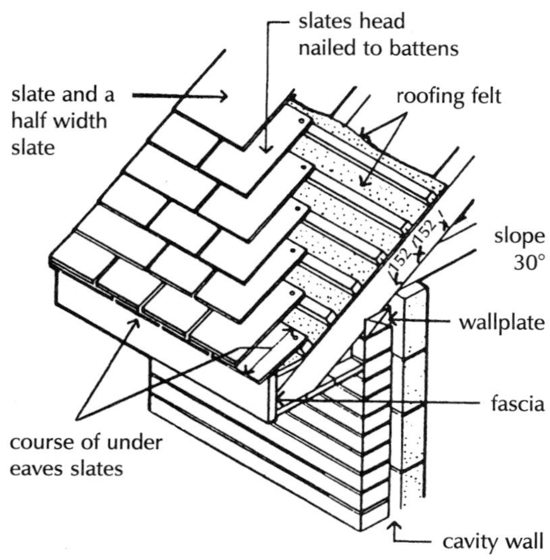

Fig. 202 Head nailed slates.

Figure 202 is an illustration of natural slates head nailed to preservative treated softwood battens with an end lap of 75 mm and the side butt joints between slates breaking joint up the slope of the roof. The roof is pitched at 25° to the horizontal, which is the least slope generally recommended for slates.

At verges and square abutments a slate and a half width slate is used in every other course to avoid using a half width slate to complete the bond.

Eaves

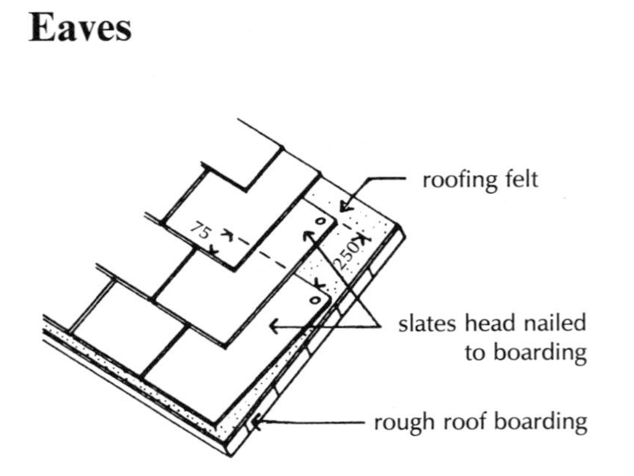

Fig. 203 Slates fixed to boards.

So that there shall be two thicknesses of slates at the eaves a course of undereaves slate is used. These slates are cut to a length equal to the gauge + lap + 25 of the slating, as illustrated in Fig. 202.

When the roof slopes are boarded to provide a wind tight, dry roof space the slates may be nailed directly through roofing felt to the boarding, as illustrated in Fig. 203.

To provide a space between the slates and the felt and boarding, down which any water that penetrates the slates may run to eaves, it is good practice to fix a system of counter battens up roof slopes, as illustrated in Fig. 184 for tiles.

Slates are usually fixed by means of nails driven through holes in the head of slates. This is the best method of fixing slates as the nail holes are covered with two thicknesses of slate so that even if one slate cracks, water will not get in. But if long slates such as 600 mm are head nailed on a shallow slope of, say, 30° or less it is possible that in a high wind the slates may be lifted so much that they snap off at the nail holes.

Centre nailed slates

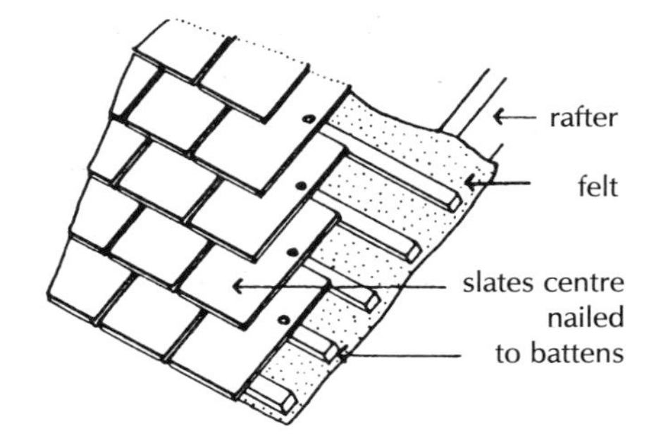

Fig. 204 Centre nailed slates.

In exposed positions on low pitch roofs it is common to fix the slates by centre nailing them to battens. The nails are not driven through holes exactly in the centre of the length of the slate, but at a distance equal to the gauge down from the head of the slate, so that the slate can double lap at tails as illustrated in Fig. 204. With this method of fixing there is only one thickness of slate over each nail hole so that if that slate cracks water may get into the roof.

The arrangement of a double course at eaves and ridge and the use of slate and a half width slates at square abutments and battens and roofing felt is the same for both head and centre nailed slates.

Ridge

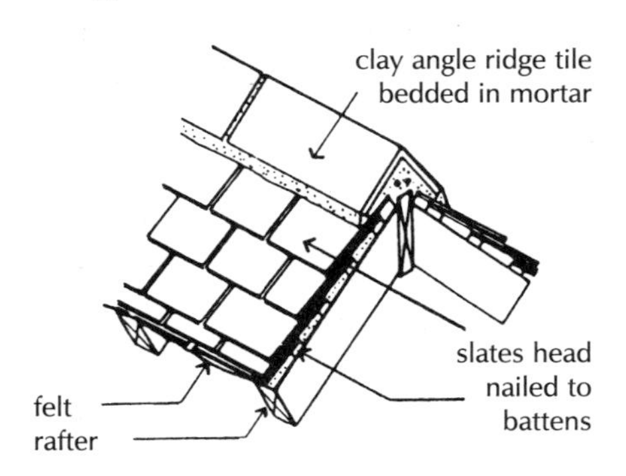

Fig. 205 Tile ridge.

Common practice is to cover the ridge, at the intersection of two slated slopes, with clay ridge tiles. So that there is a double thickness of slate a top course of slates is used at ridges. These shorter slates are usually the gauge plus lap plus 50 or 75 mm long and head nailed to a double thickness batten.

The dark colour, clay, angle ridge tiles are bedded in cement mortar on the back of the top course slates. Their ends are solidly filled with and joints pointed in cement mortar, as illustrated in Fig. 205.

A somewhat more expensive, and durable, finish is to use a sheet lead ridge capping, as illustrated in Fig. 206.

A wood roll, cut from a 50 mm square softwood section, is nailed to

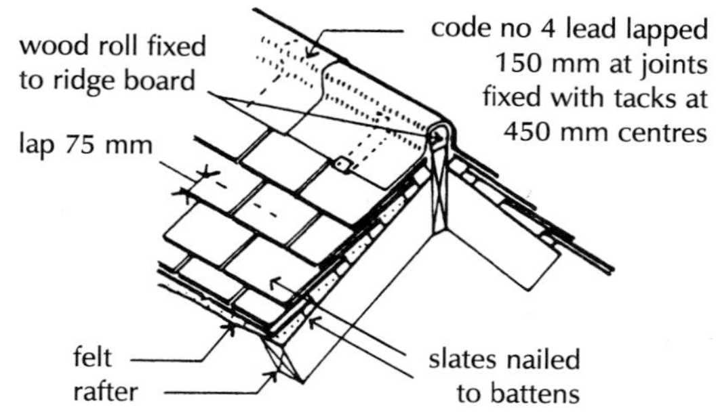

Fig. 206 Lead ridge.

the ridge board which is deeper than usual. Code No 4 sheet lead about 450 mm wide is dressed around the wood roll and down to the slates both sides.

At joints between sheets of lead capping, there is a 150 mm overlap. To prevent the wings of the sheet lead being blown up in high wind, a system of 50 mm wide strips of sheet lead is nailed to the roll under the lead at 450 mm centres. These lead tacks are turned up and around the edges of the wings of the lead capping to keep them in place.

Hips

At hipped ends of roof, slates and slate and a half width slates in every other course are cut to the splay angle up to the hip. The hip is then weathered with clay ridge tiles bedded in cement mortar or with a sheet lead capping similar to the ridge capping.

Valleys

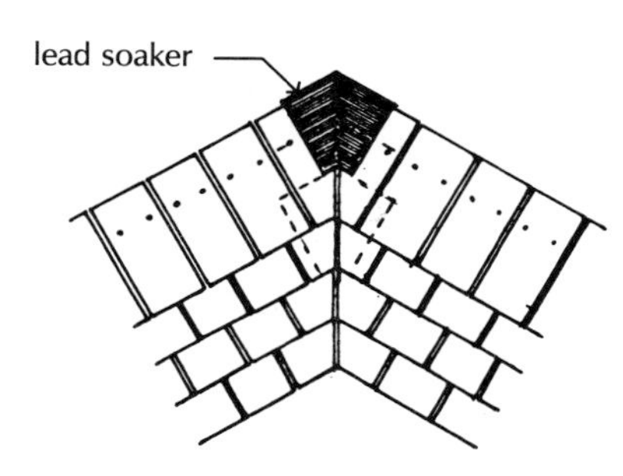

Fig. 207 Slate valley.

At the internal angle intersection of slated roof slopes the full and slate and a half width slates are splay cut up to the valley. A system of shaped lead soakers is used to weather the mitred valley. A shaped lead soaker cut from Code No 4 sheet lead is hung over the head of each course of slates. A similar soaker is hung over the head of slate courses above to overlap the soaker below, as illustrated in Fig. 207.

This involves a deal of wasteful cutting of slates. By careful cutting of slates and careful arrangement of bonding of slates a neat, mitred, watertight valley can be made.

Fixing fibre cement slates

Fibre cement slates are fixed double lap up the slope of roofs pitched at a minimum slope of 20° with a usual slope of 25° to the horizontal. The slates are fixed with the double end lap used with natural slates. The end lap may be the usual 75 mm, which is usually increased to 100 mm to provide cover for the centre nail fixing and the rivet hole.

These comparatively thin, brittle slates are centre nailed to avoid cracking the slate were it head nailed. Each slate is nailed with copper composition or aluminium nails to 38 × 25 mm or 50 × 25 mm, impregnated, sawn softwood battens nailed over roofing felt to rafters, as illustrated in Fig. 208. The gauge, spacing, of battens is calculated by the same formula used for natural slates.

To secure these thin, lightweight slates against wind uplift copper disc rivets are used. The disc of these rivets fits between slates in the under course and the tail of the rivet fits through a hole in the tail of the slate above. The tail of the rivet is bent up to hold the slate in position, as illustrated in Fig. 208.

At verges and square abutments, slate and a half width slate are used in every other course to complete the bond of slates up roof slopes.

At eaves two under eaves courses of slate are used. These cut or specially made short lengths of slate are head nailed to battens, as illustrated in Fig. 208.

At the ridge, special flanged end, fibre cement angle ridges are used. These 900 mm long ridge fittings are weathered by the overlap of the flanged ends, as illustrated in Fig. 209, and secured as illustrated with screws and plastic washers. The wings of the ridge fittings bear on the back of an under ridge course of slates without the need for cement and sand pointing.

At hips and valleys the slates are mitre cut to fit up to the hip or valley and the mitre cut slates weathered with lead soakers hung over the head of slates.

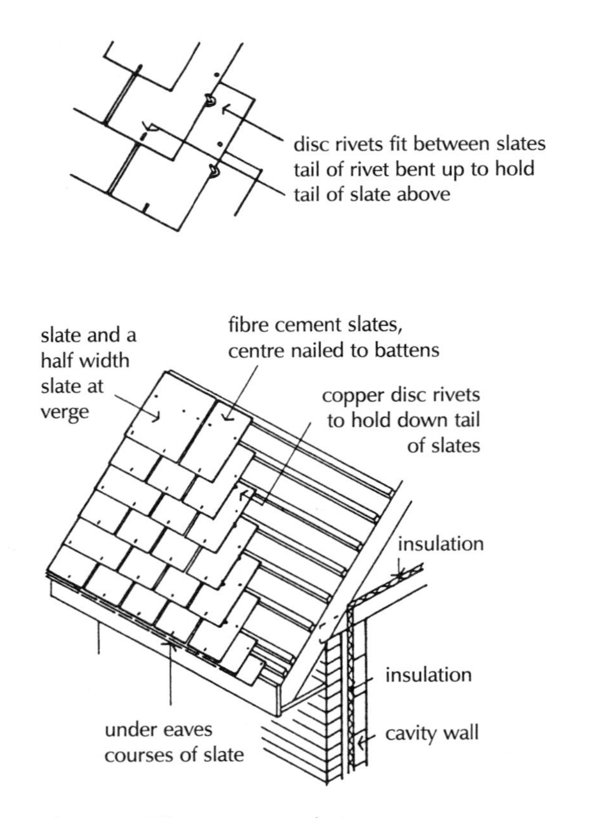

Fig. 208 Fibre cement slating.

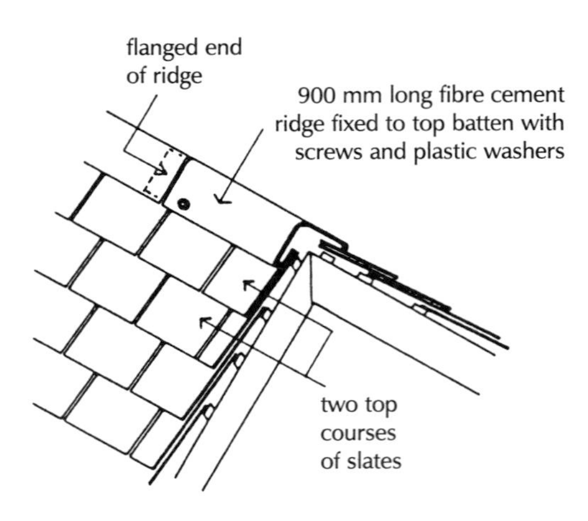

Fig. 209 Fibre cement slate ridge.

SHEET METAL COVERING TO LOW PITCH ROOFS

For many years from the middle of this century it was the fashion to construct low pitched roofs for houses, schools and other buildings. A pitched roof is generally defined as one with slopes of 10° or more to the horizontal and a low pitched roof has slopes of from 10° to 30° to the horizontal. The disadvantages of a flat roof were known and a steeply pitched roof is not always an attractive feature of small buildings. For example, a small bungalow with a pitched roof covered with tiles does not usually look attractive as the great areas of tiles dominates the lesser area of wall below. A low pitched roof is a happy compromise between flat and steeply pitched roofs. It at once looks attractive, gives reasonable insulation against loss of heat and provides roof space in which water storage cisterns can be housed.

The principal coverings for low pitched roofs are copper and aluminium strips, both of which are comparatively light in weight and therefore do not require heavy timbers to support them, and both have a useful life as a roof covering of very many years. The roofs are constructed as single slope roofs or as low pitched roofs with timber rafters pitched to a central ridge board with or without hipped ends. The construction of the rafters and their support with purlins and struts, or purlins and light timber trusses, is similar to that for other pitched roofs as previously explained.

Copper or aluminium strips 450 or 600 mm wide and up to 8.0 m long are used to cover these low pitched roofs. No drips or double lock cross welts or other joints transverse, across, the fall are used with strips up to 8.0 m long and because of this the labour in jointing the sheets is less than that required for the batten or conical roll systems used for lead sheet, consequently copper or aluminium strip coverings are cheaper. Because of the great length of each strip the fixing cleats used to hold the metal strips in position have to be designed to allow the metal to contract and expand freely.

Standing seams

Both copper and aluminium sheet are sufficiently malleable, workable, to be bent and folded without damage to the material in the formation of standing seam joints down roof slopes.

The strips of metal are jointed by means of a standing seam joint, which is a form of double welt and is left standing up from the roof as shown in Fig. 210, which is an illustration of part of a low pitched roof. The completed standing seam is constructed so that there is a gap of some 13 mm at its base which allows the metal to expand without restraint, and this is illustrated in Fig. 210. The lightweight metal strips have to be secured to the roof surface at intervals of 300 mm along the length of the standing seams. This close spacing of the fixing cleats is necessary to prevent the metal drumming due to uplift in windy weather.

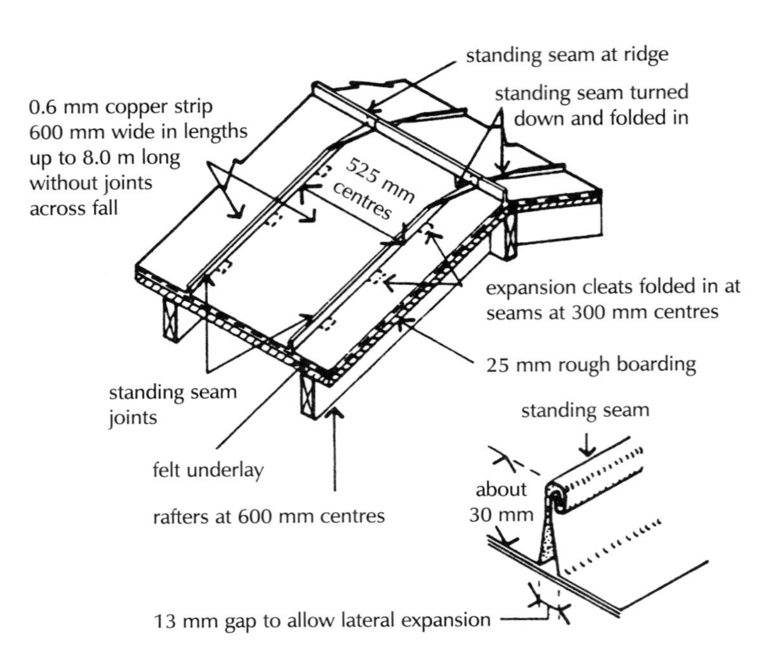

Fig. 210 Copper standing seam roofing.

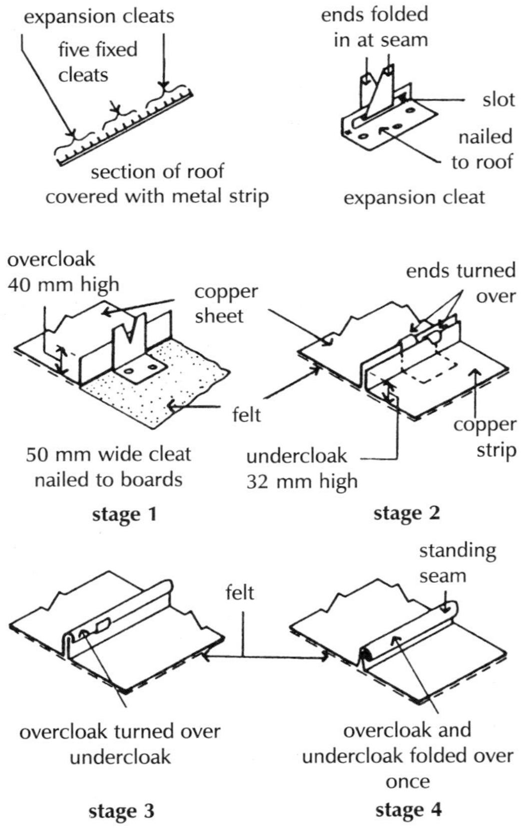

Fig. 211 Holding down cleats.

Two types of cleats are used, fixed cleats and expansion cleats. Five fixed cleats are fixed in the centre of the length of each strip and the rest of the cleats are expansion cleats. Figure 211 illustrates the arrangement of these cleats. The fixed cleats are nailed to the roof boarding through the felt underlay and Fig. 211 illustrates the formation of a standing seam and shows how the fixed cleat is folded in.

The expansion cleats are made of two pieces of copper strip folded together so that one part can be nailed to the roof and the second piece, which is folded in at the standing seam, can move inside the fixed piece. Figure 211 illustrates one type of expansion cleat used.

The ridge is usually finished with a standing seam joint, as illustrated in Fig. 212A, but as an alternative a batten roll or conical roll may be used. Whichever joint is used at the ridge, the standing seams on the slopes of the roof have to be turned down so that they can be folded in at the ridge. This is illustrated in Fig. 212A. Because copper and aluminium are generally considered to be attractive coverings to roofs, the roof is not hidden behind a parapet wall, and the roof slopes discharge to an eaves gutter, as illustrated in Fig. 212B.

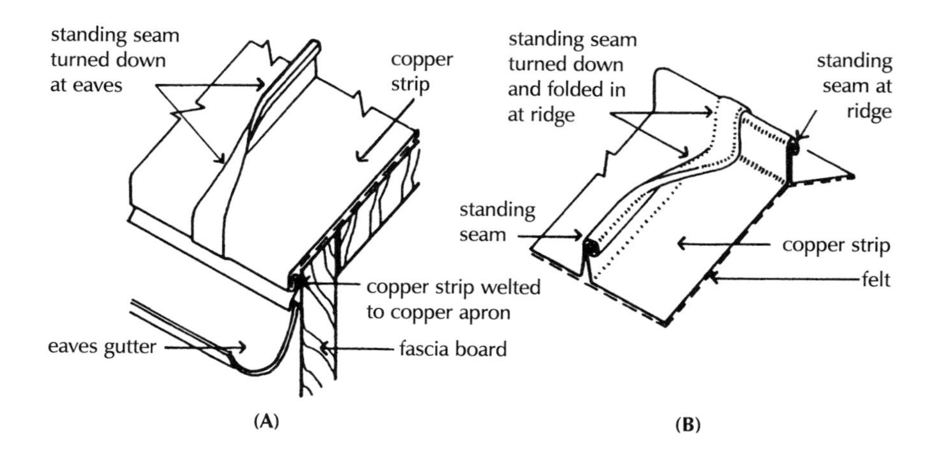

Fig. 212 (A) Ridge. (B) Eaves.

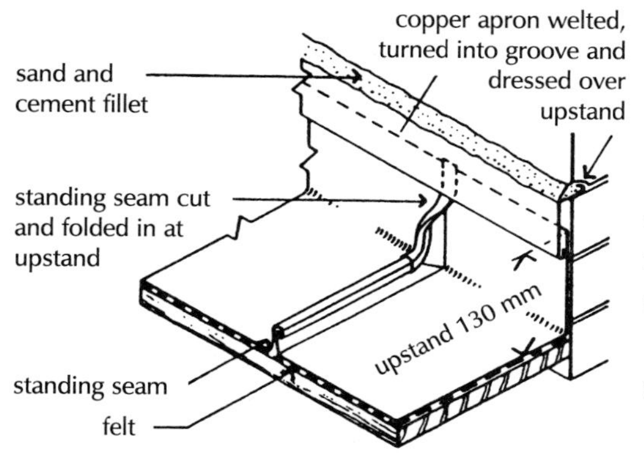

Fig. 213 Upstand.

Where the slope of the roof finishes at the parapet or wall, the strips of metal are turned up as an upstand and finished with an apron flashing. This is illustrated in Fig. 213, which also illustrates the cutting and turning down of the standing seam.

The verge of low pitched roofs can be finished with batten roll or conical roll, as illustrated for flat roof coverings.

Felt underlay

It is of importance that strip metal coverings be laid on an underlay of bitumen-impregnated roofing felt laid across the roof boards and nailed to the boards with butt side joints. The felt allows the metal strips to expand and contract without restraint.

Fire safety

The space inside a pitched roof is a void space that should be separated from other void spaces or cavities by cavity barriers that seal the junction of the cavities to prevent unseen spread of smoke and flames. The cavity in a wall is generally separated from the cavity in a pitched roof by the cavity barrier at the top of the cavity in the wall.

Separating walls, party walls, between semi-detached and terraced houses should resist the spread of fire from one house to the other by being raised above the level of the roof covering or by being built up to the underside of the roof covering so that there is a continuous fire break.

The traditional roof coverings, slate, tile and non-ferrous sheet metal, do not encourage spread of flame across their surface and are, therefore, not limited in use in relation to spread of fire between adjacent buildings.

Resistance to the passage of heat

Since the earliest requirement for thermal insulation to heated buildings, the requirements for, what is now termed, conservation of fuel and power have become increasingly stringent. The consequence of thermal insulation was at first not taken into account, so that the common sense need for moisture vapour barriers, ventilation and limits to cold draughts of air entering around ill fitting windows has been added from time to time. Most recently a standard assessment (SAP) energy rating procedure, which applies only to dwellings, has been added (see Chapter 2).

In the latest requirements in Approved Document L to the Building Regulations there should be a maximum U value of 0.2 W/m^2K for roofs to dwellings with an SAP rating of 60 or less and 0.25 W/m^2K for those with an SAP rating of over 60. For heated buildings other than dwellings a U value of 0.25 W/m^2K for roofs is required.

Two methods for the determination of the thickness of insulation required for roofs are suggested in Approved Document L. The first method is to relate the thickness of insulation required to the thermal conductivity of the insulating material to be used, ignoring the thermal resistance of the roof itself. This produces an indication of the insulation thickness required because the thermal resistance of the roof is small in comparison to that of the insulation.

As an example, if an insulant with a thermal conductivity of 0.04 W/m^2K is used and a U value of 0.25 W/m^2K is required the necessary thickness of insulation required can be read from a table as 227 mm for insulation between ceiling joists.

The second method is to make an allowance for the thermal resistance of other materials of the roof and so make some reduction in the thickness of insulation. The base thickness of insulation may be reduced by

	1 mm for roof tiles
	6 mm for roof space
	3 mm for 10 mm plasterboard
Total	10 mm reduction

The above figures are read off from a table in Approved Document L and related to an insulation with 0.04 W/m^2K conductivity and a U value of 0.25 W/m^2K for the roof.

As there is usually no advantage in using one of the thin, low conductivity, expensive insulating materials in a roof space, considerations of cost and convenience in fixing or laying are determining factors.

Insulating pitched roofs

Cold roof

The most convenient, economical and usual place for insulation in a pitched roof is either across the top of, or between, the ceiling joists. With insulation at ceiling joist level the roof is described as a cold roof.

Warm roof

A more expensive place to fix insulation is above, between or below the rafters of a pitched roof. Because the area of a pitched roof surface is greater than that of a horizontal ceiling there is a greater area to cover with insulation. The advantage of this arrangement is that the roof space will be warmed by heat rising from the heated building below and will, in consequence, be comparatively warm and dry for use as a storage place. With the insulation at roof level the roof is described as a warm roof.

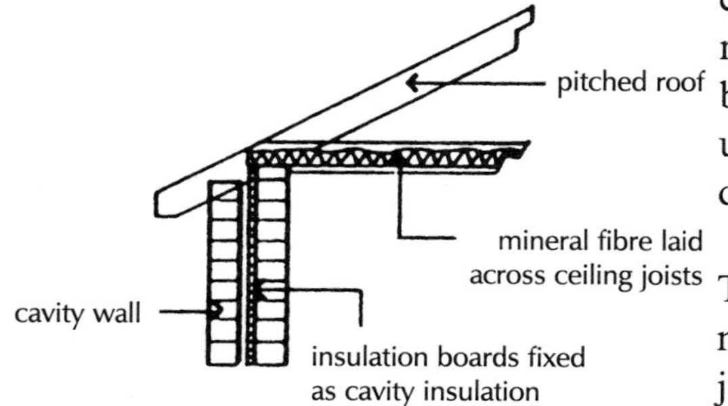

Fig. 214 Cold roof insulation.

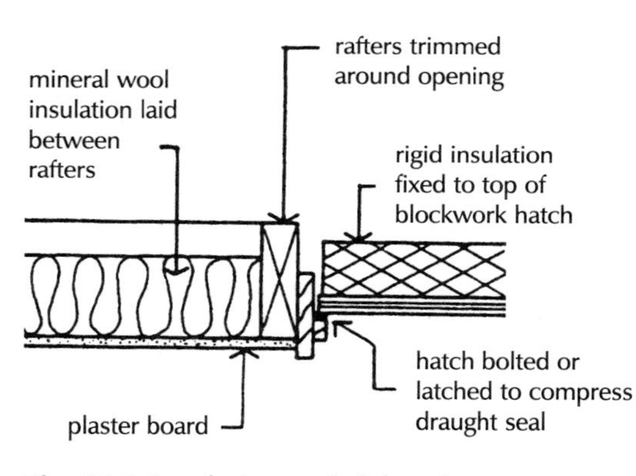

Fig. 215 Insulation to loft hatch.

The materials most used for cold roof insulation are mineral wool mats or rolls of fibre glass or rockwool spread across or between the joists or loose fill spread between the joists on top of the ceiling finish. The most straightforward way of providing a layer of insulation is to spread rolls of mineral fibre across the top of ceiling joists right across ceilings in both directions and extended up to and overlapping insulation in walls, as illustrated in Fig. 214.

So that the layer of insulation is continuous over the whole of the ceiling area it is recommended that loft hatches, giving access to roof spaces, be insulated and draught sealed, as illustrated in Fig. 215, and that where service pipes penetrate ceiling finishes some effective form of draught seal be formed.

All water carrying service pipes and water storage cisterns or tanks inside cold roof spaces must be effectively protected by a sheath of insulating material, called lagging, against the possibility of water freezing, expanding and damaging them.

With mats or rolls of mineral fibre spread across ceiling joists there is a possibility of the loose material being compressed and losing efficiency as an insulator, under walkways for access in roof spaces. Boarded access ways inside roofs should, therefore, be raised on battens above the level of the insulation.

The disadvantage of spreading or laying mineral fibre insulation or insulation boards between ceiling joists is that there will be a deal of wasteful cutting in fitting the material closely between joists. Where insulation is fixed between ceiling joists an allowance for the different thermal conductivity of wood and the insulant has to be made in calculations of thermal insulation requirements.

The insulating materials most used for cold, pitched roofs are set out in Table 10.

Table 10 Insulation materials for cold roofs.

Pitched roof cold roof	Thickness mm	U value W/m^2K
Glass fibre		
rolls laid across or between ceiling joists	60, 80, 100, 150, 200	0.04
semi-rigid batts laid across or between ceiling joists	80, 90, 100, 120, 140, 180, 200	0.04
Rockwool		
rolls laid across or between ceiling joists	80, 100, 150	0.037
granulated fibre spread between ceiling joists		0.043

Warm roof insulation

The most practical and economic place to fix insulation for a warm roof is on top of rafters where there will be the least wasteful cutting of the thin, expensive, insulation boards.

One of the comparatively thin organic insulants is used in the form of rigid boards which are used as a form of sarking, the word used to describe materials that exclude wind that would otherwise blow through tiled and slated roofs. To limit thickness, one of the materials with a low U value, such as XPS, PIR or PUR, is used.

These insulation boards are nailed directly to the roof rafters with their edges close butted or with tongued and grooved edges for a tight fit. Roofing felt is spread over the insulation boards and sawn softwood counter battens are nailed through the insulation boards to rafters to provide a fixing for tiling or slating battens.

Details of insulants for warm roofs are given in Table 11.

Condensation

A consequence of the general adoption of space heating and the requirement for thermal insulation in dwellings has been an increase in the likelihood of warm, moisture laden air condensing on cold indoor surfaces such as window glass.

The limited capacity of air to take up water in the form of moisture vapour increases with temperature. With the general demand for warm indoor temperature there has been an increase in condensation from such rooms as kitchens and bathrooms.

There have been instances of warm, moist air penetrating cracks in ceilings, due to moisture vapour pressure, and condensing to water on cold surfaces such as the underside of roofing felt. The consequent unsightly staining of ceiling finishes is due to an appreciable volume of condensate, in the form of water, running on to the ceiling.

Table 11 Insulation for warm roofs.

Pitched roof warm roof	Thickness mm	U value W/m²K
Glass fibre semi-rigid batts friction fit between rafters 1175 × 570 or 370	80, 90, 100, 120, 140, 180, 200	0.04
XPS boards fixed on top of rafters as sarking 2500 × 600	30, 40, 50, 60, 70, 80	0.025
PIR boards aluminium foil faced fixed on top of rafters as sarking 2400 × 1200	20, 25, 30, 35, 50	0.022
PUR boards glass tissue and aluminium foil faced fixed to top of rafters as sarking 1200 × 600 or 450	20, 25, 30, 35, 40, 50	0.022

XPS extruded polystyrene
PIR rigid polyisocyanurate
PUR rigid polyurethane

Where fibrous, open texture material such as glass wool is used as over ceiling insulation and warm, moist air penetrates a ceiling it may condense to water on the cold upper side of the insulation and so saturate it that the insulation property of the insulant is appreciably reduced.

Vapour barrier

In the early days when insulation was used in dwellings and condensation occurred in roof spaces the first reaction was to call for the installation of a water and moisture vapour impermeable barrier between warm, moist air below and cold surfaces above ceilings. Sheets of polythene were spread across the underside of ceiling joists, over ceiling joists under insulation or as an integral backing to plasterboard to ceilings. Efforts were made to seal the joints between sheets of polythene and plasterboards and seals around edges of ceiling and wall junction and also around pipework that penetrated ceiling finishes.

Vapour check

It was soon realised that, in spite of the most meticulous attention to detail, it was impractical to form a moisture vapour barrier. It is now accepted that a moisture vapour check is practical and the term vapour barrier has been abandoned.

The usual form of vapour check is sheets of 250 gauge polythene sheet with the edges overlapped and spread under insulation over ceiling joists with edges taped around pipes and cables that penetrate ceiling finishes. As an alternative, closed cell insulants, in the form of boards, such as extruded polystyrene, which are substantially impermeable to water vapour may be fixed across the top of ceiling joists and close side butted to serve as a vapour check.

Ventilation

A requirement from Part F to the Building Regulations is that adequate provision be made to prevent excessive condensation in a roof. The practical guidance in Approved Document F is that cold roof spaces should be ventilated to outside air to reduce the possibility of condensation.

There is no definition of what constitutes excessive, relative to condensation, or whether the roof ventilation is an alternative to a moisture vapour check, or an addition.

A more rational approach to reducing the possibility of warm, moist air penetrating a cold roof space is the recommendation for provision of both background and mechanical extract ventilation to kitchens and bathrooms contained in Approved Document F. Where such ventilation is installed and effectively used in bathrooms and kitchens below cold roofs there seems little sense in either ventilating cold roof spaces or using moisture vapour checks.

Recommendations for ventilation

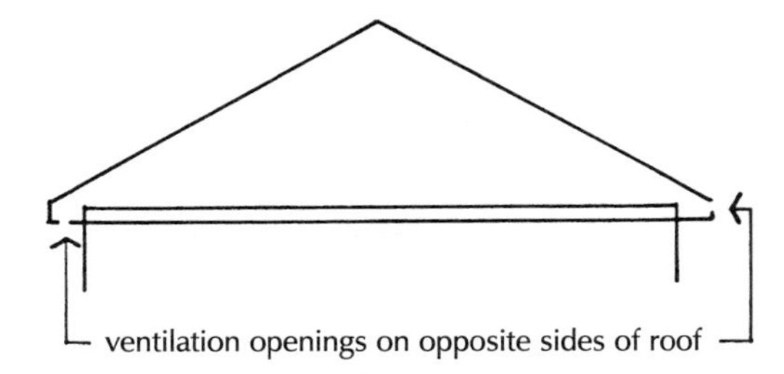

Fig. 216 Roof ventilation.

The practical guidance in Approved Document F is that cold, pitched roof spaces should have ventilation openings at eaves level to promote cross ventilation, as illustrated in Fig. 216. The openings indicated in the soffit of the eaves in Fig. 216 should have an area on opposite sides of the roof at least equal to continuous ventilation running the full length of the eaves and 10 mm wide. Plastic grilles with fine wire mesh to exclude insects are designed to fit in the soffit board of projecting eaves. Obviously there should be a clear space between the top of roof insulation and the underside of the roof covering. As an alternative to soffit ventilation the suppliers of tiles and slates offer a range of special fittings to provide roof ventilation through eaves, ridge and in roof slopes. A typical under eaves ventilator for slate is illustrated in Fig. 217. The plastic ventilator is fixed between roof rafters before the roofing felt is laid. It is fitted with insect screens over openings that provide ventilation to roofs. The roofing felt and slate are laid over the ventilators. A sufficient number of ventilators is used to provide the requisite ventilation area.

The movement of air, which is unpredictable, will vary from still, cold, to gusty, windy weather. For the system of roof ventilation proposed in Advisory Document F to be effective there would have to be an appreciable difference of, and pressure between, outside and inside. The chance of there being sufficient air pressure difference to promote ventilation, coinciding with the generation of moisture vapour laden air inside a bathroom under a cold roof are very slight. As the Building Regulations require adequate provision to prevent excessive condensation, common sense suggests that effective mechanical ventilation of bathrooms will by itself be effective in preventing excessive condensation and separate roof ventilation is unnecessary.

This is particularly true where rooms are formed in roof spaces and the ceiling follows the whole or part of the roof line. Here the guidance in Approved Document F is that the space between the roof covering and the insulation be at least 50 mm wide, as illustrated in Fig. 218. The chance of appreciable, continuous ventilation flowing through the tortuous, narrow space is slight.

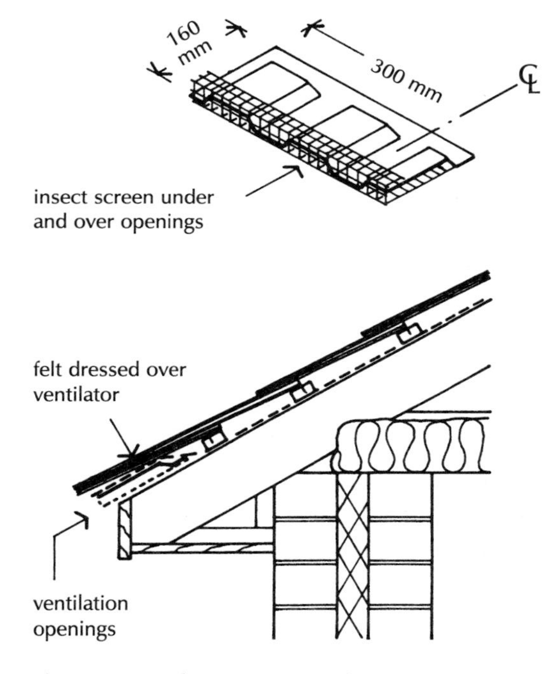

Fig. 217 Under eaves ventilator.

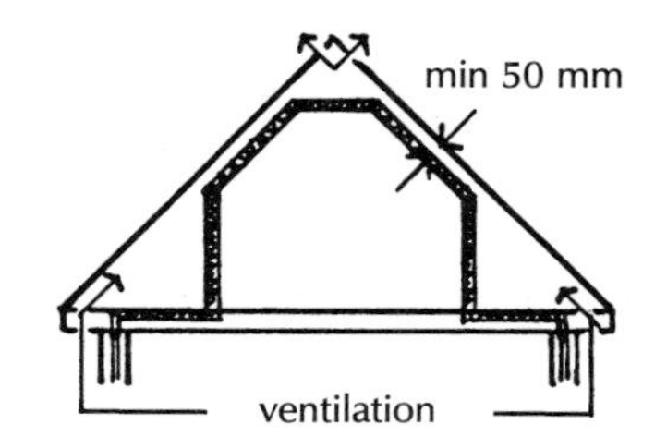

Fig. 218 Roof space ventilation.

FLAT ROOFS

History

From the earliest settlements in this country a pitched roof form as shelter from rain and wind has been common, from the early thatched houses to the traditional tile and stone slate roofed buildings. Flat roofs have not been a practical means of shelter in the rainy, northern European climate of this country. In arid, dry climates closer to the equator a flat roof has served as protection from the mainly overhead sun during the day and a roof platform on which to enjoy cooler air at night.

A flat roof was used to cover small areas of roof over the extensions, internal courts and as a cover to parts of extensive roofs. Traditional sheet metal coverings were used, lead in England and zinc in France.

Early in the twentieth century a building form, commonly known as the modern movement, adopted a severe horizontal flat roof form devoid of applied decoration. At the time the early use of reinforced concrete for floors facilitated the use of reinforced concrete as a roof platform.

Materials such as asphalt and bitumen impregnated felt, that had previously been used for road paving and temporary cover to sheds respectively, were adapted for use as covering to flat roofs.

Before the inclusion of thermal insulation in the fabric of buildings these flat roof coverings worked reasonably well for their anticipated useful life of some 20 to 30 years. The inclusion of a layer of efficient thermal insulation was a disaster.

Previously the heat generated by the overhead sun was in part dispersed from the covering down to the roof below. With insulation below, very little heat was transferred below and the covering endured the full heat of sun and chill of night. The asphalt and bitumen rapidly hardened and failed.

In consequence, flat roofs acquired and deserved a bad name to the extent that existing buildings have had pitched roofs constructed to replace flat ones.

FLAT ROOF COVERINGS

A roof is defined as flat when its weather plane is finished at a slope of 1° to 5° to the horizontal.

The shallow slope to flat roofs is necessary to encourage rainwater to flow towards rainwater gutters or outlets and to avoid the effect known as ponding. Where there is no slope or fall or a very shallow slope it is possible that rainwater may not run off the roof where deflection under load has caused the roof to sink and cause rainwater to lie as a shallow pond. This water may in time cause deterioration of a membrane and consequent failure.

The minimum falls (slope) for flat roof coverings are 1:60 for copper, aluminium and zinc sheet, and 1:80 for sheet lead and for mastic asphalt and built-up bitumen felt membranes.

To allow for deflection under load and for inaccuracies in construction it is recommended that the actual fall or slope should be twice the minimum and allowance made where slopes intersect so that the fall at the mitre of intersection is maintained.

The traditional materials that are used as a weather covering for flat roofs, lead, copper and zinc sheet, are laid as comparatively small sheets dressed over rolls at junctions of sheet down the slope of the roof and with small steps (drips) at junctions of sheet across the slope.

The size of the sheet is limited to prevent excessive expansion and contraction that might otherwise cause the sheet to tear. The upstand rolls down the slope and the steps or drips across the slope provide a means of securing the sheets against wind uplift and as a weathering to shed water away from the laps between sheets. More recently aluminium sheet has also been used.

Lead, copper and aluminium are not affected by normal weathering agents, do not progressively corrode and have a useful life of very many years. Zinc is less durable than the other non-ferrous metals used for roof covering.

Because of the cost of the material and the very considerable labour costs involved in covering a roof with one of these metals, the continuous membrane materials, asphalt and bitumen felt, came into use early this century. Asphalt is a dense material that is spread while hot over the surface of a roof to form a continuous membrane which is impermeable to water. It is generally preferred as a roof covering to concrete flat roofs. Bitumen impregnated felt is laid in layers bonded in hot bitumen to form a continuous membrane that is impermeable to water. Bitumen felt is comparatively lightweight and commonly preferred for timber flat roofs for that reason.

Both asphalt and bitumen felt oxidise and gradually become brittle and have a useful life of at most 20 years. Bitumen felt will become brittle in time and may tear due to expansion and contraction caused by temperature fluctuations. Of recent years fibre-based bitumen felts have lost favour due to premature failures and polyester-based felts are now more used for their greater resistance to tear and lower rates of oxidisation and embrittlement.

With the use of insulation under flat roof coverings the temperature fluctuations that the covering will suffer are considerable and the consequent strain on continuous coverings accelerates failures. Because of the rainfall common to northern Europe it would seem wise to avoid, where possible, the use of flat roofs, particularly for large areas of roof.

TIMBER FLAT ROOF CONSTRUCTION

The construction of a timber flat roof is similar to the construction of a timber upper floor. Sawn softwood timber joists 38 to 75 mm thick and from 97 to 220 mm deep are placed on edge from 400 to 600 mm apart with the ends of the joists built into or on to or against brick walls and partitions.

Tables in Approved Document A, giving practical guidance to meeting the requirements for dwellings of up to three storeys for single families, give sizes of joists for flat roofs related to span and loads for roofs with access only for maintenance and repair and also for roofs not limited to access for repair and maintenance.

Strutting between joists

Solid or herringbone strutting should be fixed between the roof joists for the same reason and in like manner to that used for upper timber floors.

Roof deck

Boards which are left rough surfaced from the saw are the traditional material used to board timber flat roofs. This is called rough boarding and is usually 19 mm thick and cut with square, that is plain, edges. Plain edged rough boarding was the cheapest obtainable and used for that reason. Because square edged boards often shrink and twist out of level as they dry, chipboard or plywood is mostly used today to provide a level roof deck. For best quality work tongued and grooved boards were often used.

End support of joists

If there is a parapet wall around the roof, the ends of the roof joists may be built into the inner skin of cavity walls or supported in metal hangers. The joists can bear on a timber or metal wall plate or be packed up on slate or tile slips as described for upper floors. The ends of the roof joists are sometimes carried on brick corbel courses, timber plate and corbel brackets or on hangers in precisely the same way that upper floor joists are supported. The end of roof joists built into solid brick walls should be given some protection from dampness by treating them with a preservative.

Timber firring

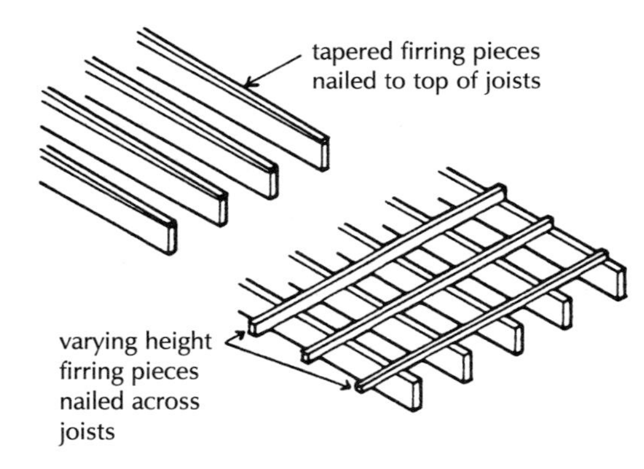

Fig. 219 Firring to timber flat roof.

Flat roofs should be constructed so that the surface has a slight slope or fall towards rainwater outlets. This slope could be achieved by fixing the joists to a slight slope but the ceiling below the roof would then also be sloping. It is usual to provide a sloping surface to the roof by means of firring pieces. These consist of either tapered lengths of fir (softwood) nailed to the top of each joist or varying depth lengths of softwood nailed across the joists, as shown in Fig. 219.

Varying depth firring is used where rough boarding is fixed as the deck for flat roof sheet metal coverings. By this arrangement the boards are fixed down the slope so that any variation in the surface of the boards, due to shrinkage or twisting, does not impede the flow of rainwater down the shallow slope.

Tapered firring is used where the roof deck is formed by chipboard or plywood sheets nailed to firring with close butted edges to form a level surface.

As an alternative to timber firring, insulation boards that are made or cut to a shallow wedge section can be used to provide the necessary shallow fall.

Sheet metal covering to timber flat roofs

Sheet metal is used as a covering because it gives excellent protection against wind and rain; it is durable and lighter in weight than asphalt, tiles or slates. Four metals in sheet form are used, lead, copper, zinc and aluminium.

Sheet lead

Lead is a heavy metal which is comparatively soft, has poor resistance to tearing and crushing and has to be used in comparatively thick sheets as a roof covering. It is malleable and can easily be bent and beaten into quite complicated shapes without damage to the sheets. Lead is resistant to all weathering agents including mild acids in rainwater in industrially polluted atmospheres. On exposure to the atmosphere a film of basic carbonate of lead oxide forms on the surface of the sheets. These films adhere strongly to the lead and as they are non-absorbent they prevent further corrosion of the lead below them. The useful life of sheet lead as a roof covering is upwards of a hundred years.

Rolled sheet lead for roof work is used in thicknesses of 1.8, 2.24 and 2.5 mm. These thicknesses are described as Code Nos 4, 5 and 6 respectively, the Code corresponding to the Imperial weight of a given area of sheet.

No sheet of lead should be larger than 1.6 m^2 so that the joints between the sheets are sufficiently closely spaced to allow the metal to contract without tearing away from its fixing. Another reason for limiting the size of sheet, which is peculiar to lead, is to prevent the sheet from creeping down the roof.

The expression creep describes the tendency of the sheet to elongate. As the temperature of the metal rises the sheet expands, but owing to its weight and poor mechanical strength it may not be able to contract fully as the temperature falls. The consequence is that the sheet gradually elongates over many years and becomes thinner and may in time let in water. It is not likely that this will happen on a flat roof with a sheet not larger than 1.6 m^2.

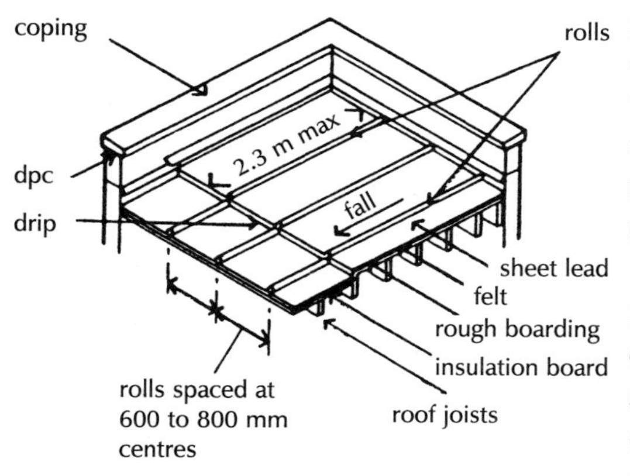

Fig. 220 Lead flat roof.

The joints across the fall of the roof are made in the form of a 50 mm drip or step down to encourage flow of water. To reduce excessive increases in the thickness of the roof due to these drips, they are spaced up to 2.3 m apart and the rolls (joint longitudinal to fall) 600 to 800 mm apart. Figure 220 illustrates part of a lead covered flat roof showing the general layout of the sheets and a parapet wall around two sides of the roof.

Wood rolls

The edges of sheets longitudinal to the fall are lapped over a timber which is cut from lengths of timber 50 mm square to form a wood roll. Two edges of the batten are rounded so that the soft metal can be dressed over it without damage from sharp edges. Two sides of the batten are slightly splayed and the waist so formed allows the sheet to be clenched over the roll. Figure 221 is an illustration of a wood roll. An underlay of bitumen impregnated felt or stout waterproof building paper is first laid across the whole of the roof boarding and the wood rolls are then nailed to the roof at from 600 to 800 mm centres. The purpose of the underlay of felt or building paper is to

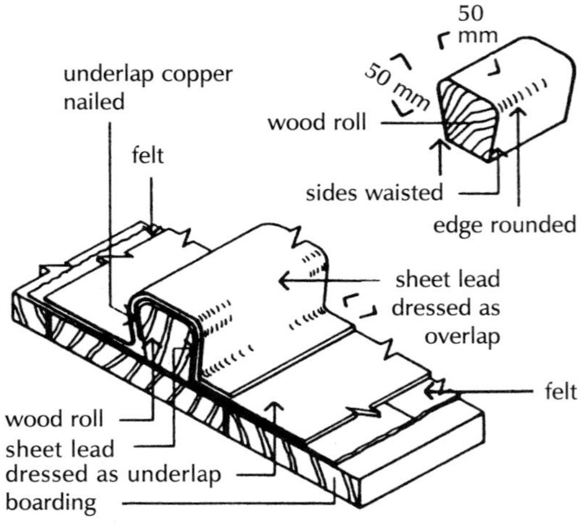

Fig. 221 Wood rolls.

provide a smooth surface on which the sheet lead can contract and expand freely.

The roof boarding on roofs to be covered with sheet lead may be fixed diagonally so that the joints between the boards are 45° to the fall. It is wasteful of timber to lay boards diagonally as the end of each board has to be cut off at 45° and boards are, therefore, laid so that they run along the fall of the roof. The reason for laying the boards either diagonally or along the fall is so that if a board shrinks and warps it will not obstruct the flow of rain off the roof or tear the lead.

The edges of adjacent sheets are dressed over the wood roll in turn. In sheet metalwork the word dressed is used to describe the shaping of the sheet. The edge of the sheet is first dressed over as underlap or undercloak and is nailed with copper nails to the side of the roll. The edge of the next sheet is then dressed as overlap or overcloak. A section through one roll is shown in Fig. 221. One edge of each sheet is dressed as underlap and nailed and the opposite edge is then dressed as overlap to the next roll. In this way no sheet is secured with nails on both sides, so that if it contracts it does not tear away from the nails.

Drips

Drips 50 mm deep are formed in the boarded roof by nailing a 50 × 25 mm fir batten between the roof boards of the higher and lower bays. The drips are spaced at not more than 2.3 m apart down the fall of the roof. The edges of adjacent sheets are overlapped at the drip as underlap and overlap and the underlap edge is copper nailed to the boarding in a cross-grained rebate, as shown in Fig. 222. An anti-capillary groove formed in the 50 × 25 mm batten is shown into which the underlap is dressed. This groove is formed to ensure that no water rises between the sheets by capillary action.

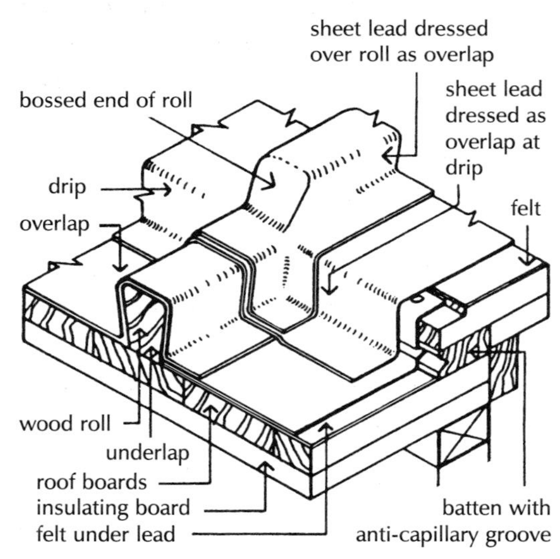

Fig. 222 Drip.

Figure 222 also shows the junction of wood rolls with a drip and illustrates the way in which the edges of the four sheets overlap. This arrangement is peculiar to sheet lead covering which is a soft, very ductile material that can be dressed as shown without damage. The end of the wood roll on the higher level is cut back on the splay (called a bossed end) to facilitate dressing the lead over it without damage.

This seemingly complicated junction of four sheets of lead, which is formed to provide a watertight overlap, encourages steady run off of rainwater and makes allowance for thermal movement of lead, can be surprisingly quickly made by a skilled plumber.

Where there is a parapet wall around the roof or where the roof is built up against a wall the sheets of lead are turned up against the wall about 150 mm as an upstand. The tops of these upstands are not fixed in any way so that the sheets can expand and contract without restraint. To cover the gap between the upstand and the wall, strips of sheet lead are tucked into a raked out horizontal brick joint, wedged in place with strips of lead sheet and then dressed down over the upstand as an apron flashing. To prevent the apron from being blown up by the wind, lead clips are fixed as shown in Fig. 223A and B, which illustrates the junction of roll and drip with upstands.

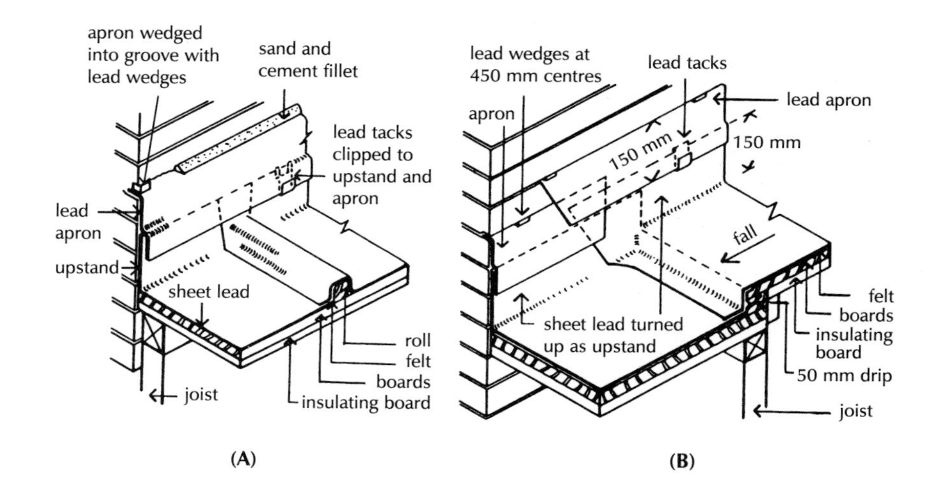

Fig. 223 (A) Junction of roll and upstand. (B) Junction of drip and upstand.

Lead gutter

Where the lead flat roof is surrounded on all sides by parapet walls it is necessary to collect the rainwater falling off at the lowest point of the roof. A shallow timber framed gutter is constructed and this gutter is lined with sheets of lead jointed at drips and with upstand and flashings similar to those on the roof itself. The gutter is constructed to slope or fall towards one or more rainwater outlets. The gutter is usually made 300 mm wide and is formed between one roof joist, spaced 300 mm from a wall, and the wall itself.

The gutter bed is supported by 50 × 50 mm gutter bearers fixed at 450 mm centres supported by 50 × 25 mm battens which are nailed to the wall and the joist to provide the necessary fall in the gutter. Gutter boards, 19 mm thick, are fixed along the length of the gutter as a gutter bed.

At drips into the gutter from the roof and at drips in the length of the gutter, the apron upstand to the parapet wall, and the apron flashing tucked into the wall are staggered and overlapped.

Figure 224 is an illustration of a lead lined gutter.

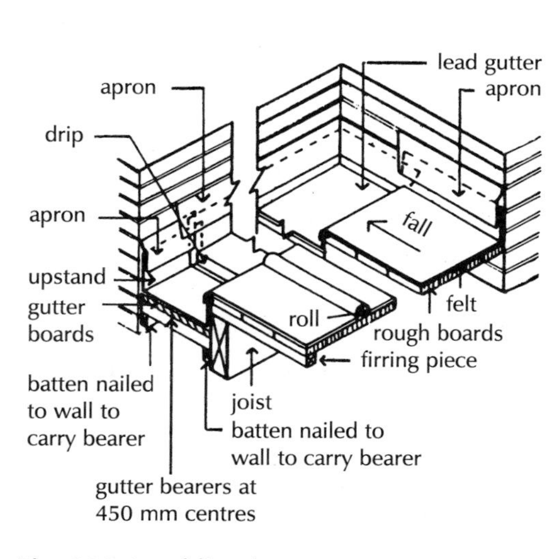

Fig. 224 Lead lined gutter.

Where the roof is surrounded by a parapet wall and a rainwater pipe can be fixed to the outside of the wall, an opening is formed in the wall as an outlet usually 225 × 225 mm square. A rainwater head is fixed to the wall below the outlet.

The sheet lead gutter lining is continued through the outlet and dressed to discharge into the rainwater head. Both upstands to the gutter lining are dressed into the sides of the rainwater outlet in the parapet wall and dressed against the wall face to form a shute to direct water into the rain water head, as illustrated in Fig. 225.

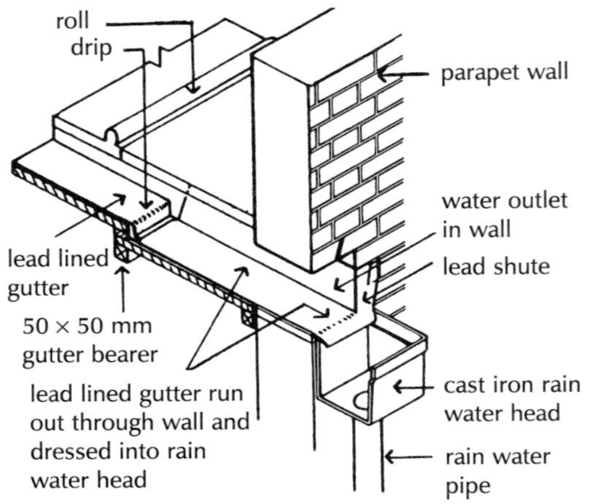

Fig. 225 Rainwater outlet.

Where it is not possible or is unacceptable for appearance sake, to fix a rainwater pipe and head externally the gutter has to be formed to discharge to a cesspool. A conventional lead lined gutter is formed to fall each side of a lead lined cesspool, as illustrated in Fig. 226.

The purpose of this lead lined box, cesspool or catchpit is as a reservoir so that during a heavy storm, when the rainwater pipe may not be able to carry the water away quickly enough, the cesspool prevents flooding of the roof.

The cesspool is formed as a box of 25 mm thick boards holed for a pipe outlet. The cesspool is 300 × 300 × 150 mm deep and supported on 50 × 50 mm bearers. The cesspool is lined with one piece of lead which is dressed to shape with upstands to be dressed up and under the lead sheet to the gutter, flat roof and upstand to parapet wall.

A lead down pipe is connected to the cesspool and run down as a down pipe or connected to a down pipe.

Where there is no parapet wall on one side of the roof it can discharge rainwater directly to a gutter fixed to a fascia board on the external face of the wall. The lead flat roof covering is dressed to discharge into the gutter, as illustrated in Fig. 227.

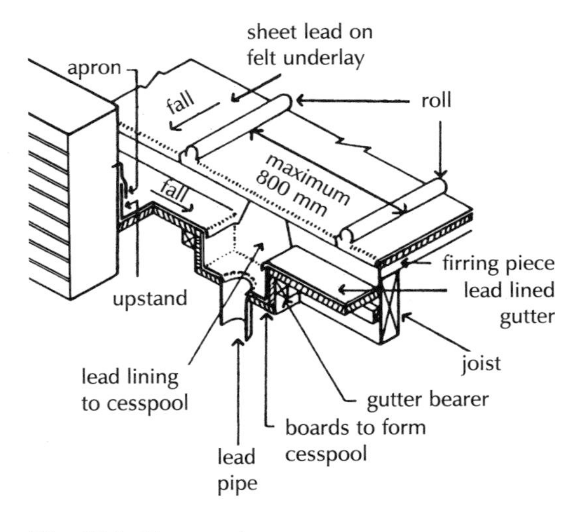

Fig. 226 Cesspool.

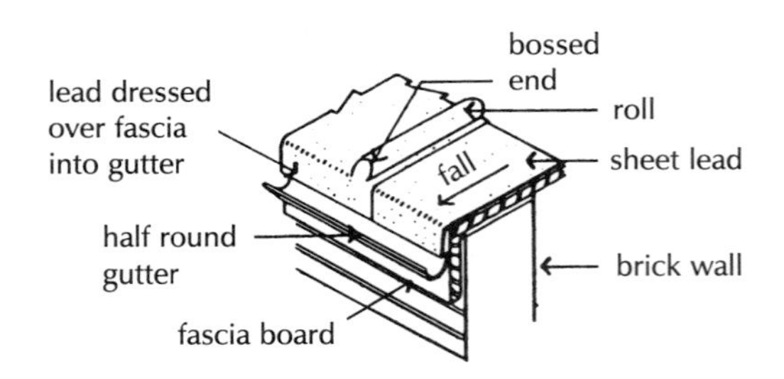

Fig. 227 Gutter.

Copper sheet

Copper is a metal which has good mechanical strength and is sufficiently malleable in sheet form that it can be bent and folded without damage. Like lead, on exposure to atmosphere a thin coat of copper oxide forms on the surface of the copper sheets, which is tenacious, non-absorbent and prevents further oxidisation of the copper below it. Copper is resistant to all normal weathering agents and its useful life as a roof covering is as long as that of lead.

The usual thickness of copper sheet for roofing is 0.6 mm. An oxide of copper forms on the surface of copper sheet and in the course of

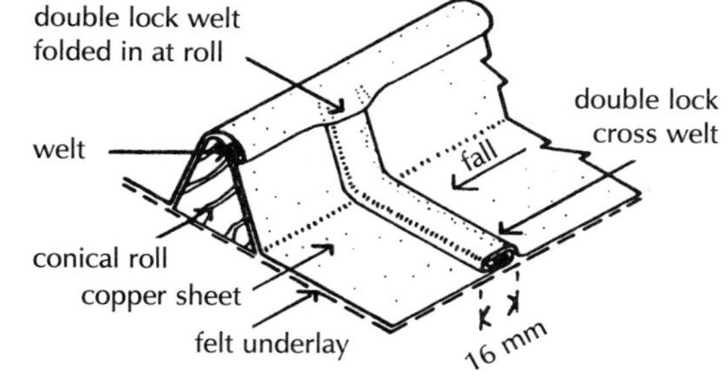

Fig. 228 Double lock welt.

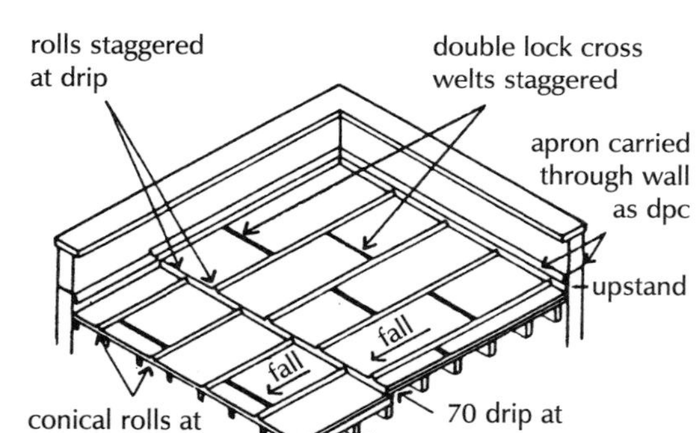

Fig. 229 Copper flat roof.

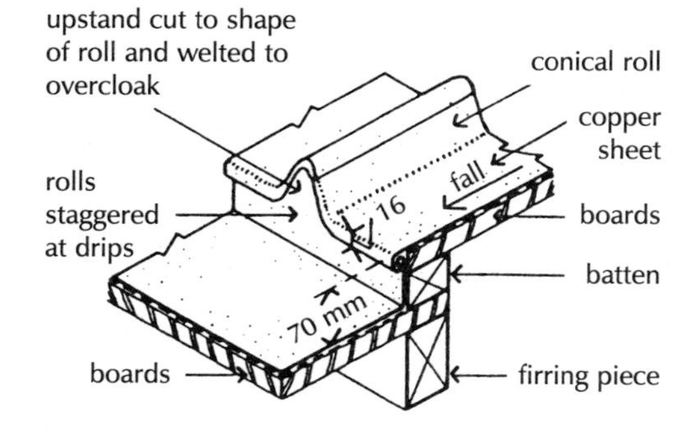

Fig. 230 Copper drip.

some few years the sheets become entirely covered with a light green compound of copper. This light green coating is described as a patina and is generally thought to give the copper sheets pleasing colour and texture. But in atmospheres heavily polluted with soot the patina is black instead of green. The patina, usually basic copper carbonate, is impervious to all normal weathering agents and protects the copper below it.

The standard sizes of sheet supplied for roofing are 1.2 m and 1.8 m × 900 mm. The minimum fall for a copper covered roof is 1 in 60 and the fall is provided by means of firring pieces just as it is for lead covered roofs.

The traditional method of fixing copper sheet to a flat roof has been by the use of wood rolls down the shallow slope, over which adjacent sheets are shaped with welted joints and drips across the slope.

Copper sheets for roofing, which cannot be shaped over rolls and drips as easily as can lead, are joined by the use of double lock welts which are a double fold at the edges of adjacent sheets. These double lock welts are used for joints of sheet over the wood rolls and at joints across the fall, as illustrated in Fig. 228. Because of the difficulty of forming a double lock welt it is necessary to stagger these joints to avoid the difficulty of forming this joint at the junction of four sheets of copper.

Figure 229 is an illustration of a copper sheet covered flat roof. The wood rolls along the fall of the roof are fixed at centres to suit the standard width of sheet, for example 750 mm. At drips the rolls are staggered to avoid welts at the junction of four sheets.

Drips are formed across the slope at from about 2 to 3 m to suit the size of standard sheet with staggered double lock cross welts between drips and drip and parapet wall. Drips are formed with a 40 × 50 mm batten which is nailed between the upper and lower roof boarding on the firring pieces on joists, as illustrated in Fig. 230.

At the drip the sheets of copper are welted with the upstand of the lower sheet shaped to the end of the roll and welted to the overcloak as illustrated in Fig. 230.

The conical rolls formed down the fall of the roof are formed around 63 × 50 mm sections of softwood which are shaped in the section of a cone with a round top. The shape is selected to facilitate shaping the metal. The wood rolls are nailed through a felt underlay to the roof boarding and also 50 mm wide strips of copper sheet at 450 mm centres as cleats. The felt underlay serves to allow the sheets to expand and contract without restraint.

The copper strips are used to restrain the sheets against wind uplift by being folded into the welted joint formed over the roll, as illustrated in Fig. 231.

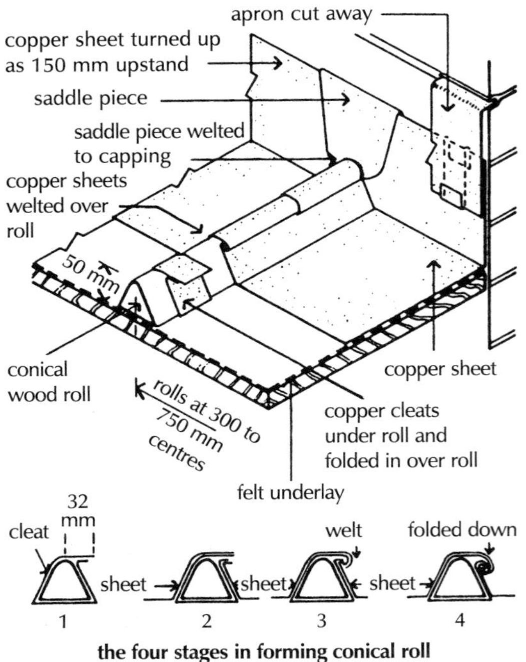

Fig. 231 Conical roll.

At upstands to parapet walls the sheets are turned up at the upstand and covered with an apron flashing which is wedged into a brick joint. At the junction of the conical roll and the upstand it is necessary to use a separate saddle piece of copper sheet that is welded to the capping and dressed up to cover the joint between adjacent upstands, as illustrated in Fig. 231.

As with lead sheet the copper covered roof may discharge to a verge gutter or to a parapet wall gutter formed in the roof and lined with copper.

As an alternative to the use of a conical roll down the slope or fall of copper roofs a batten roll may be used with drips and double lock cross welts across the fall or slope of the roof.

A wood roll with slightly shaped sides is nailed through a felt underlay and also 50 mm wide strips of copper at 450 mm centres as cleats. The sides of adjacent sheets of copper are shaped as upstand sides to the roll and the roll is covered with a separate strip of copper as a capping. The cleats are folded into the double lock welts formed between the upstand of sheets and the edges of the capping. This is the type of joint longitudinal to falls that is also used with zinc sheet roof covering.

Zinc sheet

Zinc, a dull, light grey metal, is used in sheet form as a covering for timber flat roofs.

It is the cheapest of the metals used for roofs and has been extensively used in northern European countries such as France, where it can economically be produced. In England where it has been much less used it has served as a cheap substitute for lead both as a roof covering and for flashings.

Zinc sheet is appreciably more difficult to bend and fold than copper. Being a brittle material it is liable to crack if bent or folded too closely.

As a flat roof covering the standard 2.4 m, 900 mm and 1 or 0.8 mm thick sheets are bent up to the sides of softwood batten rolls. The batten rolls are nailed at 850 mm centres through felt underlay and clips at 750 mm centres. The clips are welted over the upstand edges of the sheets.

Because of the difficulty of making a double lock welt of zinc sheet the capping is shaped to fit over the batten rolls. The lower edges of the capping are bent in, feinted, to grip the sheets.

To secure the capping, zinc holding down clips are nailed to the batten over the end of a lower capping and the end of the upper capping tucked into the fold, as illustrated in Fig. 232.

Drips are formed across the slope at 2.3 m intervals, with beaded drips and splayed cappings to roll ends, as illustrated in Fig. 233.

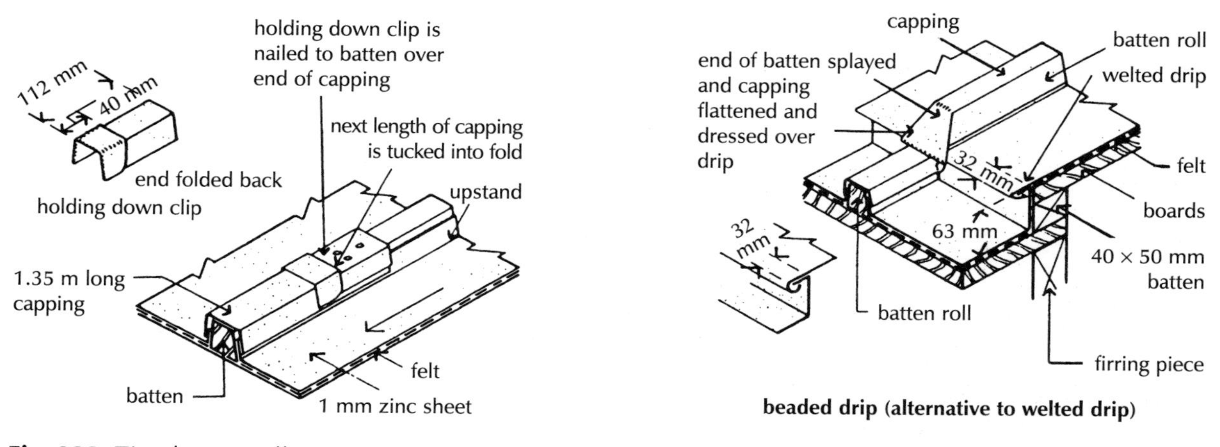

Fig. 232 Zinc batten rolls.

Fig. 233 Zinc drip.

At parapets the sheets are turned up as upstand and covered with a zinc apron that is wedged into a brick joint. The end of a batten capping is folded and flattened to cover the edges of adjacent sheets.

WATERPROOF MEMBRANES FOR TIMBER FLAT ROOFS

The two materials that have been used as a waterproof membrane for timber framed roofs are asphalt as a continuous membrane and bitumen felt laid in layers to form a membrane or skin.

Asphalt

Mastic asphalt

Asphalt, which is sometimes spelt asphalte, is described in the relevant British Standard as mastic asphalt. The material, which is soft and has a low softening (melting) point, is an effective barrier to the penetration of water.

Natural rock asphalt is mined from beds of limestone which were saturated, or impregnated, with asphaltic bitumen thousands of years ago. The rock is chocolate brown in colour and is mined in several districts around the Alps and Europe. The rock is hard and because of the bitumen with which it is impregnated it does not as readily absorb water as ordinary limestone.

Natural lake asphalt is dredged principally from the bed of a dried up lake in Trinidad. It contains a high percentage of bitumen with some water and finely divided solid material.

Asphalt is manufactured either by crushing natural rock asphalt and mixing it with natural lake asphalt, or by crushing natural limestone and mixing it with bitumen whilst the two materials are sufficiently hot to run together. The heated asphalt mixture is run into moulds in which it solidifies as it cools.

The solid blocks of asphalt are heated on the building site and the hot plastic material is spread over the surface of the roof in two layers breaking joint to a finished thickness of 20 mm. As it cools it hardens and forms a continuous, hard, waterproof surface.

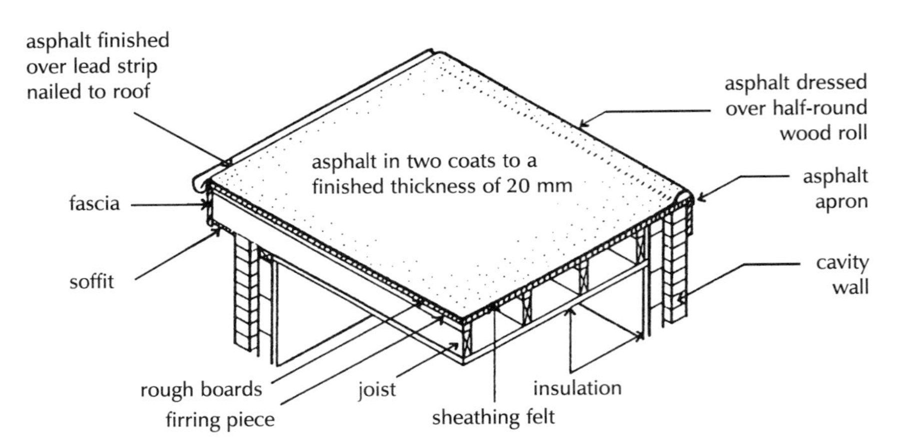

Fig. 234 Asphalt covered flat roof.

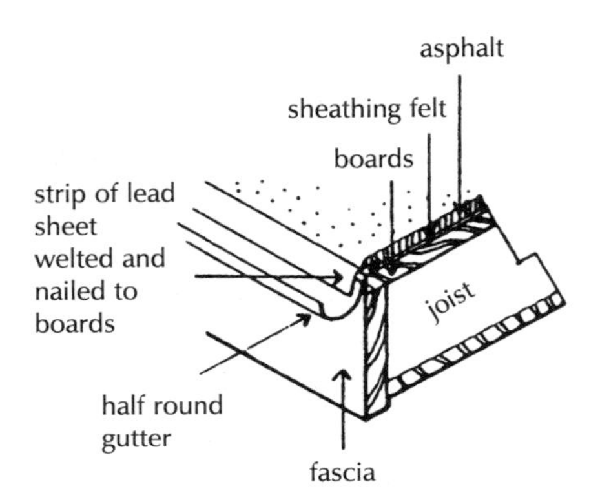

Fig. 235 Detail at eaves.

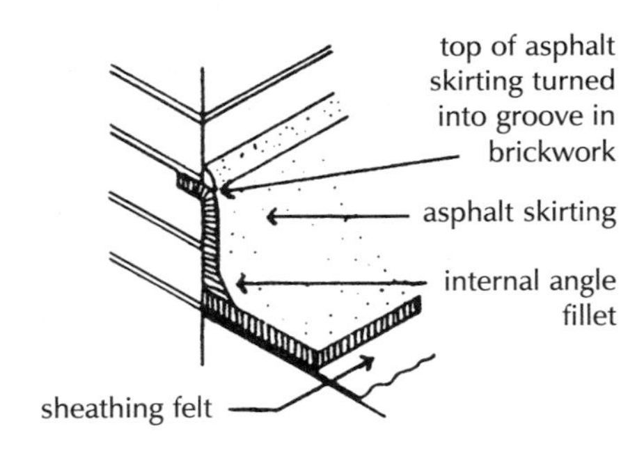

Fig. 236 Asphalt skirting.

If there is no parapet wall around the roof it is usually designed to overhang the external walls, to give them some protection, and the asphalt drains to a gutter over the strip of sheet lead which is dressed down into a gutter, as illustrated in Fig. 234, which is an illustration of a timber flat roof covered with asphalt. Figure 235 illustrates a detail of sheet lead dressed into a gutter.

If the roof has parapet walls around it, or adjoins the wall of a higher building, an asphalt skirting or upstand, 150 mm high, is formed and this skirting is turned into horizontal brick joints purposely cut about 25 mm deep to take the turn-in of the asphalt skirting, as illustrated in Fig. 236. For strengthening, an internal angle fillet is run at the junction of the flat and the asphalt skirting.

An alternative detail is to form the skirting as an upstand only, with the upstand weathered by a sheet lead apron that is wedged into a raked out brick joint and dressed down over the upstand. The advantage of this finish is that structural, thermal or moisture movements of the roof are less likely to cause cracking of the asphalt either at the internal angle fillet or at the turn in the wall.

It is essential that asphalt be laid on an isolating membrane underlay of black haired felt so that slight movements in the structure are not reflected in the asphalt membrane.

A properly laid asphalt covering to a roof will not absorb water at all and the finished surface of the asphalt can be absolutely flat as any rainwater that lies on it, due to ponding, will eventually evaporate. But it is usual to construct flat roofs so that the asphalt is laid to a slight fall, of at least 1:80, so that the rainwater drains away to a rainwater outlet or gutter.

Asphalt is a comparatively cheap roof covering and if the asphalt is of good quality and is properly laid it will have a useful life of some 20 years or more. The asphalt should be renewed about every 20 years if the roof is to be guaranteed watertight.

Built-up bitumen felt roofing

The word felt is used for fibres, such as animal hair, that are spread at random around a large, slowly rotating drum on which a mat of loosely entwined fibres, mixed with size, is built up. The mat is cut, rolled off the drum and compressed to form a sheet of felted (matted) fibres. For use as a roof surface material the felt is impregnated or saturated with bitumen.

A variety of felts is made from animal, vegetable, mineral or polyester fibre or filament for use as flat roof covering.

Sheathing and hair felts

These felts are made from long staple fibres, loosely felted and impregnated with bitumen (black felt) or brown wood tars or wood pitches (brown felt). They are used as underlays for mastic asphalt roofing and flooring to isolate the asphalt from the structure.

Fibre base bitumen felts

Fibre base bitumen felts are the original material used as covering for the pitched roofs of sheds and outhouses, in either one or two layers.

These felts are made as a base of animal or vegetable fibres that are felted, lightly compressed and saturated with bitumen. Fine granule surfaced felts may be used as a lower layer in built-up roofing and as a top layer on flat roofs that are subsequently surfaced with bitumen and mineral aggregate finish as a comparatively cheap roof finish.

Mineral surfaced fibre base felt is finished on one side with mineral granules for appearance and protection for use as a top layer on sloping roofs.

Fibre base bitumen felts have been used as a covering for small area pitched roofs. The material is either bonded to a boarded finish with bitumen in two layers, breaking joint or nailed and bonded at joints and edges for the first coat and bonded with bitumen as top coat.

The top layer of this felt covering is often surfaced with bitumen and a light coloured mineral aggregate finish for appearance sake and the benefit of reflection from the mineral. This is a cheap, perfectly satisfactory, though somewhat unattractive, weather protection to pitched roofs with a useful life of up to 20 years.

Used on flat and shallow pitch roofs the material may deteriorate quite rapidly, after a few years, once water penetrates and saturates the felt base due to oxidisation and cracking of the bitumen coat.

Reinforced fibre based felts have a layer of jute hessian embedded in the coating on one side and are used under slates and tiles where the felt is not fully supported by boarding.

Glass fibre based bitumen felts

Glass fibre based felts have for many years largely replaced natural fibre based felts as the material used for built-up felt roofing for flat roofs. The material is made from felted glass fibres that are saturated and coated with bitumen.

The felted glass fibre base forms a tenacious mat that is largely unaffected by structural, thermal and moisture movement of the roof deck and does not deteriorate due to moisture penetration.

The fine granule coated, glass fibre based felt is used as underlay and as top layer on flat roofs that are subsequently surfaced with bitumen and mineral aggregate. Mineral surfaced glass fibre based felts may be used as a top layer on flat roofs as low cost finish and on sloping roofs as a finish layer.

A perforated, glass fibre based felt is produced for use as a venting first layer on roofs where partial bonding is used.

The durability of glass fibre based felt depends principally on the bitumen with which it is saturated. In time the bitumen coating will oxidise on exposure to the radiant energy from the sun, harden and ultimately crack and let in water. A layer of insulation under the felt will appreciably increase the rise in temperature and expansion of the surface and so accelerate the cracking of the hardened top surface. A generous layer of light colour mineral aggregate dressed over the surface will, to an extent, give some protection.

The serviceable life of this weathering is of the order of 20 to 30 years.

It is generally accepted that the appearance of this roof finish is unattractive.

Polyester base bitumen felts

A polyester base of staple fibre or filament, formed by needling or spin bonding, is impregnated with bitumen. The fibre or filament base of polyester has higher tensile strength than the true felt bases. Because of this greater strength this 'felt' is better able to withstand the strains due to structural, thermal and moisture movement without rupture that a flat roof covering will suffer.

The fine granule surfaced felt is used as a base, intermediate or top layer of built-up roofing which is to be subsequently covered with bitumen and mineral aggregate finish. The mineral surface felt is for use as a top layer where there is no additional surface treatment.

Polyester base felt, which is generally used for the three layers of built-up roofing, is sometimes used as top coat to underlays of glass fibre based felt as an economy.

High performance roofing

The so called high performance bitumen coated bases are made with a polyester fabric that is coated with polymer modified bitumen. The bitumen is modified with styrene butadiene styrene (SBS) or atactic polypropylene (APP) to provide improved low temperature flexibility, improved creep resistance at high temperatures and greatly improved fatigue endurance to the bitumen.

These high performance bitumen bases, which are generally used in two layers, are more expensive than other built-up roofings and are used for their appreciably improved resistance to the strains common to flat roof coverings.

Mineral surface dressing

The finished top surface of built-up felt roof covering is often finished with a dressing of mineral aggregate spread over a bitumen coating. The purpose of this mineral dressing is to act as a reflective, protective layer of light coloured particles to reduce the oxidising and hardening effect of direct sunlight on the bitumen bonding compound and also for the sake of the appearance of this singularly unattractive covering.

Light coloured mineral aggregate, graded in size from 16 to 32 mm, is spread to a thickness of at least 12.5 mm on a layer of bitumen.

Fire safety

A requirement of the Building Regulations concerning safety is that roof coverings should not contribute to the spread of fire between adjacent buildings. To meet the requirement flat roofs covered with bitumen felt should have a surface finish of stone chippings at least 12.5 mm thick, tiles of an incombustible material, sand and cement screed or macadam.

Laying built-up bitumen felt roofing

Unlike asphalt, which is a comparatively heavy roof covering, it is necessary to provide some bond of felt to the roof deck against wind uplift and to allow some freedom for movement of the felt independent of the roof.

In the early days of the use of bitumen felt roofing it was practice to nail the first layer to the roof boarding, a practice that persists today.

Nailing may not provide sufficient freedom of movement for the felt relative to the roof deck with the consequent possibility of the felt expanding, permanently stretching to form visible ripples between fixings and so impeding run off of rainwater. Over the course of a few years the persistent wetting in troughs may cause the felt to deteriorate and let in water.

Where the surface of the roof deck for both low pitched and flat roofs is formed with rough, square edged boarding it was, and still is to an extent, practice to nail the first layer with clout nails at 150 mm staggered centres over the area of each sheet and at 50 mm centres on the centre line of all 50 mm overlaps of sheets. The reason for this is that the boards may shrink, twist and lose shape in drying out and so not provide a level base for a bitumen bond by itself.

The first layer is then covered with a further layer of felt fully bitumen bonded for short term cover or with two layers for longer term cover.

Practice today is to bond the first layer of felt to the roof deck with bitumen spread over the whole surface of the roof or more usually in a system of partial bonding with bitumen to allow some freedom of movement relative to the roof.

Bitumen felt is applied to flat roofs in three layers, the first, intermediate and top layers. The rolls of felt are spread across the roof with a side lap of 75 mm minimum between the long edges of rolls and with a head, or end, lap of at least 75 mm for felts and 150 mm for polyester based felts. To avoid an excessive build-up of thickness of felt at laps, the side lap of rolls of felt is staggered by one-third of the width of each roll between layers so that the side lap of each layer does not lie below or above that of other layers, as illustrated in Fig. 237.

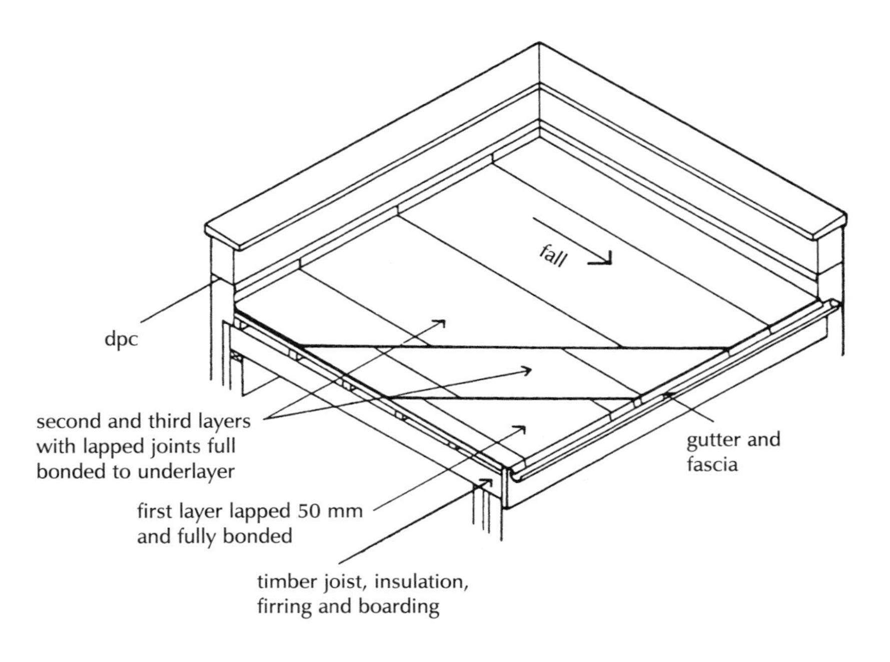

Fig. 237 Built-up felt covering to timber flat roof.

Laying felt roofing

Full bond

The traditional method of fixing felt roofing is by fully bonding the felt to the roof deck and the layers of felt to one another by the 'pour and roll' technique in which hot bitumen is poured on the roof deck and the rolls of felt are continuously rolled out as the bitumen is poured. The pour and roll method of bonding is used for all three layers in the full bonding method and for the two top layers in the partial bond method. The purpose of the bitumen is initially to bond the first layer to the roof deck against wind uplift and then to bond the succeeding layers to each other and form a watertight seal at overlaps of rolls of felt.

The majority of timber framed flat roofs today are formed with a deck of sheets of plywood or particle board, for economy in labour and to provide a stable, level surface for built-up bitumen felt roofing.

The joints between the boards are first covered with strips of self-adhesive tape and the three layers of felt are fully bonded to the deck or a perforated first layer of felt is laid loose over the deck. This first layer is of fibre glass based felt which is perforated with holes. The intermediate layer is then fully bonded in bitumen to the perforated layer so that the hot bitumen runs through the perforations to bond with the deck as a form of partial bonding. The final layer is then fully bonded in bitumen to the intermediate layer.

The full bond method is used where appreciable wind uplift is anticipated in exposed positions and the perforated bottom layer where wind uplift is low in sheltered positions that will also allow some freedom of movement of the felt relative to the roof.

Partial bond

Where it is anticipated that wind uplift will be moderate it is usual to use a perforated, glass fibre based felt as a first layer to allow some movement of the deck independent of the felt covering. The perforated first layer is laid loose over the deck and the intermediate layer is fully bonded to it. The hot bitumen bed will penetrate the perforated bottom layer to effect a partial bond to the deck. The top layer is then fully bonded to the intermediate layer.

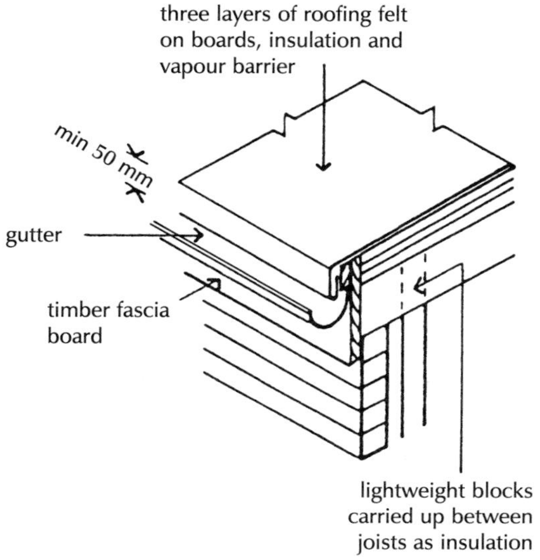

three layers of roofing felt on boards, insulation and vapour barrier

min 50 mm

gutter

timber fascia board

lightweight blocks carried up between joists as insulation

Fig. 238 Eaves gutter.

On a roof deck covered with insulation boards the method of bonding the first layer depends on the composition of the insulation. The majority of insulation boards either have a surface to which hot bitumen can be applied or are coated to assist the bond of bitumen. On polyurethane and polyisocyanurate boards, provision should be made for the escape of gases generated by the use of hot bitumen, by using a perforated venting first layer of felt that is laid loose over the boards.

Built-up bitumen felt roof coverings should be laid to a shallow fall so that rainwater will run off to a gutter. The most straightforward way of draining these flat roofs is by a single fall to one side to discharge directly to a gutter fixed right across one side of the roof, as illustrated in Fig. 238. Here the roof is enclosed inside a parapet wall on three sides.

The roof joist ends are carried over the wall to provide a fixing for a softwood timber fascia board which supports a gutter, as illustrated in Fig. 238. To direct the run off of water into the gutter a strip of felt is nailed to the fascia board, welted and then turned up on the roof and bonded between the first and intermediate layers of felt.

Because felt is difficult to welt without damage a strip of sheet lead may be used as an alternative to the felt strip.

For continuity of insulation a lightweight insulation inner leaf of wall or cavity insulation is carried up to the level of the roof insulation.

At the junction of a felt covered flat roof and a parapet or a wall the top two layers of three-ply bitumen felt are dressed up the wall over an angle fillet as an upstand 150 mm high. The top of the felt upstand is covered with a lead flashing which is tucked into a raked out brick joint and dressed over the upstand, as illustrated in Fig. 239.

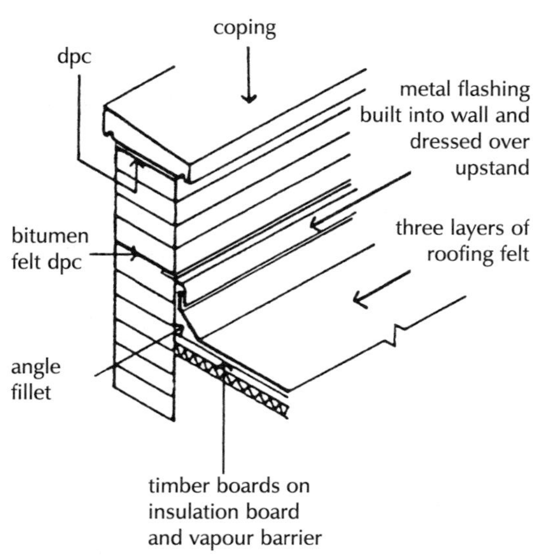

Fig. 239 Upstand and flashing.

Single-ply roofing

For details of single-ply, polymeric roof membranes see Volume 3.

Inverted or upside down roof

An effective layer of insulation laid or fixed below flat roof coverings will appreciably increase the temperature fluctuations that a roof covering will suffer between hot, overhead sun and cold nights. Without the insulation the extremes of temperature are reduced by transmission through the covering to the roof structure.

Bitumen felt coverings are particularly affected by direct sunlight which causes oxidisation of the bitumen which hardens and becomes brittle and liable to crack.

The most rational place for the layer of insulation on a flat roof, particularly those covered with bitumen felt and, to a lesser extent, asphalt, is on top of the covering to reduce temperature fluctuations.

In this form of construction one of the organic closed cell insulation boards such as those made of coalesced glass beads, which do not absorb water and are sufficiently dense to support the overlay of stones or paving slabs, is used.

The roof covering of bitumen felt, asphalt or one of the single-ply membranes described in Volume 3 is laid on the roof deck with upstands carried up to 150 mm above the finished level of the loading stones or paving slabs. The roof covering may be laid to slight falls to rainwater outlets or laid flat to outlets. The insulation boards are laid, butt side jointed, on top of the roof covering.

To protect the insulation against wind uplift the roof is covered with a layer of mineral aggregate, stones, of sufficient depth to hold the insulation in place. Where the flat roof serves as a terrace, paving stones are used as a loading coat, as illustrated in Fig. 240. The paving stones are laid, open jointed, on a filter layer of dry sand or mineral fibre mat as a bed and filter for rainwater.

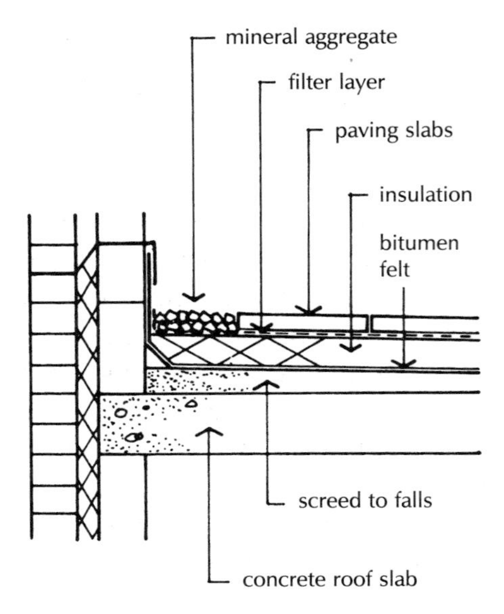

Fig. 240 Inverted or upside down roof.

Rain falling on the roof will in part be retained by the stone or paving slab covering and the filter layer, and will in part drain through the insulation to the covering. During dry periods much of the rain retained by the stones or paving slabs will evaporate to air.

For access to clear any blockages that might occur in rainwater outlets it is wise to provide a margin, around the edge of the roof, of loose aggregate over a narrow strip of insulation that can be lifted to clear blockages.

The disadvantage of the inverted roof is that where leaks occur in the covering the whole top layer may have to be removed. Because of the protection afforded by the two top layers, faults in the covering are less likely than with coverings laid on the insulation.

Concrete flat roofs

Reinforced concrete flat roofs for small buildings are constructed in the same way as reinforced concrete floors with hollow beam, beam and filler blocks or in situ cast concrete slab, reinforced to support the self-weight of the roof, rain, snow and wind pressure or uplift. Roofs are designed to provide support for access for maintenance or for use as a terrace. The roof is supported by external walls with intermediate support as necessary from internal loadbearing walls or beams.

To provide a smooth level surface ready for the roof covering the concrete or concrete topping may be power floated. More usually a cement and sand screed is used as a finish which can be spread and finished to the level or levels necessary to provide a fall to drain rainwater off the roof. A screed can be finished to one way, two way or four way falls to rainwater outlets with the necessary currents at intersections of falls.

When insulation of roofs first became necessary to meet the early requirements of the Building Regulations it was quite common practice to use lightweight screeds composed of lightweight aggregate and cement to provide the necessary surface and serve as thermal insulation to meet the early modest requirements for insulation.

All wet, cement mix materials take time to thoroughly dry out and this is singularly true of lightweight aggregate screeds which may, during the drying period, absorb rainwater and so increase the drying out time. A consequence was that water drying out from these screeds, trapped below asphalt and bitumen felt coverings, would expand to moisture vapour due to direct sunlight and raise blisters and possibly rupture the weather membrane. The remedy was to provide complicated ventilation paths and systems of ventilation to relieve moisture vapour pressure.

With the current more stringent requirements for conservation of fuel the use of lightweight aggregates, which provide only modest insulation, has been largely abandoned.

With the use of a layer of insulation under flat roof coverings it is sometimes considered necessary to provide ventilation against the possibility of water drying out from cement concrete and screeds penetrating the insulation, expanding to moisture vapour and so rupturing the weather covering.

Asphalt roof covering

The conventional wisdom is that asphalt is best suited as a covering to concrete and felt to timber roofs. The sense is that the comparative stability and freedom from shrinkage movement of concrete is best suited to the heavy, inflexible nature of asphalt, whereas the lighter, more flexible nature of built-up bitumen felt is better suited to a timber roof.

Hot asphalt is spread over a layer of loose laid sheathing felt on a dry screed finish in two layers to a finished thickness of 20 mm and covered with a dusting of dry, fine, sharp sand to absorb the 'fat' of the neat asphaltic bitumen that is worked to the surface by hand spreading. On a screed laid to falls and currents the asphalt drains to

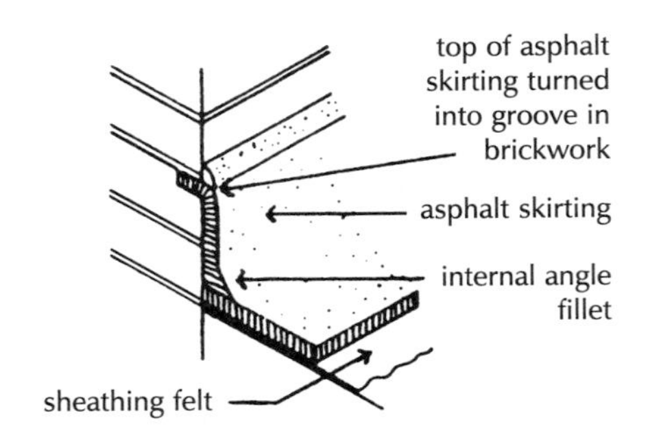

Fig. 241 Asphalt skirting.

outlets formed in parapet walls as illustrated in Fig. 242. The asphalt skirting is dressed into the outlet in the parapet over a lead chute dressed down over a rainwater head.

A reinforcing, internal angle, fillet of asphalt is formed at the junction of the flat roof and the parapet wall and the 150 mm high asphalt skirting. The top of the skirting may be turned into a groove cut in a horizontal brick joint, as illustrated in Fig. 241.

As an alternative the asphalt skirting may be run up the face of the wall by itself or over plastic vents placed at intervals to provide ventilation for moisture vapour. The top of the skirting is then covered with a sheet lead flashing, tucked into and wedged in a raked out brick joint and dressed down over the upstand asphalt skirting.

The spacing and size of skirting vents and the need and frequency of vents in the surface of the asphalt is indeterminate. First or second hand knowledge of past failures of asphalt by blisters may dictate the assumption of the need for vents rather than soundly based need.

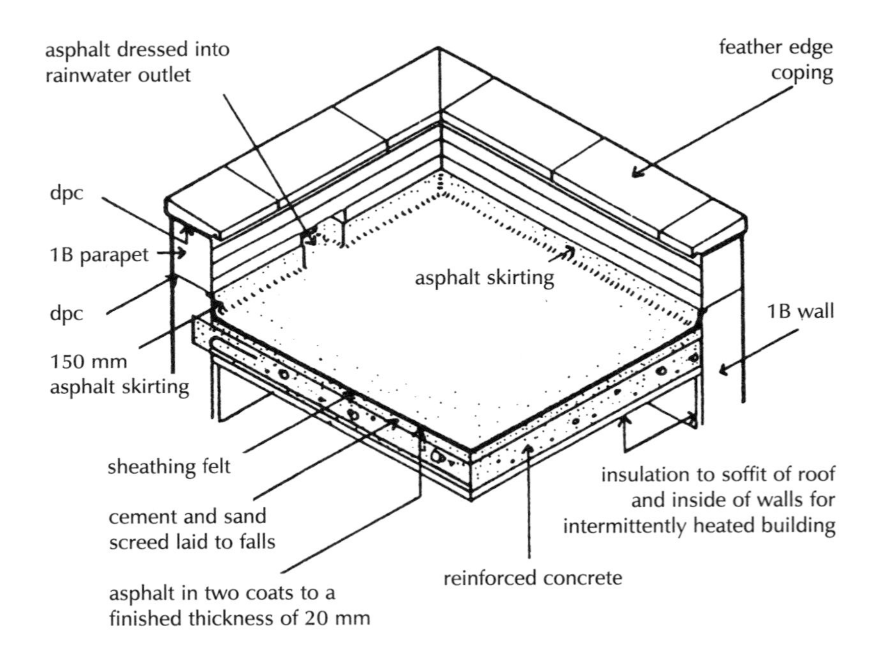

Fig. 242 Asphalt covering to concrete flat roof.

Built-up bitumen felt covering

As a precaution against the possibility of water drying out from screeded concrete roofs and rising to the surface under bitumen felt coverings, expanding and causing blisters, it is usual to use the partial bonding method.

With the partial bonding method the first layer of felt is bonded to the surface of the screed with perimeter and intermediate strips of bitumen with 180 mm wide vents between the strips of bitumen to allow moisture vapour to vent to perimeter and central vents.

The surface of the screed is first coated with a bitumen primer to improve the bond of bitumen to screed. Perimeter strips of bitumen 450 mm wide are spread around the roof with 150 mm wide vents at intervals for moisture vapour to vent to the perimeter. Strips of bitumen are spread over the body of the roof, as illustrated in Fig. 243.

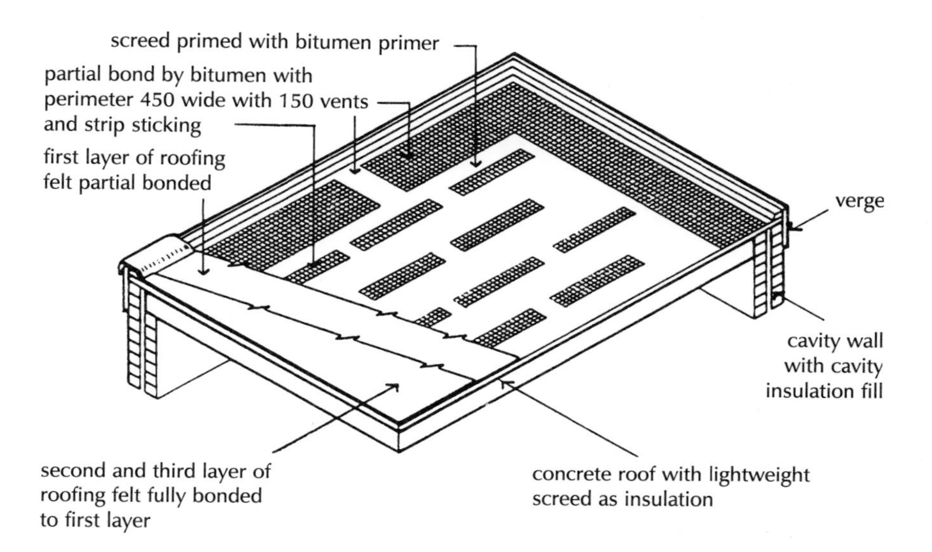

Fig. 243 Partial bond of felt roofing to concrete.

The principal adhesion of the felt covering, against wind uplift, is effected by the perimeter bonding. The size and spacing of the strips of bitumen is chosen as a matter of judgement between the need for adhesion to keep the felt covering flat and the assumed need to provide ventilation paths for moisture vapour pressure.

The first layer of felt is rolled out on to the hot bitumen and the intermediate and final layers of felt are rolled out and fully bonded and lapped as described for laying built-up bitumen felt roofing on timber.

At verges the intermediate and final layers of felt can be shaped over a splayed wood block and then covered with a strip of felt that is welted to a timber batten and turned over on to the roof, as illustrated in Fig. 244.

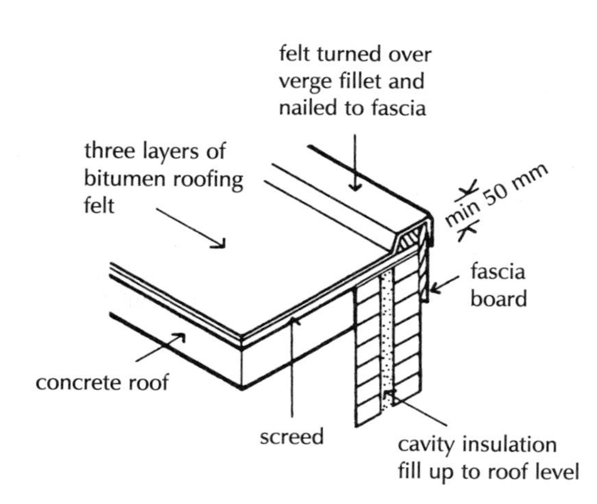

Fig. 244 Verge.

Where there is a parapet wall around the roof the intermediate and top layers of felt are turned up against the wall some 150 mm and covered with a lead sheet flashing which is turned into and wedged in a raked out brick joint.

Because the end laps of rolls of felt are made to overlap down the slope or fall of a roof it is difficult to form cross falls and currents to slopes of roof to a parapet wall rainwater outlet; it is usual to drain felt roofing to one continuous verge gutter.

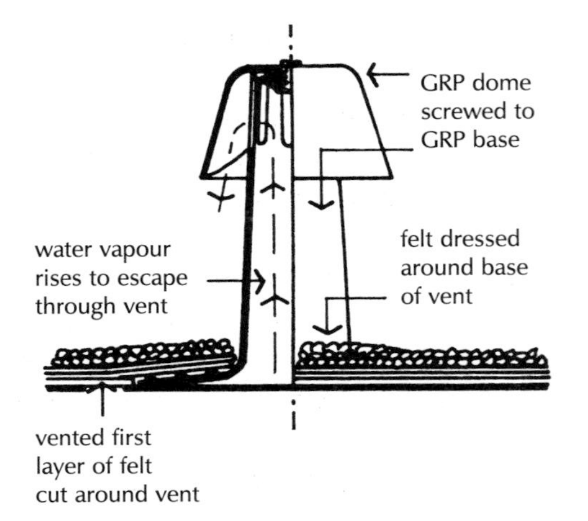

Fig. 245 Ventilator.

Where a built-up bitumen felt covering is laid to falls to parapet wall outlets it is impossible to avoid an untidy build-up of overlaps at oblique cuts of felt.

The usual method of providing ventilation for moisture vapour pressure is plastic ventilators that are fixed behind the felt upstand to parapet walls and to the felt overlap at verges. The parapet ventilators are covered with the apron flashing.

On larger roofs, where it is deemed necessary to provide additional ventilators in the surface of the roof, glass fibre reinforced plastic vents are fixed in the roof at about 6 m centres. These ventilators, illustrated in Fig. 245, are fixed to the roof and the intermediate and top layers of felt are cut and bitumen bonded around the vents.

Conservation of fuel and power

The requirement contained in the Building Regulations for the conservation of fuel and power is that 'Reasonable provision shall be made for the conservation of fuel and power in buildings by limiting the heat loss through the fabric of the building'. In Approved Document L to the regulations is practical guidance on meeting the regulations. There is no obligation to adopt the guidance contained in the Approved Document.

One additional requirement in the Building Regulations, that applies only to new dwellings, is that the calculation by a standard assessment procedure (SAP) of energy rating for the new dwelling should be submitted to the building control body. There is no obligation to achieve a particular SAP rating providing reasonable provision is made to conserve fuel and energy.

Resistance to the passage of heat

The resistance of flat roofs to the transfer of heat is poor. To achieve improved resistance to the transfer of heat Approved Document L advises the inclusion of a material with good resistance to heat transfer, insulation, in the construction of the roof.

As a practical, sensible measure of the maximum transfer of heat that should be allowed to conserve fuel and energy Advisory Document L suggests standard U values for roofs of 0.2 W/m^2K for dwellings, with an SAP rating of 60 or less, 0.25 W/m^2K for those with an SAP rating of over 60 and a U value of 0.25 W/m^2K for all other buildings.

The U value is a measure of how much heat will pass through one square metre of a structure when the air temperatures on either side differ by one degree, expressed as watts per square metre per degree temperature difference as W/m^2K. The U value, coefficient of thermal transmittance, is used as a convenience to determine the transfer of heat through a structure made of various materials.

Most of the materials used in the construction of flat roofs, separately or together, provide insufficient resistance to the transfer of heat to meet the requirement of the Building Regulations. It is necessary, therefore, to build in or fix some material with high resistance to heat transfer to act as a thermal insulation.

The most practical position for a layer of insulation for a flat roof is on top of the roof structure either under or over the weathering cover. In this position the insulation boards can be fixed or laid without undue wasteful cutting to provide insulation for the roof structure and utilise the heat store capacity of a concrete roof to provide some heat during periods when the heating is turned off.

As an alternative to fixing insulation on top of a flat roof it may be fixed between the ceiling joists of a timber roof so that the thermal resistance of the timber joists combines with the greater resistance of the insulation material. Tables in Approved Document L provide details of the thickness of insulation required for common insulating materials for a range of U values for insulation between joists.

The disadvantage of fixing insulation between the joists of timber flat roofs is the wasteful cutting necessary to fit the material between the joists, the labour necessary to support or wedge the material in position and the need for oversize electrical cables run in the insulation.

Thermal bridge

For efficient insulation against heat loss through the fabric of a building it is advisable to unite the system of insulation used in walls with that used in roofs and minimise those parts of the construction that provide a low resistance path to transfer of heat across, what have been called, thermal bridges.

The extent to which a detail of construction will act as a thermal bridge will depend on the difference in thermal resistance of the thermal bridge, and that of the adjoining construction. Where the difference is gross an appreciable concentration of condensation may appear on the colder surface of the bridge and cause unsightly stains and mould growth and adversely affect materials such as iron and steel.

To reduce the thermal bridge effect at the junction of a concrete roof built into a cavity wall and the inner leaf carried up as a parapet, the cavity insulation should be carried up the cavity face of a concrete block inner leaf to minimise the possibility of condensate stains on the soffit of the concrete roof at the intersection of ceiling and wall, as illustrated in Fig. 246.

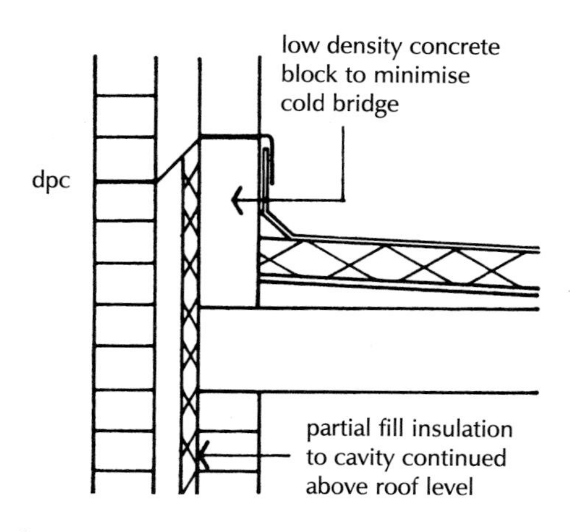

Fig. 246 Junction of roof and wall.

Similarly the cavity insulation of a wall should be carried up to unite with the roof level insulation of a timber flat roof to minimise condensation stains at the ceiling and wall junction, as illustrated in Fig. 247.

In buildings such as offices and places of assembly where the building is intermittently heated with high temperature radiant hea-

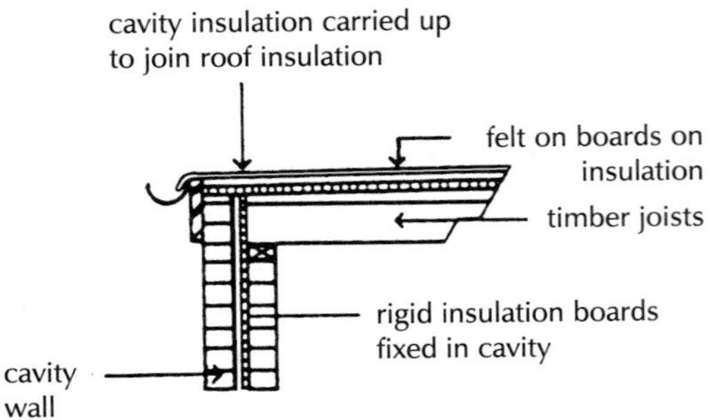

Fig. 247 Wall insulation joined to roof insulation.

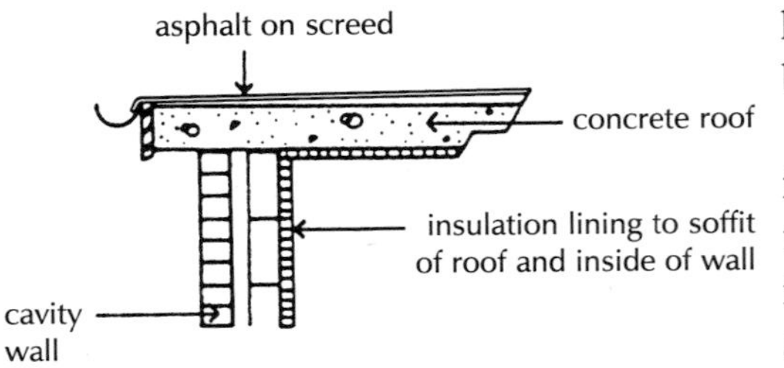

Fig. 248 Internal insulation.

ters it is often practice to fix the insulation to the fabric on the inside face of walls and ceilings in the form of insulation backed plasterboard, as illustrated in Fig. 248. In this way the radiant heat, when first turned on, immediately heats inside instead of expending some of its energy on heating the fabric.

Where a flat roof projects beyond the face of an external wall as a projecting verge to provide some protection for the wall below, or for the sake of appearance, it may not be practical to unite the wall insulation with that of the roof. In consequence there may be a path for the more rapid transmission of heat than in the surrounding fabric.

Heat may well be more readily conducted from inside through the concrete roof to outside than elsewhere in the detail illustrated in Fig. 249. To limit the thermal bridge effect in the construction illustrated it has been suggested that insulation is fixed behind a timber soffit board and behind a timber fascia board, as illustrated in Fig. 249.

The edge and soffit insulation will reduce the external surface area for thermal bridging but leave a possible path from inside to outside through the concrete and the external leaf to outside. The probability is that the extended thermal bridge of the concrete roof will provide adequate thermal resistance by itself and the edge and soffit insulation are unnecessary.

Similarly with the overhanging verge of a timber flat roof it has been suggested that edge and soffit insulation be used as illustrated in Fig. 250. The probability here is that the extended thermal bridge from inside, through the roof to outside, is in all likelihood sufficient to provide adequate thermal resistance, taking into account the

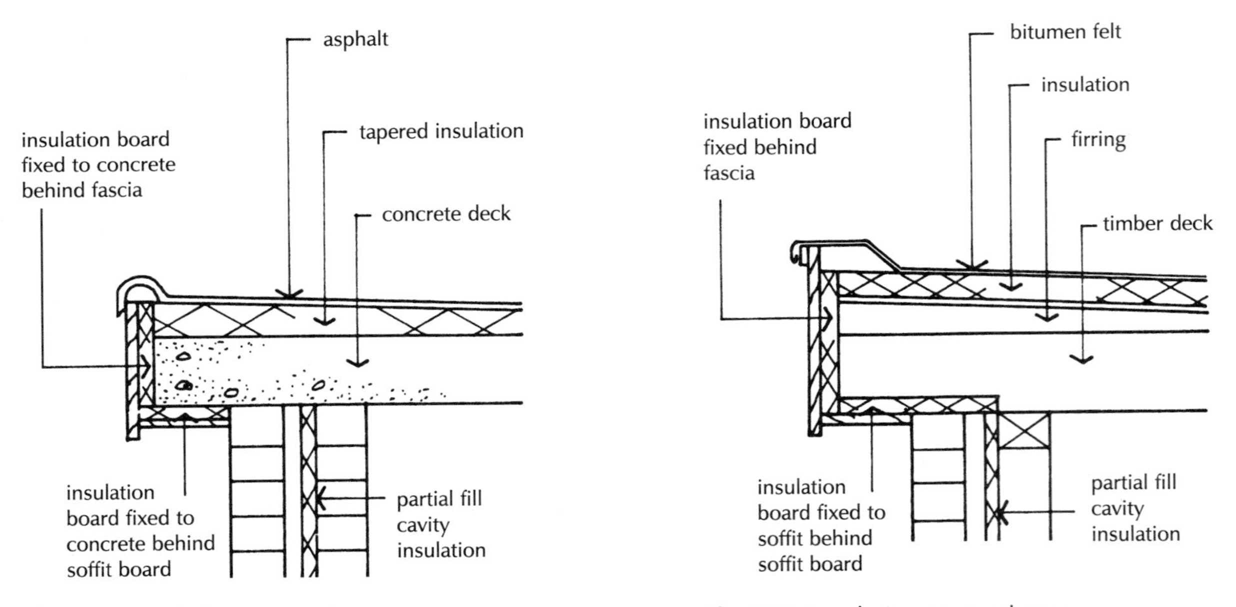

Fig. 249 Insulation to overhang.

Fig. 250 Insulation to overhang.

thermal resistance of the timber joists, fascia and soffit boards and air between joists where the roof space is not ventilated.

Insulation materials

To provide the required thermal resistance for a flat roof one of the semi-rigid insulation boards is used for roof level insulation. The cheapest material that is used is rockwool slabs either of uniform thickness or cut to provide falls for roof drainage, particularly for asphalt finishes.

Expanded and extruded polystyrene boards are the cheapest of the inorganic materials made in boards of uniform thickness or tapered for falls to roofs with expanded polystyrene.

The three inorganic material boards PIR, PUR and PUR have the lowest U value. They are the more expensive of the materials and are faced with glass fibre tissue for protection and as a finish impermeable to moisture vapour so that they may be used without the need for a moisture vapour check.

Some details of these materials are set out in Table 12.

Ventilation

The requirement of Part F to the Building Regulations is that 'adequate provision be made to prevent excessive condensation in a roof or a roof void above an insulated ceiling'.

Condensation is the effect caused by the reduction in temperature of warm, moisture vapour laden air coming into contact with a cold surface on which water is deposited as condensation. Where the condensation is excessive it may so saturate open texture insulation that it appreciably reduces the thermal resistance of the insulation and may cause unsightly damp stains and mould growth on decorated surfaces.

To minimise the extent of such an effect it has been practice to form a moisture vapour check on the warm side of insulation to reduce the penetration of moisture vapour to insulation. This practice, which is in all probability unnecessary, persists today.

The principal sources of warm, moist air are enclosed bathrooms and kitchens. The current requirement of Part G to the Building Regulations is that there shall be both natural and mechanical ventilation to bathrooms and kitchens to extract moist, warm air to outside. The combination of effective extract ventilation to the sources of warm, moist air, the use of moisture vapour checks and the use of moisture vapour impermeable insulation has reduced the likelihood of excessive condensation occurring.

There is a recommendation in Approved Document F to the Building Regulations that where there is a likelihood of excessive condensation in roof voids above insulated ceilings, the void space or spaces in this cold roof form of construction should be ventilated to outside air.

Table 12 Insulation materials.

Flat roof warm roof	Thickness mm	U value W/m^2K
Rockwool		
slab 1200 × 60	30, 40, 50, 60, 70, 80, 90, 100	0.036
slab cut to falls 1200 × 890	30, 40, 50, 60, 70, 80, 90, 100	0.036
Cellular glass		
board 1200 × 600	30, 35, 40, 50, 60, 70, 80	0.042
EPS		
board uniform in thickness or tapered	from 20 up in 5 increments	0.034
XPS		
board 1250 × 600	50, 75, 90, 100, 120	0.025
PIR		
board faced with glass fibre tissue 1200 × 750	35, 40, 50	0.022
PUR		
board bitumen impregnated glass fibre tissue faced 1200 × 750	26, 32 35, 40, 50, 75	0.022
PUR		
board glass tissue fibre faced 2400 × 1200	25, 30, 35, 50, 55	0.022

EPS expanded polystyrene
XPS extruded polystyrene
PIR rigid polyisocyanurate
PUR rigid polyurethane

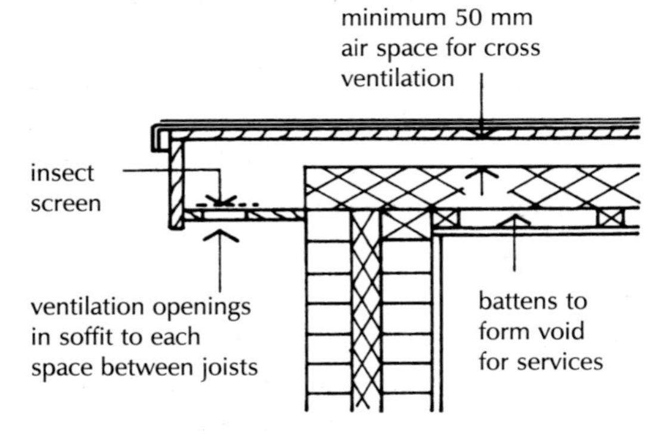

Fig. 251 Ventilation.

In the unlikely event of excessive condensation occurring it is recommended that there be formed a clear air space of at least 50 mm above the insulation. This space should be ventilated by continuous strips at least equal to continuous strips 25 mm wide running the full length of eaves on opposite sides of the roof. The ventilating openings are formed in the soffit of the overhang by plastic ventilators fitted with insect screens.

Where the insulation is laid or fixed between the joists there should be a clear space above the top of the insulation and underside of the roof of at least 50 mm for air to circulate across the roof from opposite sides, as illustrated in Fig. 251. So that electric cables are not run in the insulation the ceiling is fixed to battens to provide a space in which to run cables.

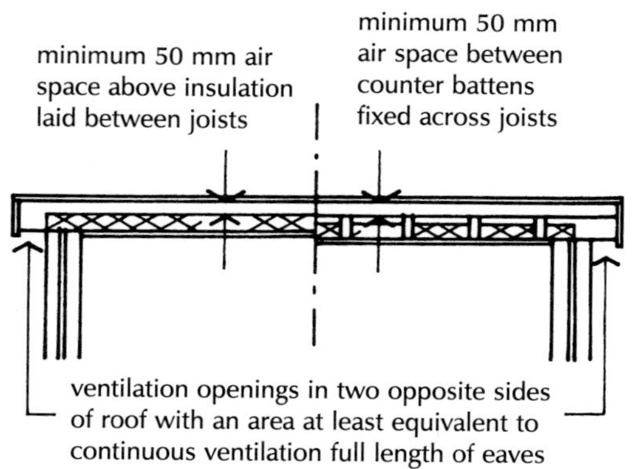

Fig. 252 Ventilation.

Where rolls of mineral fibre are laid between joists to the extent that there is not a clear 50 mm space between joists above the insulation, it is necessary to fix timber counter battens across the ceiling joists to provide the recommended 50 mm, minimum ventilation space as illustrated in Fig. 252.

Because of unpredictable, variable weather, it is uncertain whether the ventilation suggested will be effective in providing appreciable air movement to cause ventilation. Unless there is some realistic likelihood of there being excessive condensation in a timber flat roof, in spite of ventilation to bathrooms and kitchens, there seems little sense in the ventilation proposed.

PARAPET WALLS

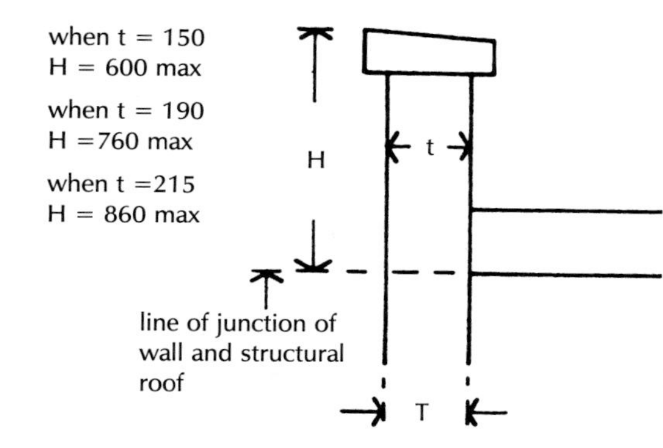

Fig. 253 Solid parapet wall.

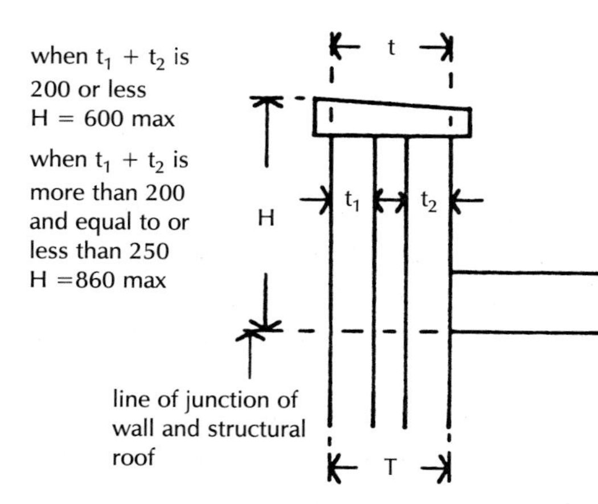

Fig. 254 Cavity parapet wall.

External walls of buildings are raised above the level of the roof as parapet walls for the sake of the appearance of the building as a whole. Parapet walls which are exposed on all faces to driving rain, wind and frost are more liable to damage than external walls below eaves level.

Because parapet walls are freestanding their height is limited in relation to their thickness for the sake of stability. Approved Document A, giving practical guidance to the requirements of the Building Regulations for houses and other small buildings, sets limits to the thickness and height of solid parapet walls, as illustrated in Fig. 253. Where the height (H) of the parapet walls is not more than 600 mm it should be not less than 150 mm thick, where H is 760 mm the thickness should be not less than 190 mm thick and where it is not more than 860 mm the thickness should not be less than 215 mm.

Where an external cavity wall is carried up as a parapet wall, as illustrated in Fig. 254, the limits of height (H) are not more than 600 mm where the combined thickness of the two leaves ($t_1 + t_2$) is equal to or less than 200 mm and 860 mm where the combined thickness of the two leaves ($t_1 + t_2$) is greater than 200 mm and equal to or less than 250 mm.

To protect the top surface of a parapet wall which is exposed directly to rain, it is essential that it should be covered or capped with some dense material to prevent rain saturating the wall. Natural stone was commonly used for this purpose as it is at once protective and decorative. The stones which are described as coping stones are cut so that they have a sloping surface when laid which is described as weathering. The stones usually project some 50 mm or more each side of the parapet wall so that rainwater running from them drips clear of the face of the wall. Three common sections employed for coping stones are shown in Fig. 255.

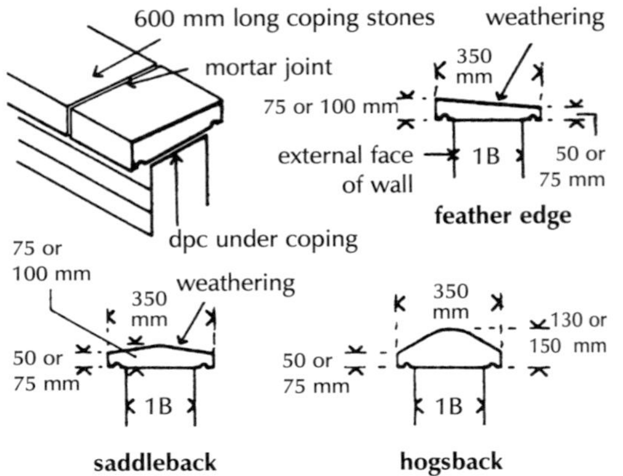

Fig. 255 Coping stones.

To encourage rainwater to run off the coping stones, clear of the wall faces, it is practice to cut semi-circular grooves in the underside of the overhang edges of the stones so that water runs off the extreme drip edges of the stones. Featheredge copings are laid so that the weathered top surface slopes towards the roof to minimise staining fair face external brick faces.

The stones are bedded in cement mortar on the parapet wall and butt end joints between stones are filled and pointed in cement mortar.

For economy, cast stone copings are used instead of natural stone copings. These stones are made as artificial stone with a core of concrete faced with a mixture of crushed stone particles and cement. The surface of cast stone may soon show irregular, unsightly staining.

Coping stones are usually in lengths of 600 mm with the joints between them filled with cement mortar. In time the mortar between the joints may crack and rainwater may penetrate and saturate the parapet wall below. If frost occurs, the parapet wall may be damaged. To prevent the possibility of rainwater saturating the parapet through the cracks in coping stones it is common practice to build in a continuous dpc of bituminous felt, copper or lead below the stones, as illustrated in Fig. 255.

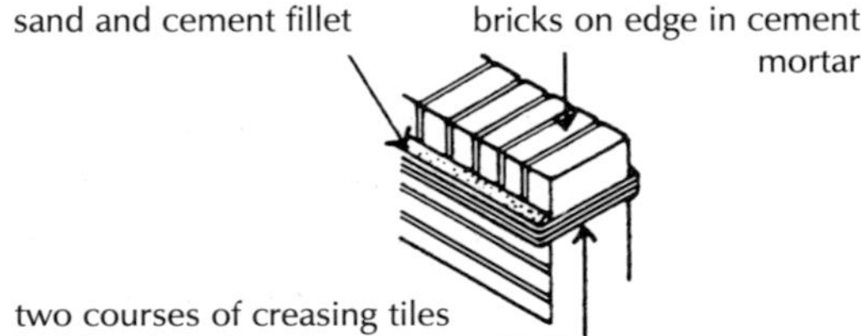

Fig. 256 Brick capping.

Another method of capping parapet walls is to form a brick on edge and tile creasing capping. This consists of a top course of bricks laid on edge, and two courses of clay creasing tiles laid breaking joint in cement mortar, as illustrated in Fig. 256. The bricks of the capping are laid on edge, rather than on bed, because many facing bricks have sand faced stretcher and header faces. By laying the bricks on edge only the sanded faces show, whereas if the bricks were laid on bed, the bed face which is not sanded would show. Also a brick on edge capping looks better than one laid on bed. Creasing tiles are made of burned clay and are usually 265 mm long by 165 mm wide and 10 mm thick. The tiles are laid in two courses breaking joint in cement mortar.

The tiles overhang the wall by 25 mm to throw water away from the parapet below. A weathered fillet of cement and sand is formed on top of the projecting tile edges to assist in throwing water away from the wall. Two courses of good creasing tiles are generally sufficient to prevent water soaking down into the wall and no dpc is usually necessary under them. Parapet walls should be built with sound, hard, well burned bricks which are less liable to frost damage than common bricks. The bricks should be laid in cement mortar mix 1 cement to 3 of sand.

Parapet wall dpc

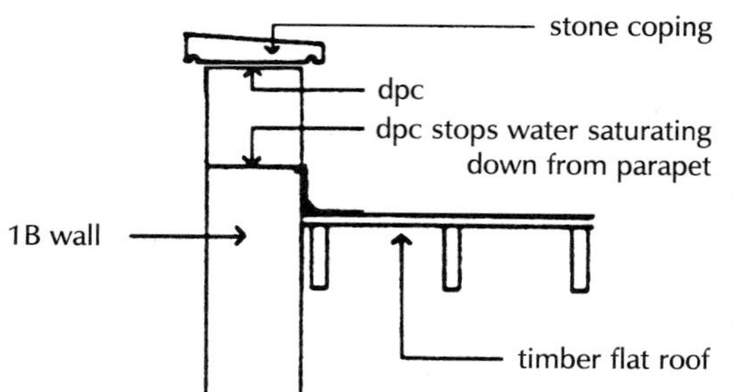

Fig. 257 Solid parapet.

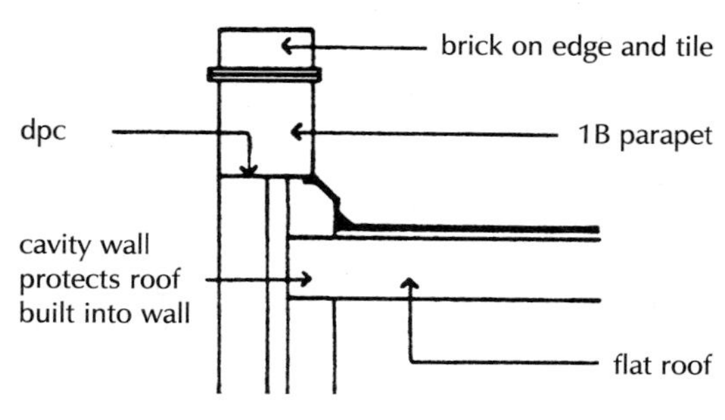

Fig. 258 Cavity parapet wall.

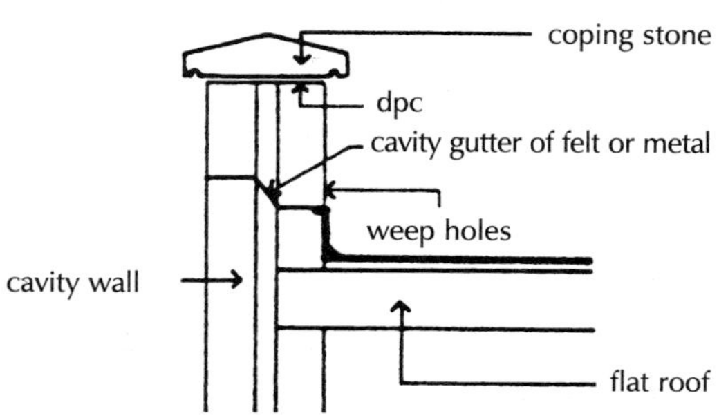

Fig. 259 Cavity carried above roof.

It has been practice for some years to form a horizontal dpc in parapet walls at the level of the top of upstand aprons and flashings to flat roof coverings. The purpose of this dpc is as insurance against the possibility of moisture in the exposed parapet wall penetrating down the wall to the roof itself. In reasonably sheltered positions this dpc is probably unnecessary in a wall built of sound bricks.

A parapet dpc to a solid brick parapet wall is illustrated in Fig. 257.

To provide protection to the roof structure it is usual to extend a cavity external wall up to the level of the top of the upstand of the roof covering to flat roofs as illustrated in Fig. 258. In this construction cavity insulation can be continued up to the level of insulation under roof coverings. A horizontal dpc is usually built in at the level that the solid brick parapet wall is raised on the cavity external wall, as illustrated in Fig. 258.

A cavity external wall is sometimes continued up as a parapet wall, as illustrated in Fig. 259. The only advantage of this arrangement is to continue the normal stretcher bond of external brickwork up to the parapet instead of having to change from stretcher to English or Flemish bond in a solid parapet, for the sake of appearance. Otherwise a cavity parapet serves no useful purpose and is probably an inconvenience as rainwater may penetrate to the cavity, particularly in exposed positions.

As a precaution against water penetrating to the cavity in a parapet wall it is practice to form a cavity dpc and tray which is continued across the cavity and built in one course lower in the roof side, as illustrated in Fig. 259. To collect and drain any water that may enter the cavity, weep holes are formed with raked out vertical joints in brickwork so that water runs down on to the roof.

The material best suited to this cavity dpc and tray is sheet lead that can more easily be dressed to shape across the cavity and over the upstand of the roof covering. An advantage of the lead dpc is that it will make a thinner, less ugly, horizontal joint on the external face of the wall than would a thicker felt dpc.

Index